WATERPLANTS OF NEW SOUTH WALES

WATER RESOURCES COMMISSION N.S.W.

WATERPLANTS
OF NEW SOUTH WALES

G. R. Sainty and S. W. L. Jacobs

Photographs by R. F. Brayne

with major contributions on

BIOLOGICAL CONTROL by K. L. S. Harley

FRESHWATER ALGAE by Valerie May

CARP, CRAYFISH AND WATER RAT by K. Shearer

EUTROPHICATION by Kathleen H. Bowmer

National Library of Australia
Cataloguing-in-Publication entry

Sainty, G. R. (Geoff R.).
    Waterplants of New South Wales
    Bibliography.
    Includes index.
    ISBN 0 7240 3730 6.
    1. Aquatic plants—New South Wales.
    I. Jacobs, S. W. L. (Surrey W.L.).
    II. New South Wales Water
    Resources Commission. III. Title.
    574.92'09944

First published 1981 by
WATER RESOURCES COMMISSION
NEW SOUTH WALES
Copyright © Water Resources Commission
New South Wales
All rights reserved

Design and Production by
George Bambagiotti

Edited by
Kevin Jeffcoat

Typesetting in Century Schoolbook by
Dalley Photocomposition
Ultimo, New South Wales

Colour Reproductions by
Mansfield Reproductions Pty Ltd
Chiswick, New South Wales

Printed in Australia on Geisha Satin Art by
C. C. Merritt Pty Ltd
Lakemba, New South Wales

Bound by
Stanley Owen & Sons Pty Ltd
Alexandria, New South Wales

Frontispiece: *Melaleuca quinquenervia* with
Casuarina glauca on the banks of Myall Lake.

CONTENTS

ACKNOWLEDGEMENTS

During the past seven years many specialists in various fields have freely given information that has been essential in compiling this book. For this help we acknowledge the Director and Staff, particularly *Karen Wilson,* of the Royal Botanic Gardens and National Herbarium, Sydney; *Stuart Bill, Bill Graham,* State Rivers and Water Supply Commission, Victoria; *Charles Julian,* Water Resources Commission, Queensland; *Jim Swain,* Ciba Geigy; *Peter Dunk,* Ministry of Conservation, Victoria; *Helen Aston,* National Herbarium, Melbourne; *Barry Fox,* University of New South Wales; *Frank Eady,* Monsanto; *Mike Barrett,* Imperial Chemical Industries; *Penny Wade, Stan McFadzean,* Dupont; *David Mitchell, John Adeney, Barrie Steer, Mike Julian, Helena Hicks,* CSIRO; *Don Buckmaster,* Fisheries and Wildlife, Victoria; *Gerry Smith, Ken Shaw, Ken Armstead, John Duchatel, Max Dickens, Chris Ripper, John Hardie, Ian Kelly, Jim Neville,* Water Resources Commission, New South Wales; *Roger Carolin,* University of Sydney; *Barbara Briggs,* National Herbarium, Sydney; *Ian Bolton; Gus Crosby; Kathy Reynolds; Leslie McCullough; Mike Perri,* Department of National Development, Canberra; *Steve Lillyman,* Department of Lands & Surveys, Western Australia; *Bill Fischer,* University of California, Fresno, U.S.A.; *Bill Maier,* Department of Natural Resources, Florida, U.S.A.; *James McGehee,* U.S. Corps of Army Engineers, Florida, U.S.A.; *Orrie Pastega,* Clarence River County Council Flood Mitigation Authority, for section on flood mitigation drains; *Cedric Jones,* Water Resources Commission, New South Wales, for the seed section; *Jim Thorman,* for sketches of plants; *Alan Heritage,* CSIRO and *Ari van der Lelij,* Water Resources Commission, New South Wales, for section on iron and manganese bacteria; *Laurie Jakel* and *Trevor Miles* of State Rivers and Water Supply Commission, Victoria, and *Kathleen Bowmer* for assistance with the section on aquatic plant management; *Keith Leamon,* Department of Agriculture, Victoria, and *Brian Smith,* National Museum of Victoria, for section on Little Basket Shell; *Brenda Prenzel,* for manuscript checking and typing; *Betty Jacobs,* for typing and glossary drawings; *David Mackay* for some of the drawings. We also acknowledge those people without whose help this book would not have been published.

PHOTO CREDITS

Stephen Hogg: Iron bacteria, *Echinochloa telmatophila.* Keith Leamon: *Corbiculina australis.* Charles Julian: *Salvinia molesta.* Geoff King: *Montia australasica.* Ann Williams: *Halophila ovalis, Posidonia australis.* Karen Wilson: *Ipomoea diamentensis, Bacopa monniera.* Peter Room: *Monochoria cyanea.* Trevor Miles: *Sium latifolium, Hydrocleys nymphoides.*
Biological Control: R. P. Bradbury, J. E. Broadbent, J. W. Forno, J. Green, M. H. Julien, A. D. Wright.

INTRODUCTION

Waterplants are species which thrive in water or on wet land.

Most waterplants cause no problems and besides their obvious aesthetic and horticultural values, most play an important part in the ecosystem. They provide food and cover for animals and generally improve water quality and stabilise soil and sand. However, a few are serious weeds of rice, and irrigation and drainage systems. In lakes and creeks, dense growth of submerged and floating plants can interfere with boating and swimming and adversely affect fish population. Floating pieces can bank up at structures, impeding flow and causing flooding. Some species will taint water and reduce availability in farm dams and small reservoirs.

The plants described and photographed in this book include both common and potentially troublesome species in New South Wales.

Irrigationists and those involved in the supervision and control of waterways and lakes, should be able to identify the common waterplants and be aware of uncommon or potentially troublesome species.

The early recognition of an unwelcome waterplant in a system can result in isolated outbreaks being eradicated by simple and economic measures, whereas if left unchecked a major, costly and continuing program of control may be necessary.

This book is designed as a practical adjunct to botanical texts, the main Australian work being Aston's "Aquatic Plants of Australia". The range covered here includes grasses, sedges, sea grasses and a contributed section on common freshwater algae. In addition there is a section describing insects introduced to Australia to control noxious waterplants. Freshwater animals known to cause damage to irrigation works have also been included. Separate chapters on biological control and eutrophication have been added to complement the section on aquatic plant management.

Plants recorded as troublesome in aquatic situations throughout Australia have been described. A number of exotic species not occurring in Australia and known to be serious weeds overseas have also been included.

HOW TO USE THIS BOOK

The plants have been arranged in alphabetical order of families commencing with the Alismataceae and concluding with Zosteraceae. The exceptions are the algae which are grouped together under "Algae" and mangroves which are grouped under "Mangroves". The genera are arranged alphabetically within the family and the species arranged alphabetically within the genus.

The arrangement is based on the scientific name of the plant. The standardised common name has been included when available and, in some instances, alternative common names are given. The standardised common names are listed by Hartley (1979) in "A Checklist of Economic Plants in Australia". Common names are listed in the index.

The species are described and, in most cases, the description is accompanied by one or more photographs. In some families, for example the grasses and the sedges, extra information is supplied on other not truly aquatic species or genera which could be confused with the aquatic species.

The botanical description is followed by notes on growth biology, habitat, economic significance and, in the case of a few serious weeds, control measures. Distribution in New South Wales is indicated in the small map on each page. This map is divided into New South Wales botanical subdivisions (see page 44). The scales used in the photographs are in centimetres and millimetres.

If you do not know the scientific name of the plant and you cannot find an appropriate common name in the index then you will need to first identify the plant.

If you have limited knowledge of botany turn to the synoptic or pictorial key (Key No. 3 on page 36). Plants here are grouped according to their shape and growth, e.g. submerged feathery leaves, floating leaves. Decide which group, then check your specimen against the descriptions and photographs.

If you have some knowledge of botany you may be able to use either Key No. 1 (page 13) or Key No. 2 (page 25). Key No. 1 is based on floral characters and requires fairly complete specimens but is the most accurate of the keys.

Key No. 2 relies more on vegetative parts and characters more likely to be present on incomplete specimens.

Because of the greater variability of vegetative parts, Key No. 2 may be less accurate than Key No. 1. Any results obtained from the keys should be checked against the photographs and descriptions.

HOW TO USE A BOTANICAL KEY

A key is normally used as an aid for the identification of plants unknown or only partly known to the user.

For many reasons the identity and name of a plant obtained solely by use of a key may be incorrect, incomplete or both. When identifying any unknown plant with the aid of keys, always check the description of each name arrived at by means of the key to ensure that there is reasonable agreement between the characters observed in the unknown plant and those provided in the description of the plant it is presumed to be. When a marked discrepancy exists between plant and description, a misidentification has probably occurred.

A key is an artificial device whereby a choice is provided between (usually two) contradictory propositions about the unknown plant, resulting in the acceptance of one and the rejection of the other(s). In some cases, however, this may not be a clear-cut choice.

For example, find an appropriate name for this shape:

First choose between choice A and AA. If AA fits, then choose between D and DD, and so on. This is a dichotomous indented key.

```
A    A shape with 3 sides .............................B
     B    A shaded shape...........................SHADED TRIANGLE
     BB   A non-shaded shape....................C
          C    Contains dots .....................SPOTTED TRIANGLE
          CC   No dots present .................PLAIN TRIANGLE
AA   A shape with 4 sides .............................D
     D    A shaded shape...........................SHADED SQUARE
     DD   A non-shaded shape....................E
          E    With dots ...........................SPOTTED SQUARE
          EE   Without dots ......................PLAIN SQUARE
```

In a dichotomous bracketed (non-indented) form, this key would appear as:

A	A shape with 3 sides	B
AA	A shape with 4 sides	D
B	A shaded shape	SHADED TRIANGLE
BB	A non-shaded shape	C
C	Contains dots	SPOTTED TRIANGLE
CC	No dots present	PLAIN TRIANGLE
D	A shaded shape	SHADED SQUARE
DD	A non-shaded shape	E
E	With dots	SPOTTED SQUARE
EE	Without dots	PLAIN SQUARE

As you can see, these two types are quite similar but, with longer keys, the bracketed non-indented key uses much less space.

Although this is a very simple key it illustrates how, by choosing one of the two alternatives at each step, it is possible to name the shape as a SHADED SQUARE.

This type of key is called a dichotomous key as there are only two choices at each step.

It should be emphasised that a key like this is finite, i.e., there are a limited number of alternatives offered and if a shape not catered for in the key is encountered, e.g.

it may be possible to arrive at a result (PLAIN TRIANGLE) but comparison with a description would indicate the discrepancy.

In some cases the choices are not quite so obvious and it may be necessary to follow both lines available until one series of choices becomes either not applicable or the final description does not match.

In order to use a botanical key, certain terms need to be known, so reference to a glossary of botanical terms is necessary in most cases. "The Language of Botany" by C. Debenham (available from the Society for Growing Australian Plants) is a suitable reference. Glossaries are often included in floras, for example "Flora of the Sydney Region" by Beadle, Evans and Carolin; see also "Aquatic Plants of Australia" by H. I. Aston. A glossary is included on page 532 of this book.

There are two different types of keys used in this book. Most of the keys are bracketed or non-indented. They are mainly dichotomous, but the first choice in Key No. 2 to the families and genera is a multiple choice. Key No. 3 is a synoptic type. A synoptic or pictorial key provides a summary of the characters in a single table, it gives broad groupings but often results in a list of possible answers which require checking against descriptions or photographs. The photographs themselves can also be regarded as a pictorial key. All of these keys have advantages and disadvantages and, with the major keys at least, some variation is supplied in arrangement and key characters to help overcome the problems that inevitably develop.

In many cases it is possible to identify plants to the species level with this book. In other cases (e.g. Gramineae (grasses), Cyperaceae (sedges)) information is only provided to identify plants accurately to the genus level (although the photographs may allow you to go a little further in a *few* cases). This is done because reliable distinctions at the species level in such groups can often only be made by experienced botanists using microscopic characters.

The generic name is adequate for most practical purposes. There is too great a risk of misidentification at the species level.

In all such cases, providing adequate specimens can be supplied, it should be possible to obtain a reliable identification by forwarding a specimen to the appropriate State Herbarium.

PLANT SPECIMENS
collection for identification

Persons forwarding specimens to a Herbarium for identification should follow the directions given below.

1. All specimens sent should be as perfect as possible. They should consist of a small branch or portion of the stem, 20-30 cm long, showing the leaves in position together with flowers and/or fruits. Owing to the large number of plant species it is difficult, sometimes impossible, to determine specimens from leaves alone. In the case of small plants and grasses the whole plant should be sent.

2. Specimens should be numbered and a second set of specimens with corresponding numbers should be kept. The correspondent will be supplied with a list of the numbered names and other information desired, but the specimens will not be returned, except under special circumstances.

3. The date and place of collection should be given, as well as details of the growth habit, height, flower colour, habitat, type of soil, and, in the case of trees, a description of the bark.

4. It is desirable to press the specimens between newspaper sheets and to place them between two sheets of cardboard when forwarding. Large wet and bulky specimens such as Waterlilies, *Nymphaea* spp., require the newspaper to be changed to speed drying and prevent mould growth. Specimens should never be wrapped in plastic or damp paper or forwarded in a jar of water. If pressing is difficult then specimens can be preserved in an aqueous solution containing 70 per cent methylated spirit or a similar preservative.

5. If rare specimens are sent, or specimens of which more material is required for the Herbarium, the correspondent is expected to supply the material in return for the information received.

6. All specimens should be forwarded for identification to the appropriate State Herbarium, and an explanatory letter should be sent under separate cover. Specimens from New South Wales should be forwarded to the Director, National Herbarium, Royal Botanic Gardens, Sydney 2000.

KEY 1
Primarily floral characters

Key to Major Divisions

1. Fertile plants not producing true seeds or flowers. **2.**

1a. Fertile plants producing seeds and flowers. **4.**

2. Female sex organs single celled and/or not surrounded by a layer of sterile cells; zygotes never developing into multicellular embryos while still within the female sex organs. **Algae**

2a. Sex organs multicelled, surrounded by a layer of sterile cells; zygotes developing into multicelled sporophytes while still attached to the gametophyte. **3.**

3. Mature sporophyte always parasitic on gametophyte and with an undifferentiated vascular system. **Bryophyta**

3a. Mature sporophyte independent of gametophyte and with a differentiated vascular system. **Pteridophyta**

4. Germinating plants with a single cotyledon; leaves with parallel venation; floral structures basically in whorls of 3 although 1 or more may be absent in any particular whorl; vascular bundles in stem often scattered or in 2 or more rings. **Monocotyledons**

4a. Germinating plants usually with 2 cotyledons (rarely 1); leaves with reticulate venation; floral structures basically in whorls of 4 or 5 (or more) although 1 or more may be absent in any particular whorl; vascular bundles in stem usually in a single ring. **Dicotyledons**

Algae

1. Cells lack true nucleus surrounded by membrane; photosynthetic pigments diffused through peripheral cytoplasm.
Myxophyceae or Cyanophyceae 2.

1a. Cells with a nucleus surrounded by a nuclear membrane; photosynthetic pigments localised in definite chromatophores (chloroplasts) or absent. **3.**

2. Plants filamentous, often occurring as coils of cells floating in water. *Anabaena*

2a. Plants non-filamentous, usually occurring as clumps of cells in mucilage, the clumps floating in water. *Anacystis*

3. Cell walls consist of two overlapping halves, containing silicon (persistent after treatment with sodium hypochloride solution).
Bacillariophyceae (Diatoms)
3a. Cell walls not highly silicified, present or absent. **4.**

4. Food storage as paramylon, never as starch; plants unicellular often motile; cell wall frequently absent; chloroplasts often absent.
Euglenineae (Euglenoids)
Wall lacking except in spore. ***Euglena***
4a. Food storage usually as starch; grass-green chloroplasts present; plants uni- or multicellular. **Chlorophyceae or Isokontae 5.**

5. Plant with a main central axis with whorls of branches arising at intervals; reproductive structures elaborate, occurring at nodes and often appearing as black or orange spheres; plants often encrusted with lime.
6.

5a. Plants without whorls of branches along the main axis; reproductive structures comparatively simple. **7.**

6. Oogonium structure with 10 cells at apex; whorls of branches also present on secondary branches. ***Nitella***

6a. Oogonium structure with 5 cells at apex; whorls of branches absent on secondary branches. ***Chara***

7. Plant tubular, many cells forming a complete boundary of a hollow tube. ***Enteromorpha***
7a. Plants various, but not forming a tissue surrounding a hollow tube.
8.

8. Individual cells joined to form an open network. ***Hydrodictyon***
8a. Plants filamentous, single cells joined end to end; filaments branched or unbranched. **9.**

9. Each cell with a spirally arranged chromatophore (chloroplast); filaments unbranched. ***Spirogyra***
9a. Chromatophores peripheral, either solitary and reticulate, not spiral *or* numerous and discoid. ***Cladophora***

Bryophyta (Mosses and Liverworts)

1. Gametophytes dorsiventrally differentiated, externally simple (differentiated into leaves and stems, but then not radially symmetrical, in some non-aquatic groups); sex organs formed from superficial cells on the dorsal side of the thallus (terminal cells in some non-aquatic species); sporophyte consisting solely of a capsule in the aquatic species.
Hepaticae (Liverworts)
(Ricciaceae, *Riccia*)

1a. Gametophyte with a transitory prostrate stage bearing erect sexual branches which continue growing as independent plants; branches differentiated into stems and leaves and usually radially symmetrical; sex organs terminal; sporophyte consisting of (a) foot and capsule or (b) foot, seta and capsule.
Musci (Mosses)

Pteridophyta (Ferns and Fern allies)

1. Plants with rhizomes or bases attached to the bottom even when covered by water; roots in mud.
2.

1a. Plants floating when water present; roots or root-like leaves suspended in water.
3.

2. Rhizome elongate, leaves arising at intervals along the rhizome; leaves submerged and/or floating, clover-like (4-lobed) or linear; sporangia borne in woody sporocarps produced at or near the base of the stipes.
Marsileaceae
(*Marsilea, Pilularia*)

2a. Rhizome corm-like, bearing a dense rosette of stiff linear leaves; sporangia borne on the adaxial surface near the base of the leaves.
Isoetaceae
(*Isoetes*)

3. Leaves pinnately divided; sporangia borne on abaxial leaf surface of emergent leaves.
Parkeriaceae
(*Ceratopteris*)

3a. Leaves simple (but rhizome branching) or simply lobed; sporangia axillary or borne on submerged leaves.
4.

4. Mature plants usually less than 5 cm long; individual leaves usually less than 2 mm long, alternate; leaves in two rows; roots produced along rhizome.
Azollaceae
(*Azolla*)

4a. Mature plants more than 5 cm long; individual leaves usually over 1 cm long, opposite; leaves in 3 rows, lower row much dissected and root-like.
Salviniaceae
(*Salvinia*)

Angiosperms
(Flowering Plants)
Monocotyledons

1. Ovary inferior.   **2.**
1a. Ovary superior.   **10.**

2. Ovules solitary; flowers bisexual, inflorescence large, open, terminal or sub-terminal panicles; emergent freshwater aquatics.   **Marantaceae (Thalia)**

2a. Ovules numerous; flowers usually unisexual (bisexual in *Ottelia*), spatheate; female (or bisexual) flowers solitary or few together, often axillary; submerged marine or submerged or emergent freshwater aquatics.   **Hydrocharitaceae 3.**

3. Mature leaves differentiated into blade and petiole.   **4.**
3a. Mature leaves not differentiated into blade and petiole.   **6.**

4. Plants marine or estuarine, rhizomatous; flowers and leaves submerged unless exposed by low tide.   ***Halophila***
4a. Plants growing in fresh water; young leaves may be submerged but mature leaves and flowers floating or emergent.   **5.**

5. Stoloniferous; petioles with stipules; flowers less than 3 cm across, unisexual.   ***Hydrocharis***
5a. Caespitose, stolons absent; petioles without stipules; flowers usually more than 3 cm across, bisexual.   ***Ottelia***

6. Leaves in a basal rosette, long and grass-like, mostly more than 4 cm long.   ***Vallisneria***
6a. Leaves produced along the stem, not grass-like, less than 4 cm long.   **7.**

7. Leaves alternate, recurved; petals about the same size as sepals.   ***Lagarosiphon***
7a. Leaves whorled, not recurved; petals about the same size as, or larger than sepals.   **8.**

8. Petals much larger than sepals; leaves mostly in whorls of 4-5.   ***Egeria***
8a. Petals about the same size as sepals; leaves in whorls of 3-8.   **9.**

9. Leaves usually in whorls of 3.   ***Elodea***
9a. Leaves usually in whorls of more than 3.   ***Hydrilla***

10. Perianth absent or present (by definition) as lodicules or hypogonous bristles or scales; flowers small, grouped into spikelets subtended by glumes.   **11.**
10a. Perianth mostly present, if absent then the flowers not grouped into spikelets subtended by glumes.   **13.**

11. Fruit dehiscent; carpels 3 or more, free, on a carpophore; stamens mostly 1. **Centrolepidaceae *(Centrolepis)***
11a. Fruit indehiscent; ovary apparently 1-locular, sessile; stamens mostly more than 1. **12.**

12. Flowers subtended by a single bract and variously arranged into spikelets; leaves mostly with closed sheaths and mostly without ligules (*Gahnia* has ligules). **Cyperaceae** (see separate generic key)
12a. Flowers enclosed by a lemma and palea and these variously arranged into characteristic spikelets which are mostly subtended by sterile glumes; leaves mostly with open sheaths and ligules present although sometimes much reduced. **Gramineae** (see separate generic key)

13. Plants very small or minute (up to 1 cm long but several may be joined together) floating on or below the surface; no obvious differentiation into leaves and stems; simple roots present or absent; flowers minute, unisexual, spatheate. **Lemnaceae (*Lemna, Spirodela, Wolffia*)**
13a. Plants all much larger than 1 cm long and differentiated ± obviously into stem (may be only as flower stem) and leaf; plants free-floating or rooted; roots present; flowers unisexual or bisexual, ± spatheate.

14.

14. Carpels more than one, free or partially fused (but may be mostly fused in fruit). **15.**
14a. Carpels fused or solitary. **25.**

15. Flowers bracteate; carpels numerous (6-∞), emergent aquatics. **Alismataceae 16.**
15a. Flowers ebracteate; carpels mostly fewer than 6 (up to 16 in *Ruppia polycarpa*); plants submerged or emergent. **19.**

16. Carpels spirally arranged; flowers unisexual, lower flowers female or bisexual, upper flowers male. ***Sagittaria***
16a. Carpels in 1 whorl; flowers bisexual. **17.**

17. Carpels roughly triangular with a long beak, united along central axis. ***Damasonium***
17a. Carpels rounded on the back, not triangular, not united along central axis but often closely packed. **18.**

18. Carpels usually more than 15; back of carpel ridged but without warty projections. ***Alisma***
18a. Carpels usually fewer than 15; back of carpel usually with warty or other projections. ***Caldesia***

19. Inner perianth whorl yellow, showy, delicate; flowers more than 1 cm across. **Butomaceae *(Hydrocleys)***
19a. Inner perianth whorl not showy, nor particularly delicate; flowers less than 1 cm across. **20.**

20. Flowers axillary, solitary, sessile or pedicellate; submerged aquatics, often in saline waters.　　　　　　　　　　　　　**Zannichelliaceae 21.**
20a. Flowers 2 or more on a branched or simple spike-like inflorescence; submerged or emergent aquatics.　　　　　　　　　　　　　**22.**

21. Carpels enclosed by a spathe-like membranous perianth with or without a slit; leaves mainly opposite and with free membranous stipules; carpels ± warty on curved back; male flowers without a perianth.
　　　　　　　　　　　　　　　　　　　　　　　Zannichellia
21a. Carpels subtended by 3 membranous perianth segments; leaves mostly alternate, with distinct sheaths; carpels mostly smooth on the back; male flowers with a perianth.　　　　　　　　*Lepilaena*

22. Flowers 2 per inflorescence on a reduced spike; individual carpel stalks elongating in fruit; submerged aquatics.　　　　**Ruppiaceae**
　　　　　　　　　　　　　　　　　　　　　　　(Ruppia)
22a. Flowers usually more than 2 per inflorescence; individual carpel stalks not elongating in fruit.　　　　　　　　　　　**23.**

23. Leaves without sheathing bases (but bases expanded), arising from tuber more than 3 cm long when mature, usually buried deeply in mud; carpels mostly 3 (-6); floating leaves with expanded blades and obvious cross-veins joining parallel veins; plants with submerged leaves, ± floating leaves and emergent inflorescenses.　　　　**Aponogetonaceae**
　　　　　　　　　　　　　　　　　　　　　　　(Aponogeton)
23a. Leaves with sheathing bases; tubers (really turions) if present, much smaller than 3 cm and present as overwintering organs; carpels 4 or more; if floating leaves produce expanded blades then these without obvious cross-veins.　　　　　　　　　　　　　　**24.**

24. Leaves produced all along the stem; stamens 4; carpels 4, free; leaves submerged or floating, floating leaves differing from submerged leaves; inflorescences emergent.　　　　　　　　**Potamogetonaceae**
　　　　　　　　　　　　　　　　　　　　　　　(Potamogeton)
24a. Leaves all basal; stamens 6 or more (but not all necessarily fertile); carpels mostly 4, fused in lower half; leaves emergent but some may be floating, all similar; inflorescence emergent.　　**Juncaginaceae 25.**

25. Perianth segments 6; stamens 6; carpels united below the middle; leaves elliptical in cross-section.　　　　　　　　*Triglochin*
25a. Perianth segments absent (or 2, depending on interpretation); stamens 12; carpels free near top; leaves triangular in cross-section.
　　　　　　　　　　　　　　　　　　　　　　　Maundia

26. Carpels solitary.　　　　　　　　　　　　　　**27.**
26a. Carpels more than 1, fused.　　　　　　　　　　**32.**

27. Flowers axillary or in open inflorescences. **28.**

27a. Flowers in dense spike-like or globular inflorescences or arranged along one side of a flattened axis. **29.**

28. Flowers axillary, solitary, unisexual; submerged estuarine or freshwater plants. **Najadaceae** *(Najas)*

28a. Flowers in an open bracteate inflorescence, bisexual; submerged estuarine or marine aquatics. **Posidoniaceae** *(Posidonia)*

29. Inflorescence spicate on one side of a flattened axis enclosed within a spathe; flowers unisexual, male flowers alternating with female flowers on the inflorescence; submerged estuarine or marine aquatics. **Zosteraceae 30.**

29a. Inflorescence of dense spike-like or globular clusters; flowers distributed all around the axis; flowers unisexual, male and female flowers separated on the inflorescence; plants emergent freshwater aquatics. **31.**

30. Erect stems all fertile; transection of middle of rhizome or stem internode with 2 cortical vascular bundles; retinacula obtuse. *Zostera*

30a. Both sterile and fertile erect stems present; transection of middle of rhizome or stem internodes with 4-8 cortical vascular bundles; retinacula acute. *Heterozostera*

31. Flowers in remote dense globular clusters; female flowers in one or more clusters below similar but smaller clusters of male flowers. **Sparganiaceae** *(Sparganium)*

31a. Flowers in dense tightly packed spike-like inflorescences; female flowers below, male flowers above ± separated by a short piece of stem. **Typhaceae** *(Typha)*

32. Inflorescence a spadix with naked (no perianth) unisexual flowers, female flowers below, male above; the spadix subtended by a ± showy spathe longer than the spadix. **Araceae 33.**

32a. Inflorescence not a spadix and not subtended by a spathe; perianth present; flowers bisexual or unisexual. **34.**

33. Plants free-floating; inflorescence not obvious. *Pistia*

33a. Plants caespitose, rooted with perennial rhizomes; inflorescence obvious, spathe white. *Zantedeschia*

34. Perianth dry and scarious; flowers bisexual or unisexual.

Juncaceae
(Juncus)

34a. Perianth obvious and showy (petaloid); flowers bisexual.   **35.**

35. Stamen 1; flowers yellow; leaves tapering gradually from the base; emergent aquatic.

Philydraceae
(Philydrum)

35a. Stamens 6; flowers blue; leaf blade expanded; plants either free— floating or emergent.   **Pontederiaceae 36.**

36. Fruit one seeded; ovary 1-locular; plants erect, emergent.

Pontederia

36a. Fruit many-seeded; ovary 3-locular.   **37.**

37. Plants free-floating when water present; petioles inflated; corolla tube well developed.

Eichhornia

37a. Plants decumbent, often growing out over water but normally rooted in mud covered by shallow water; petioles not obviously inflated; petals fused only near base.

Monochoria

Dicotyledons

1. Ovary superior, receptacle enlarged and shaped like an inverted cone with individual carpels sunk into the tissue; leaves all basal, peltate; aquatic with floating and/or emergent leaves and emergent flowers.

Nelumbonaceae
(Nelumbo)

1a. Ovary superior or inferior; plants various but without the enlarged, inverted cone-shaped receptacle with individual carpels sunk into the tissue.   **2.**

2. Carpels free, 2 or more, each with separate styles and stigmas.   **3.**

2a. Carpels solitary or ovary of 2 or more fused carpels.   **6.**

3. Leaves all alternate, divided or lobed, never peltate, submerged or emergent.

Ranunculaceae
(Ranunculus)

3a. Leaves all opposite or submerged leaves only opposite and/or floating, alternate, peltate leaves present.   **4.**

4. All leaves opposite, simple; flowers white; leaves ± succulent.

Crassulaceae
(Crassula)

4a. Submerged leaves opposite, much divided; floating leaves (if present) alternate, peltate, entire; flowers white, pink or red.

Cabombaceae 5.

5. Dissected submerged leaves common, peltate floating leaves rarely formed. ***Cabomba***

5a. Dissected submerged leaves absent, floating peltate leaves common, often with mucilage on under surface. ***Brasenia***

6. Ovary superior. **7.**
6a. Ovary inferior or half inferior. **34.**

7. Petals present and free. **8.**
7a. Petals present and at least partially fused or petals absent. **12.**

8. Leaves all basal, arising from a corm or a horizontal rhizome, floating or emergent; sepals 4, 5, or 6; petals numerous; stamens numerous. **Nymphaeaceae 37.**

8a. Leaves opposite, whorled or alternate, spread along the stem. **9.**

9. Leaves alternate. **10.**
9a. Leaves opposite or whorled. **11.**

10. Sepals 4; petals 4; stamens 6. **Brassicaceae** **(Rorippa)**

10a. Sepals 2; petals 5; stamens 5. **Portulacaceae** **(Montia)**

11. Leaves opposite; sepals 3; petals 3; stamens 3. **Elatinaceae** **(Elatine)**

11a. Leaves whorled, specialised insect traps present; sepals 5; petals 5; stamens 5; submerged free-floating aquatic. **Droseraceae** **(Aldrovanda)**

12. Petals present and at least partially fused. **13.**
12a. Petals absent. **25.**

13. Woody trees or shrubs, mangroves. **14.**
13a. Herbs or scramblers, not woody. **15.**

14. Leaves opposite. **Avicenniaceae** **(Avicennia)**

14a. Leaves irregular. **Myrsinaceae** **(Aegicerus)**

15. Leaves opposite or whorled or all basal on mature plants. **16.**
15a. Leaves alternate, immature plants may be tufted. **21.**

16. Flowers regular (actinomorphic); leaf blade well developed, shortly petiolate or sessile. **Lythraceae** **(Lythrum)**

16a. Flowers irregular (zygomorphic) though sometimes only slightly so, but then the leaf blade usually not well developed, passing gradually into the elongated petiole. **Scrophulariaceae 17.**

17. Stamens 2.   **18.**
17a. Stamens 4.   **19.**

18. Flowers solitary; calyx 3-lobed.   *Glossostigma diandrum*
18a. Flowers in axillary racemes; calyx 4-lobed.   *Veronica*

19. Stigmas 2, free or mostly free.   *Bacopa*
19a. Stigma 1, entire or lobed.   **20.**

20. Stigma spathulate, entire, longer than the stamens.
  Glossostigma elatinoides
20a. Stigma sub-capitate, entire or slightly 2-lobed, subequal to the
stamens.   *Limosella*

21. Specialised insect traps developed on submerged divided leaves; plants
free-floating, usually below the surface, inflorescence emergent.
  Lentibulariaceae
  (Utricularia)
21a. No specialised insect traps present; plants rooted, leaves floating or
emergent; inflorescence emergent.   **22.**

22. Petals united for most of their length, 5 lobes at tip; fruit with 4 or
fewer seeds.   **Convolvulaceae**
  (Ipomoea)
22a. Petals fused only at base with 5 lobes longer than the tube; fruit
3—many-seeded.   **23.**

23. Calyx 2-lobed; capsule usually 3-seeded.   **Portulacaceae**
  (Montia)

23a. Calyx 5-lobed; capsule usually many-seeded.
  Menyanthaceae 24.

24. Inflorescence of few—many pedicellate flowers from a short common
axis much shorter than the pedicels; fruit dehiscing irregularly.
  Nymphoides
24a. Inflorescence an open panicle with the main axis very much longer
than the individual floral pedicels; fruit a capsule with 4 distinct valves.
  Villarsia

25. Woody trees or shrubs.   **26.**
25a. Herbs.   **29.**

26. True leaves present and blade well developed but plants
deciduous.   **27.**
26a. True leaves absent.   **28.**

27. Winter-deciduous trees or large shrubs.   **Salicaceae**
  (Salix)

27a. Permanently deciduous shrubs with tangled stems, leaves present
only on new growth.   *Muehlenbeckia*

28. Photosynthetic tissue in non-articulated stems; shrubs.
Muehlenbeckia

28a. Photosynthetic tissue in cylindrical ultimate articulated branchlets with a ring of small membranous "teeth" at each node; trees or shrubs.
Casuarinaceae
(Casuarina)

29. Leaves alternate; plants with sheathing stipules (ochreae).
Polygonaceae 30.

29a. Leaves opposite or whorled; plants without sheathing stipules.
32.

30. Perianth segments 6.
Rumex

30a. Perianth segments 5.
31.

31. Flowers unisexual; shrub with intricately branched stems, often leafless or new growth only with leaves.
Muehlenbeckia

31a. Flowers bisexual; robust herbs, ± woody at the base, always leafy.
Polygonum

32. Leaves whorled, much divided; plants submerged, mostly free-floating; flowers submerged.
Ceratophyllaceae
(Ceratophyllum)

32a. Leaves opposite, entire; plants floating, submerged or emergent.
33.

33. Leaves less than 1.5 cm long; flowers without perianth, unisexual.
Callitrichaceae
(Callitriche)

33a. At least some leaves more than 1.5 cm long; perianth segments 5, flowers bisexual.
Amaranthaceae
(Alternanthera)

34. Petals free.
35.

34a. Petals at least partially fused or petals absent.
41.

35. Trees with papery bark.
Myrtaceae
(Melaleuca)

35a. Herbs or shrubs.
36.

36. Leaves all basal, arising from either a corm or horizontal rhizome; leaves floating or emergent, ± round, simple, clearly differentiated from the petiole; sepals 4, 5, or 6; petals numerous; stamens numerous.
Nymphaeaceae 37.

36a. Leaves scattered along a stem, if arising from a horizontal rhizome then not floating nor clearly differentiated into blade and petiole and/or flowers with 5 sepals, 5 petals and 5 stamens.
38.

37. Sepals 4; ovary semi-inferior; petals spreading, showy.

Nymphaea

37a. Sepals 5 or 6; ovary superior; petals neither showy nor spreading.

Nuphar

38. Sepals 4; petals 4; stamens 8. **Onagraceae (Ludwigia)**

38a. Sepals 5; petals 5; stamens 5. **Apiaceae 39.**

39. Leaves linear, septate. *Lilaeopsis*
39a. Leaves peltate or compound, neither linear nor septate. **40.**

40. Leaves peltate. *Hydrocotyle*
40a. Leaves compound. *Sium*

41. Petals absent. **Haloragaceae (Myriophyllum)**

41a. Petals present and at least partially fused. **42.**

42. Trees with papery bark; inflorescence a spike. **Myrtaceae (Melaleuca)**

42a. Herbs; inflorescence a head of small flowers. **Compositae 43.**

43. Plants erect; flowers blue. *Aster*
43a. Plants prostrate or scrambling; flowers yellow or white. **44.**

44. Leaves opposite. *Eclipta*
44a. Leaves alternate. *Cotula*

KEY 2
Primarily vegetative characters

1a. Trees, if shrubs then growing on margins of estuaries or along areas of tidal influence. **2.**
1b. Herbs or shrubs rooted in mud, leaves mostly emergent, flowers emergent, rarely with some leaves floating. **6.**
1c. Herbs, plants free-floating, on or below the surface. (See also 1f.) **37.**
1d. Herbs, plants rooted in mud, leaves floating, flowers (if present) emergent, often some leaves submerged. **49.**
1e. Herbs, plants normally rooted in mud, most leaves completely submerged, some leaves may be emergent or floating when plant flowering, sometimes flowering above surface. Detached plants may survive for a considerable period and appear to be floating. **58.**
1f. Herbs or shrubs, plants normally not rooted in water, subjected to periodic inundation or plants normally rooted on bank or margins and growing out over water surface. **83.**

2a. Trees with papery bark. *Melaleuca*
2b. Trees or shrubs with grey non-papery bark. **3.**

3a. True leaves absent, photosynthetic tissue in cylindrical ultimate branchlets. *Casuarina*
3b. True leaves present (unless in winter for deciduous trees). **4.**

4a. Trees deciduous growing near fresh water, not in areas subjected to tidal inundation. *Salix*
4b. Trees evergreen, growing in areas subject to tidal inundation. **5.**

5a. Leaves opposite. *Avicennia*
5b. Leaves alternate or irregular. *Aegiceras*

6a. Plants grass-like *or* plants with long, narrow leaves, many times longer than wide. **7.**
6b. Plants not grass-like; leaf blades only, at most, a few times longer than wide. **16.**
7a. Plants grass-like; leaves flat *or* cylindrical; flowers without true perianth, *and* subtended by one or more bracts (glumes, lemmas or paleas) *and* organised into characteristic spikelets. **8.**
7b. Plants not grass-like *or,* if grass-like, then flowers with a true perianth (hard and scarious or petaloid) *or* flowers not arranged in spikelets. **9.**

8a. Flowers subtended by a single bract and variously arranged into spikelets; leaves mostly with closed sheath and mostly without ligules (*Gahnia* has ligules). **Cyperaceae** (see separate generic key)

8b. Flowers enclosed by a lemma and palea and these variously arranged into characteristic spikelets which are mostly subtended by sterile glumes; leaves usually with open sheaths and ligules present although often much reduced. **Gramineae** (see separate generic key)

9a. Flowers or flower clusters in an open panicle, perianth dry and scarious; leaves or pseudo-leaves often cylindrical, sometimes flat. *Juncus*

9b. Flowers with various forms of perianth, but not dry and scarious *and* arranged in an open panicle. **10.**

10a. Stems and often also basal leaf margins hairy, plants bifacial. *Philydrum*

10b. Neither stems nor leaf-bases hairy. **11.**

11a. Leaves septate cylindrical constricted slightly at each septum, plants stoloniferous. *Lilaeopsis*

11b. Leaves not septate. **12.**

12a. Leaves inflated and spongy when fresh. **13.**

12b. Leaves neither inflated nor spongy. **14.**

13a. Leaves inflated, elliptical in cross-section. *Triglochin*

13b. Leaves inflated, triangular in cross-section. *Maundia*

14a. Inflorescence open, flowers on clearly visible pedicels, inflorescence axis visible between flowers. *Sagittaria*

14b. Flowers arranged in dense elongated or globular heads; individual flower pedicels not visible; inflorescence axis visible between clusters but not between individual flowers. **15.**

15a. Inflorescence dense, spike-like; male flowers above, female below, ± separated by a piece of stem. *Typha*

15b. Inflorescence of dense globular clusters arranged along the axis; upper clusters male, lower clusters female. *Sparganium*

16a. Leaves compound and/or lobed. **17.**

16b. Leaves simple or absent. **19.**

17a. Leaves lobed or palmately divided, rarely pinnate; flowers yellow, solitary. *Ranunculus*

17b. Leaves pinnate, flowers white in a raceme or umbel. **18.**

18a. Crushed leaves with a "carroty" smell; plants only rarely growing in water; flowers white; inflorescence an umbel.   ***Sium***
18b. Crushed leaves with a "cress" smell; plants mostly growing in water; flowers white; inflorescence a raceme.   ***Rorippa***

19a. Leaves all basal.   **20.**
19b. Leaves arranged along the stem (basal leaves may also be present), or leaves absent.   **28.**

20a. Leaves almost circular, with or without a radial slit, petiole attached near centre.   **21.**
20b. Leaves not almost circular and/or petiole attached near the margin.
                                                                  22.

21a. Leaves peltate, without a radial slit.   ***Nelumbo***
21b. Leaves with a radial slit.   **Nymphaeaceae**

22a. Plants without horizontal rhizomes or stolons.   **23.**
22b. Plants with horizontal rhizomes or stolons.   **25.**

23a. Leaves without an obvious midrib, longitudinal veins are similar, arising from blade-petiole junction.   ***Caldesia***
23b. Leaves with an obvious midrib, thicker and larger than the lateral vascular bundles; lateral veins arising from base only, or from the base and further along the vein.   **24.**

24a. Lateral veins arising only from blade/petiole junction.
                                                            Damasonium
24b. Lateral veins arising from blade/petiole junction and from further along the midrib, especially on larger mature leaves.   ***Alisma***

25a. Leaf blades less than 10 cm long; plants less than 30 cm tall when mature.   ***Hydrocharis***
25b. Leaf blade more than 10 cm long; mature plants usually more than 50 cm tall.   **26.**

26a. Expanded leaf blades with longitudinal midrib only, lateral veins parallel to each other, arising at an angle to the midrib, all along the midrib, and curving towards the margin; flowers blue.   ***Thalia***
26b. Expanded leaf blades with or without a well defined midrib but with several other longitudinal veins arising from near the blade/petiole junction; flowers white or blue.   **27.**

27a. Longitudinal veins usually less than 10 (always less than 20); midrib well defined; cross-veins obvious; flowers white.   ***Sagittaria***
27b. Longitudinal veins numerous, usually more than 20; midrib not well defined; cross-veins not visible; flowers blue.   ***Pontederia***

28a. Leaves with sheathing stipules (ochreae); leaves alternate or absent.
Polygonaceae 29.
28b. Leaves opposite or alternate, without sheathing stipules but leaf
bases may be winged and half encircling the stem.   **31.**

29a. Plants shrubby, with numerous stiff, intertwining branches; leaves
absent, or present only on new growth.   *Muehlenbeckia*
29b. Plants ± herbaceous; leaves always well developed and obvious;
plants not with numerous stiff intertwining branches.   **30.**

30a. Inflorescence large, usually at least half the height of the plant;
perianth segments 6, 3 small outer segments.   *Rumex*
30b. Inflorescence small, usually very much less than half the height of
the plant, often axillary; perianth segments 5, equal.   *Polygonum*

31a. Leaves opposite.   **32.**
31b. Leaves alternate.   **34.**

32a. Plants small, usually much less than 30 cm high; flowers white;
leaves united at base, ± succulent.   *Crassula*
32b. Plants taller than 30 cm; flowers pink or blue; leaves not united at
base, not succulent.   **33.**

33a. Inflorescence a terminal spike, leaf margins entire.   *Lythrum*
33b. Inflorescence an axillary raceme, leaf margins usually serrulate.
Veronica

34a. Leaves without obvious petioles, either the lamina sessile or the leaf
linear to spathulate. Flowers pink, blue, or purple or white.   **35.**
34b. Leaves with obvious petioles. Flowers yellow.   **36.**

35a. Leaf lamina sessile, without a winged base. Flowers blue, pink or
purple.   *Lythrum*

35b. Leaves linear or spathulate with a winged base half encircling the
stem. Flowers white.   *Montia*

36a. Flowers in a large panicle; leaves smooth, glabrous, without stipules
at base.   *Villarsia*
36b. Flowers solitary, axillary; leaves either smooth with globular stipules
at base and/or stems and leaves hairy.   *Ludwigia*

37a. All or most of the vegetative parts of the plant, except for the roots or
root-like structures, above the water surface.   **38.**
37b. All or most of the vegetative parts of the plants below the water
surface.   **44.**

38a. Leaves whorled, plant without a horizontally elongated leafy axis.
39.
38b. Plants with an elongated leafy axis, but plants may be very small.
42.

39a. Plants very small usually less than 1 cm (always less than 2 cm) long.
Lemnaceae
39b. Plants always bigger than 5 cm in at least one dimension.	**40.**

40a. Leaves divided; ferns, flowers never produced, reproduction by spores.	*Ceratopteris*
40b. Leaves simple; flowering plants.	**41.**

41a. Petioles large, inflated, obvious; flowers blue, obvious.
Eichhornia
41b. Petioles not inflated, leaves tapering gradually to base; flowers not obvious, white to light green.	*Pistia*

42a. Plants small, less than 5 cm long; leaves alternate, usually less than 2 mm long.	*Azolla*
42b. Plants more than 5 cm long; leaves opposite.	**43.**

43a. Leaf surface without hairs.	*Alternanthera*
43b. Leaf surface covered with hairs.	*Salvinia*

44a. Plants without an elongated vegetative axis, not clearly differentiated into leaves and stems.	**45.**
44b. Plants with an elongated vegetative axis, clearly differentiated into leaves (often very specialised) and stems.	**47.**

45a. Plants composed of a repeatedly dichotomously branched thallus; roots absent; plants never flowering, reproducing by spores but aquatic forms normally sterile.	*Riccia*
45b. Plants not composed of a repeatedly dichotomously branched thallus; if plants thallose then branching lateral; roots present or absent; plants reproducing by flowers or spores.	**46.**

46a. Plants thallose, branching lateral; flowers minute, dorsal; roots present or absent.	*Lemnaceae*
46b. Plants various, never thallose; often filamentous, one or few-cellular, tubular or spherical; flowers never produced, reproduction by spores; roots never present.	(*see separate key page 13*)	**Algae**

47a. Leaves alternate but 2 lobed from base; specialised insect traps present.	*Utricularia*
47b. Leaves whorled; specialised insect traps present or absent.	**48.**

48a. Specialised insect traps present and obvious.	*Aldrovanda*
48b. Specialised insect traps absent; leaves dichotomously forked, segments linear-cylindrical.	*Ceratophyllum*

49a. Petioles attached near middle of blade; blades ± circular with one or more radial slits.	**50.**
49b. Petioles attached on the margins of the blade; blades broad-lanceolate, elliptic or ovate.	**56.**

50a. Leaves entire.   **51.**
50b. Leaves with one or more radial slits.   **53.**

51a. Leaves all basal; petioles with stiff, short spines.   ***Nelumbo***
51b. Leaves arranged along the stem; petioles without spines, often
covered with mucilage.   **Cabombaceae 52.**

52a. Submerged leaves present, much divided, quite different from
floating leaves; floating leaves small, usually less than 5 cm long.
                                                                          Cabomba
52b. Submerged leaves absent, all leaves the same, large, usually more
than 5 cm long.   ***Brasenia***

53a. Leaves with 4 radial slits, i.e. composed of 4 leaflets, clover-like.
                                                                          Marsilea
53b. Leaves with one radial slit.   **54.**
54a. Fresh petioles with obvious transverse septa; leaves arising from base
and/or in groups along the stem; flowers yellow, 2-5 cm across.
                                                                          Hydrocleys
54b. Petioles without obvious transverse septa; if flowers yellow then
very much smaller.   **55.**

55a. Leaves all basal, arising from a corm or buried horizontal rhizome;
petals numerous.   **Nymphaeaceae**
55b. Leaves arising along floating horizontal stolon; petals 5.
                                                                          Nymphoides

56a. Leaves attached along trailing stem; leaves dimorphic, submerged
leaves different from floating leaves.   ***Potamogeton***
56b. Leaves all basal; leaves either all similar or dimorphic.   **57.**

57a. Plants forming deeply buried tubers; flowers numerous per
inflorescence.   ***Aponogeton***
57b. Plants not forming tubers, root system superficial; flowers solitary.
                                                                          Ottelia

58a. Submerged leaves divided or deeply lobed.   **59.**
58b. Submerged leaves simple, entire.   **61.**

59a. Submerged leaves pinnately divided; emergent leaves pinnately
divided or entire.   ***Myriophyllum***
59b. Submerged leaves dichotomously or trichotomously divided.
                                                                                   60.

60a. Submerged leaves opposite or whorled; floating leaves (if present)
peltate.   ***Cabomba***
60b. Submerged leaves alternate.   ***Ranunculus***

61a. Leaves all basal.   **62.**
61b. Leaves arising along stem.   **69.**

62a. Plants of marine, estuarine or brackish habitats.   **63.**
62b. Plants of freshwater habitats.   **66.**

63a. Leaves elliptical.   *Halophila*
63b. Leaves elongated, linear, grass-like.   **64.**

64a. Leaves less than 0.4 cm wide.   *Ruppia*
64b. Leaves more than 0.4 cm wide.   **65.**

65a. Leaves more than 1 cm wide.   *Posidonia*
65b. Leaves 0.5-1 cm wide.   *Zostera*

66a. Leaf blades stiff $\pm$ cylindrical; plants reproducing by spores, flowers never produced.   *Isoetes*
66b. Leaf blades soft, flattened, grass-like; plants flower when mature but often absent in submerged immature growth.   **67.**

67a. Stolons present.   *Vallisneria*
67b. Stolons never produced.   **68.**

68a. Plants producing a well-buried corm, leaves arising from the apex of the corm.   *Aponogeton* (immature)

68b. Plants not producing a corm, roots superficial, arising from near leaf bases.   *Ottelia* (immature)

69a. Plants not producing flattened leaf blades; branches in whorls; secondary branches $\pm$ in whorls; often encrusted with lime; plants not flowering, reproducing by spores.   **Characeae 70.**
69b. Plants producing flattened leaf blades but often flat, narrow, linear and grass-like; rarely lime encrusted; mature plants flowering.   **71.**

70a. Secondary branches with whorls of branches.   *Nitella*
70b. Secondary branches without whorls of branches.   *Chara*

71a. Leaves alternate.   **72.**
71b. Leaves opposite or whorled.   **78.**

72a. Leaves with no sheath below or stipular sheath above junction of blade and stem; leaf bases not stem-clasping; blades recurved.   *Lagarosiphon*
72b. Leaves with either a leaf-sheath below and/or stipular sheath above junction of blade and sheath *or* leaf bases stem-clasping.   **73.**

73a. Leaves with stipular sheath (or ligule) extending above junction of blade and sheath; $\pm$ sheath below *or* base of blade stem-clasping.   **74.**

73b. Leaves with sheath below junction of blade and sheath, no stipular sheath or ligule above but auricles may be present.   **75.**

74a. Leaf blades mostly more than 1 mm wide. *(Potamogeton pectinatus may have leaves less than 1 mm wide)*; perennials with rhizomes, often with tough stems; leaves with stipular sheaths *or* base of blade stem clasping.   ***Potamogeton***
74b. Leaf blades mostly less than 1 mm wide; usually delicate annuals (in New South Wales); stems weak; leaves with stipular sheaths.
Zannichelliaceae *(Zannichellia, Lepilaena)*

75a. Plants marine or estuarine only.   **76.**
75b. Plants of freshwater or brackish (*Ruppia* rarely estuarine) habitat.
77.

76a. Plants with leaves mostly basal in vegetative stage, flowering culms with leaves along stem; middle of stem and rhizome internodes with 2 cortical vascular bundles.   ***Zostera***
76b. Plant with leaves on erect fertile and sterile stems except during winter; middle of stem and rhizome internodes with 4 or more cortical vascular bundles.   ***Heterozostera***

77a. Leaf blade base with two short auricles.   ***Ruppia***
77b. Leaf blade base without auricles.   ***Scirpus***

78a. Leaves opposite; leaf blade expanded.   **79.**
78b. Leaves whorled or leaves linear, without expanded blade.   **80.**

79a. Leaves with a midrib, lateral veins branching off at acute angle to margin, each ending in a minute gland; stipules at leaf base.
Elatine
79b. Leaves with one or more longitudinal, more or less parallel veins, arising from base of central vein and re-uniting below tip; no marginal glands; stipules absent.   ***Callitriche***

80a. Leaf bases with a short, expanded sheath-like base.   ***Najas***
80b. Leaves sessile, whorled, without expanded sheath-like bases.
81.

81a. Leaf margins strongly serrate; leaves in whorls of 3-8; whorls usually well spaced along stem.   ***Hydrilla***
81b. Leaf margins entire or minutely serrate; leaves in whorls of 3-6; whorls usually crowded, especially near tip of stem.   **82.**

82a. Leaves mostly in whorls of 3.   ***Elodea***
82b. Leaves mostly in whorls of 4-5.   ***Egeria***

83a. Plants grass-like *or* plants with long, narrow leaves many times longer than wide.   **84.**
83b. Plants not grass-like; leaf blades only at most a few times longer than wide.   **87.**

84a. Flowers with true perianth; perianth members dry and scarious; leaves or pseudo-leaves round or flat. **Juncaceae**
(Juncus)
84b. Flowers without a true perianth, subtended by one or more bracts and variously arranged into characteristic spikelets. **85.**

85a. Carpels 2 or more, free, inserted at different levels on a gynophore; plants small; leaves filiform; in or near perennial or ephemeral swampy areas. **Centrolepidaceae**
(Centrolepis)
85b. Carpels connate, apparently single; leaves flat or cylindrical; plants small to large, habitats various. **86.**

86a. Flowers subtended by a single bract and variously arranged into spikelets; leaves mostly with closed sheath and mostly without ligules (*Gahnia* has ligules); leaves flat or cylindrical. **Cyperaceae**
(see separate generic key)
86b. Flowers enclosed by a lemma and palea and these variously arranged into characteristic spikelets which are mostly subtended by sterile glumes; leaves usually with open sheaths and ligules present although often much reduced. **Gramineae**
(see separate key)

87a. Shrubs with divaricate green branches; leaves permanently deciduous, young branches only with leaves. **Muehlenbeckia**
87b. Herbs or if shrubby, then plants never permanently deciduous. **88.**

88a. Leaves compound or deeply (more than half the width of entire part) divided. **89.**
88b. Leaves simple. **92.**

89a. Leaves with bases sheathing the stem, i.e. the leaf bases extend more than half way around the circumference of the stem. **90.**

89b. Leaves without sheathing bases but petiole base may be expanded. **91.**

90a. Leaves compound; flowers white; crushed leaves with carroty smell. **Sium**
90b. Leaves lobed; flowers yellow. **Cotula**

91a. Leaves pinnate, mature leaves usually with three or more pairs of leaflets, flowers white. **Rorippa**
91b. Leaves variously lobed, divided or rarely pinnate but then usually with only two pairs of leaflets. Flowers mostly yellow, when white leaves finely di- or trichotomously divided into fine segments. **Ranunculus**

92a. Leaves all basal.   **93.**
92b. Leaves arranged along a prostrate or variously erect stem.   **95.**

93a. Plants small with small fibrous roots; leaves less than 10 cm long, often less than 5 cm long.   ***Glossostigma, Limosella***
93b. Plants with thick tuberous rootstocks and leaves more than 10 cm long.   **94.**

94a. Plants with large, glossy sagittate leaves; inflorescence subtended by a large showy white spathe.   ***Zantedeschia***
94b. Leaves not glossy, more or less lanceolate or broad lanceolate, not sagittate; inflorescence red or green, not subtended by a showy spathe but not present when leaves only basal.   ***Rumex***

95a. Leaves opposite.   **96.**
95b. Leaves alternate.   **103.**

96a. Leaves with obvious petioles.   **97.**
96b. Leaves without obvious petioles, either sessile or the blade tapering gradually to the junction with stem.   **99.**

97a. Leaves usually more than 5 cm long; plants growing on bank or growing out over water; inflorescence white, axillary.
                                                                     Alternanthera
97b. Leaves usually less than 5 cm long; inflorescence or flowers axillary, white, yellow or inconspicuous.   **98.**

98a. Leaf margin entire; flowers solitary, axillary, inconspicuous.
                                                                           Elatine
98b. Leaf margins serrate; inflorescence axillary, yellow or white.
                                                                           Eclipta

99a. Leaf blades tapering gradually to junction with stem.   **100.**
99b. Leaves sessile.   **101.**

100a. Sepals and petals absent; flowers unisexual, subtended by bracteoles.   ***Callitriche***

100b. Sepals and petals present; flowers bisexual.   ***Glossostigma***
                                                                          Limosella

101a. Leaves barely flattened, $\pm$ cylindrical, joined together at the base with a small sheath; flowers white.   ***Crassula***

101b. Leaf blades flattened, without sheathing bases; flowers blue, pink, purple or white.   **102.**

102a. Leaves usually less than 2 cm long; flowers solitary, axillary, blue, pink or white.   ***Bacopa***
102b. Leaves usually more than 2 cm long; inflorescence a terminal leafy spike; flowers pink, purple or slightly bluish.   ***Lythrum***

103a. Leaves peltate. *Hydrocotyle*
103b. Leaves not peltate, various. **104.**

104a. Leaves with a sheathing stipular base (ochreae). **105.**
104b. Leaves without ochreae. **106.**

105a. Perianth segments 6, 2 whorls of unequal size. *Rumex*
105b. Perianth segments 5, subequal. *Polygonum*

106a. Stems prostrate or trailing. **107.**
106b. Stems erect or decumbent. **109.**

107a. Leaves with parallel venation and sheathing leaf base; flowers blue. *Monochoria*
107b. Leaves with reticulate venation and with non-sheathing leaf bases; flowers yellow, pink or white. **108.**

108a. Leaf bases hastate, sagittate, cordate or truncate, blade roughly triangular; stipules absent; flowers pink or almost white. *Ipomoea*
108b. Leaf bases attenuate; blades approximately elliptical; 2 small swollen stipules at base of petiole; flowers yellow. *Ludwigia*

109a. Shrubs, often cold deciduous; plants hairy; flowers yellow. *Ludwigia*
109b. Herbs but may be slightly woody at base. **110.**

110a. Flowers in compact heads (a daisy) which are arranged in a terminal leafy panicle; flower heads blue or white; plants usually erect. *Aster*

110b. Flowers solitary, axillary; flowers blue or purple; plants usually decumbent or weakly ascending. *Lythrum*

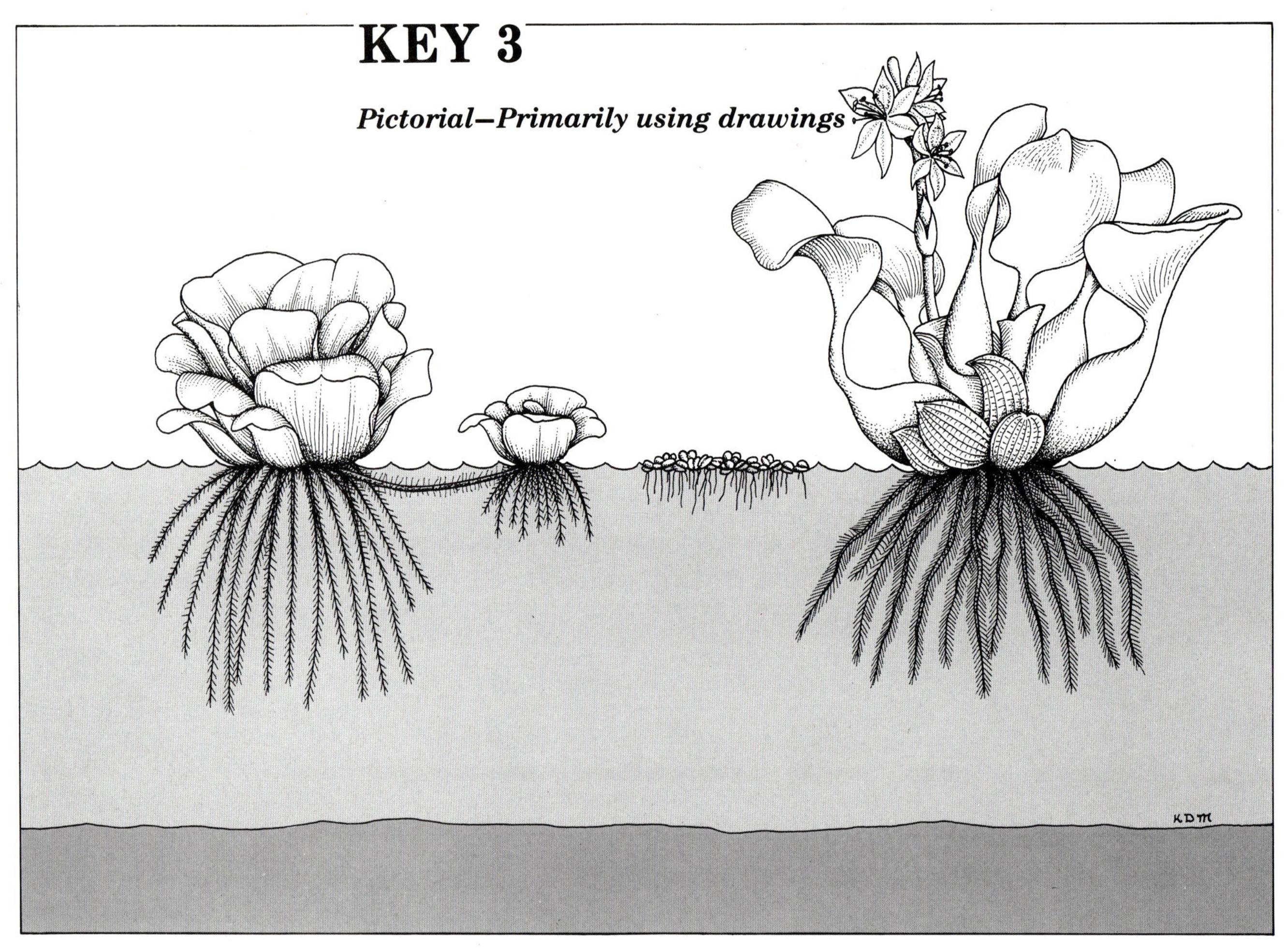

Free Floating—Unattached

Ceratopteris	*339*
Azolla	*71*
Eichhornia	*357*
Lemna	*273*
Pistia	*66*
Salvinia	*403*
Spirodela	*273*
Wolffia	*273*

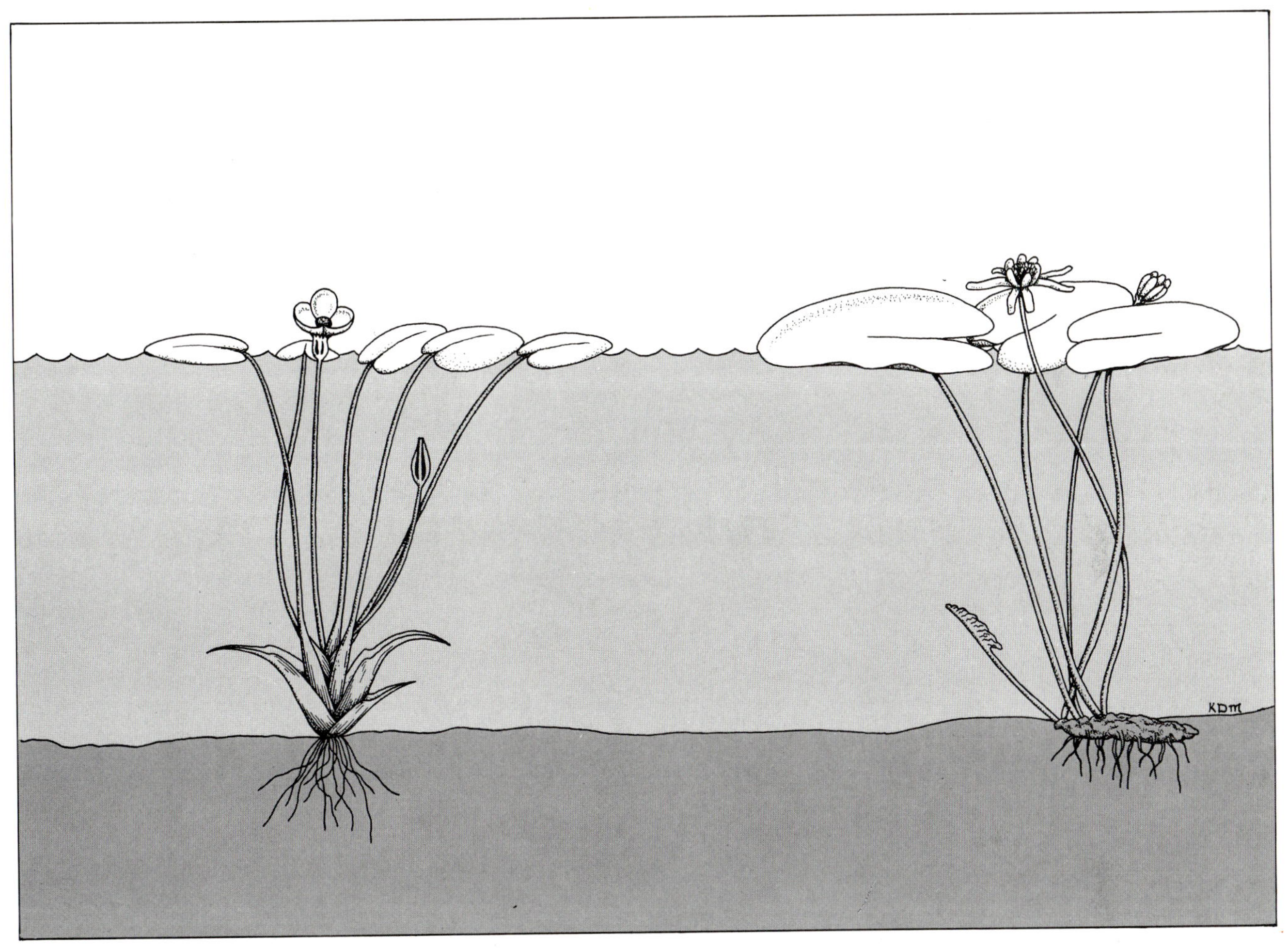

Surface Floating—Attached

Aponogeton	*62*
Brasenia	*79*
Cabomba	*81*
Callitriche	*82*
Damasonium	*49*
Marsilea	*296*
Nuphar	*323*
Nymphaea	*325*
Nymphoides	*301*
Ottelia	*247*
Potamogeton javanicus	*371*
Potamogeton tricarinatus	*378*

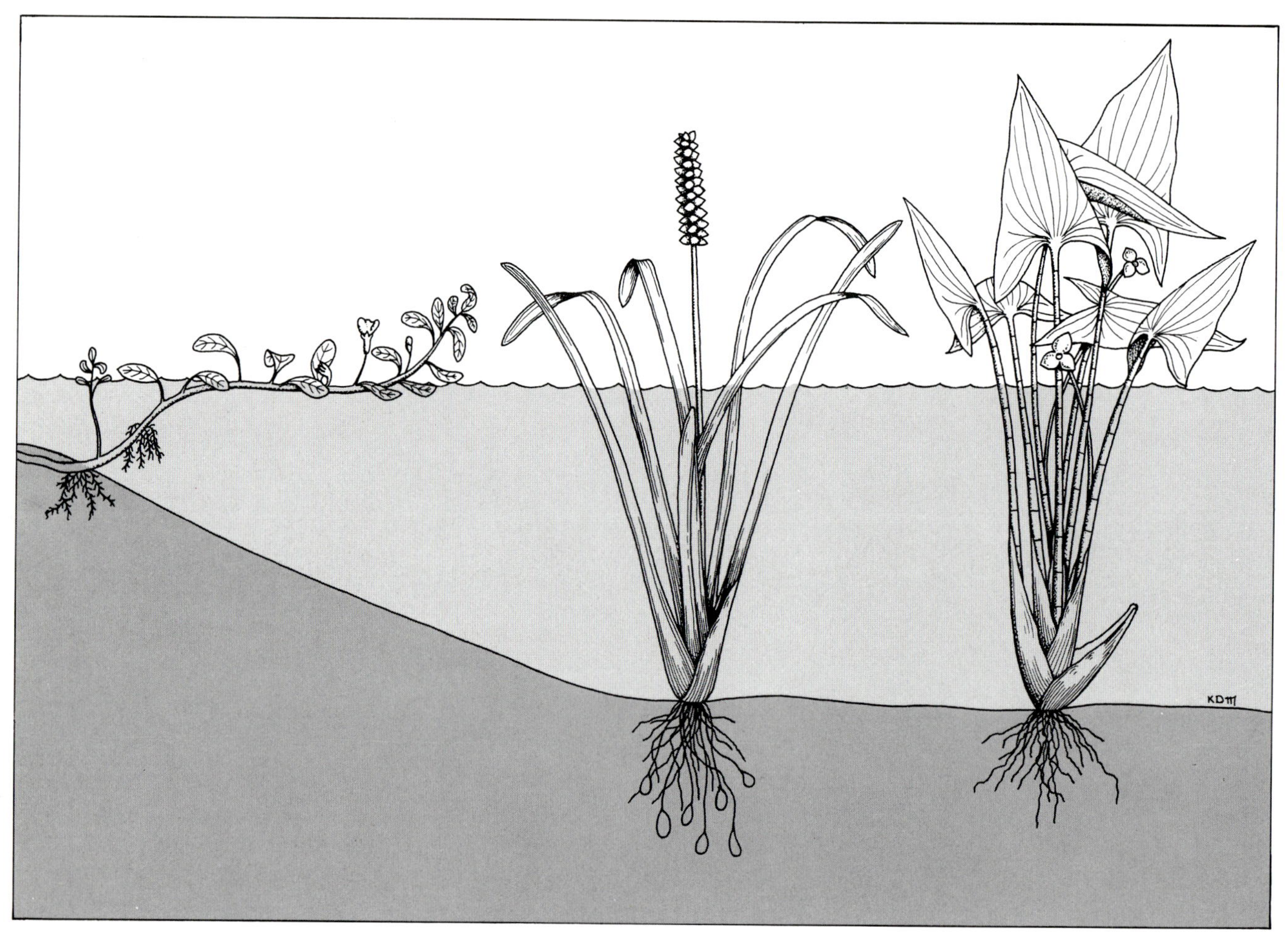

Floating and/or Emergent Leaves

Surface runner			Strap leaves			Broad leaves	
Alternanthera	54		*Maundia*	267		*Damasonium*	49
Ipomoea	99		*Triglochin*	268		*Hydrocharis*	243
Ludwigia	334					*Hydrocleys*	77
Monochoria	361					*Hydrocotyle*	57
Polygonum	345					*Marsilea*	296
						Monochoria	361
						Nelumbo	321
						Nuphar	323
						Sagittaria	50

Mostly Submerged Feathery Leaves—Some Emergent

Some emergent leaves-pinnate		Divided leaves-opposite or in whorls		Divided leaves-alternate on stem	
Myriophyllum		*Aldrovanda*	153	*Ranunculus trichophyllus*	389
M. aquaticum	219	*Ceratophyllum*	89	*Utricularia*	277
M. elatinoides	221	*Ceratopteris*	339		
M. propinquum	223	*Chara*	450		
M. spicatum	225	*Myriophyllum*	229		
M. verrucosum	229	*Riccia*	391		

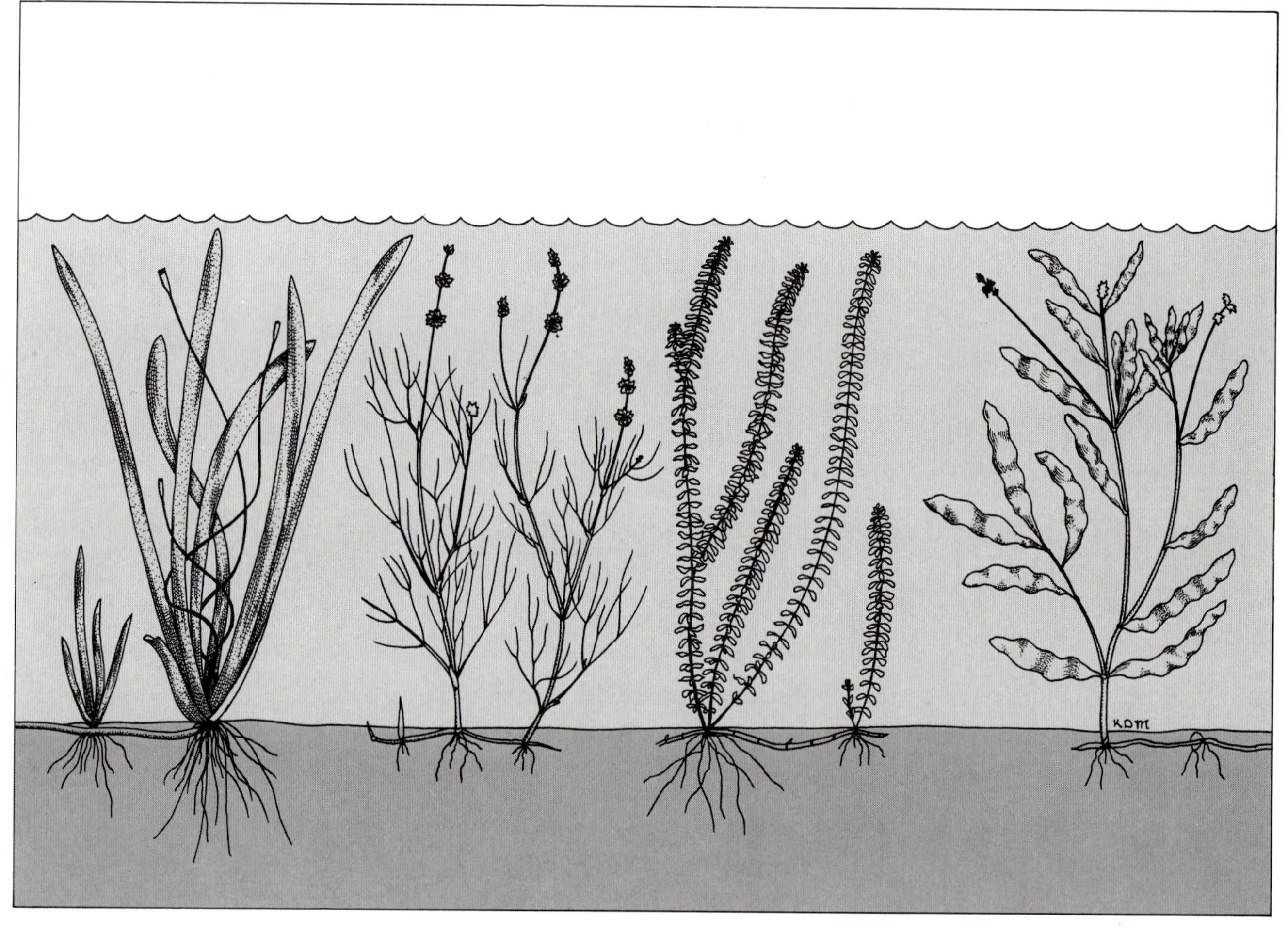

Submerged Leaves

Strap—usually more than 3mm wide		Narrow—usually less than 3mm wide		Entire leaves in whorls		Other submerged leaf types	
Halophila	*236*	*Eleocharis pusilla*	*133*	*Egeria*	*230*	*Crassula*	*100*
Heterozostera	*429*	*Isoetes*	*253*	*Elodea*	*232*	*Elatine*	*155*
Ottelia	*247*	*Lepilaena*	*421*	*Hydrilla*	*238*	*Halophila*	*236*
(submerged juvenile)		*Najas tenuifolia*	*319*			*Lagarosiphon*	*244*
Posidonia	*366*	*Potamogeton*				*Marsilea angustifolia*	*295*
Triglochin	*268*	*pectinatus*	*374*			*Najas marina*	*317*
Vallisneria	*249*	*Ruppia*	*394*			*Potamogeton crispus*	*369*
Zannichellia	*425*	*Scirpus fluitans*	*143*			*Potamogeton*	
Zostera	*426*					*perfoliatus*	*377*
						Potamogeton ochreatus	*372*
						Sium	*61*
						Veronica	*413*

Emergent Plants

Thin, cylindrical or grasslike leaves

Baumea	106	Juncus	225
Brachiaria	169	Leersia	190
Carex	110	Lepironia	141
Cyperus	115	Leptochloa	195
Diplachne	171	Panicum	196
Echinochloa	172	Paspalum	200
Eclipta	97	Phalaris	205
Eleocharis	127	Phragmites	207
Eragrostis	181	Polypogon	211
Gahnia	138	Potamophila	213
Glyceria	182	Pseudoraphis	215
Hemarthria	187	Scirpus	142
Isachne	189	Sparganium	415

Broad leaves

Alisma	46	Pontederia	363
Aster	92	Ranunculus	380
Cotula	95	Rorippa	392
Eclipta	97	Rumex	352
Ludwigia peruviana	337	Sagittaria	51
Lythrum	281	Sium	61
Monochoria	361	Thalia	293
Nuphar	323	Veronica	413
Polygonum	345	Villarsia	309
		Zantedeschia	69

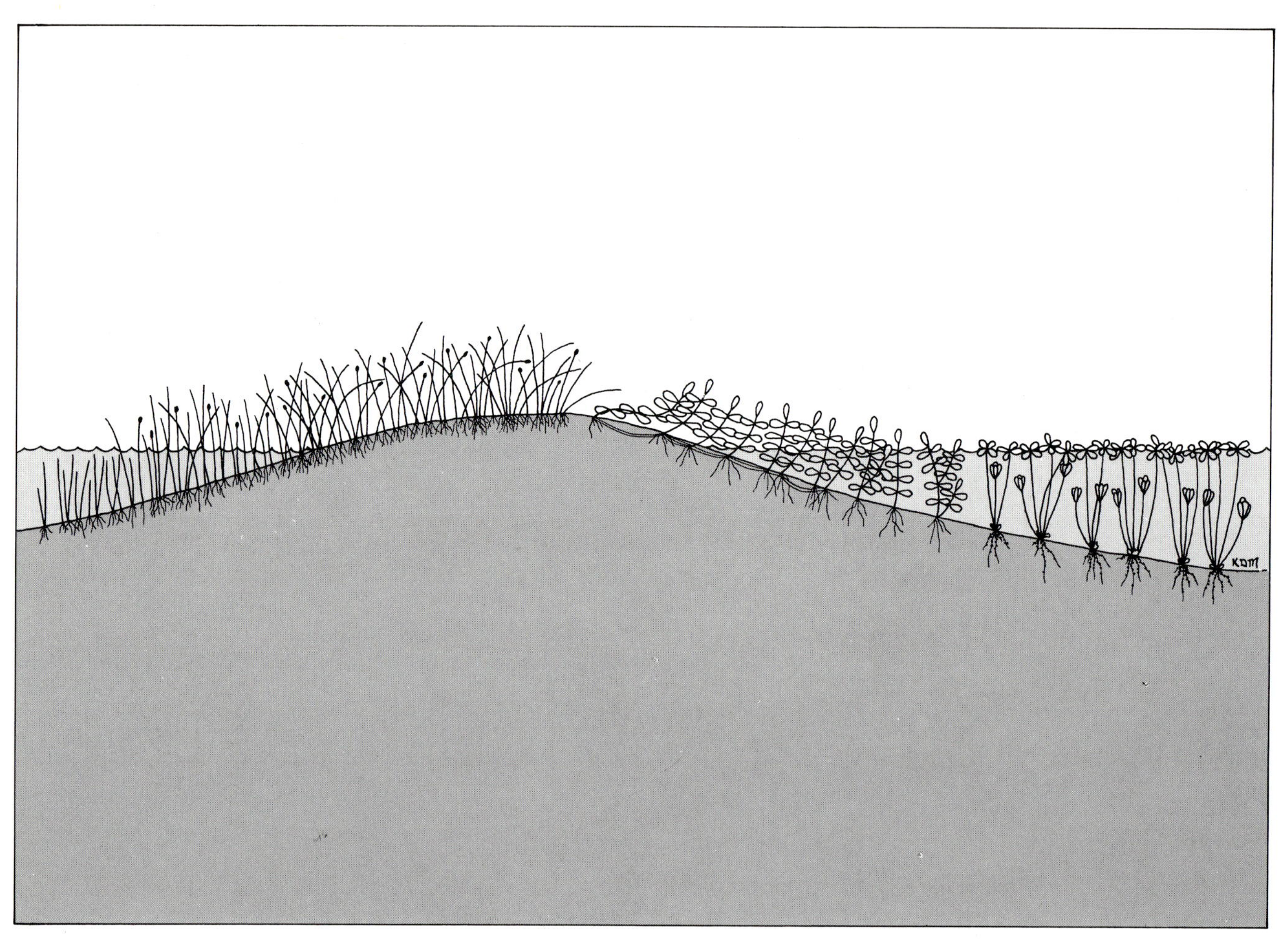

Low Profile—Small—Submerged or on Mud

Bacopa	*407*
Chara	*451*
Crassula	*100*
Eleocharis pusilla	*133*
Glossostigma	*409*
Limosella	*411*
Montia	*365*
Nitella	*451*
Triglochin striata	*271*

Trees

Casuarina	85
Mangroves	285
Melaleuca	314
Salix	399

Shrubs

Ludwigia peruviana	337
Mangroves	285
Muehlenbeckia	343

BOTANICAL SUBDIVISIONS OF NEW SOUTH WALES

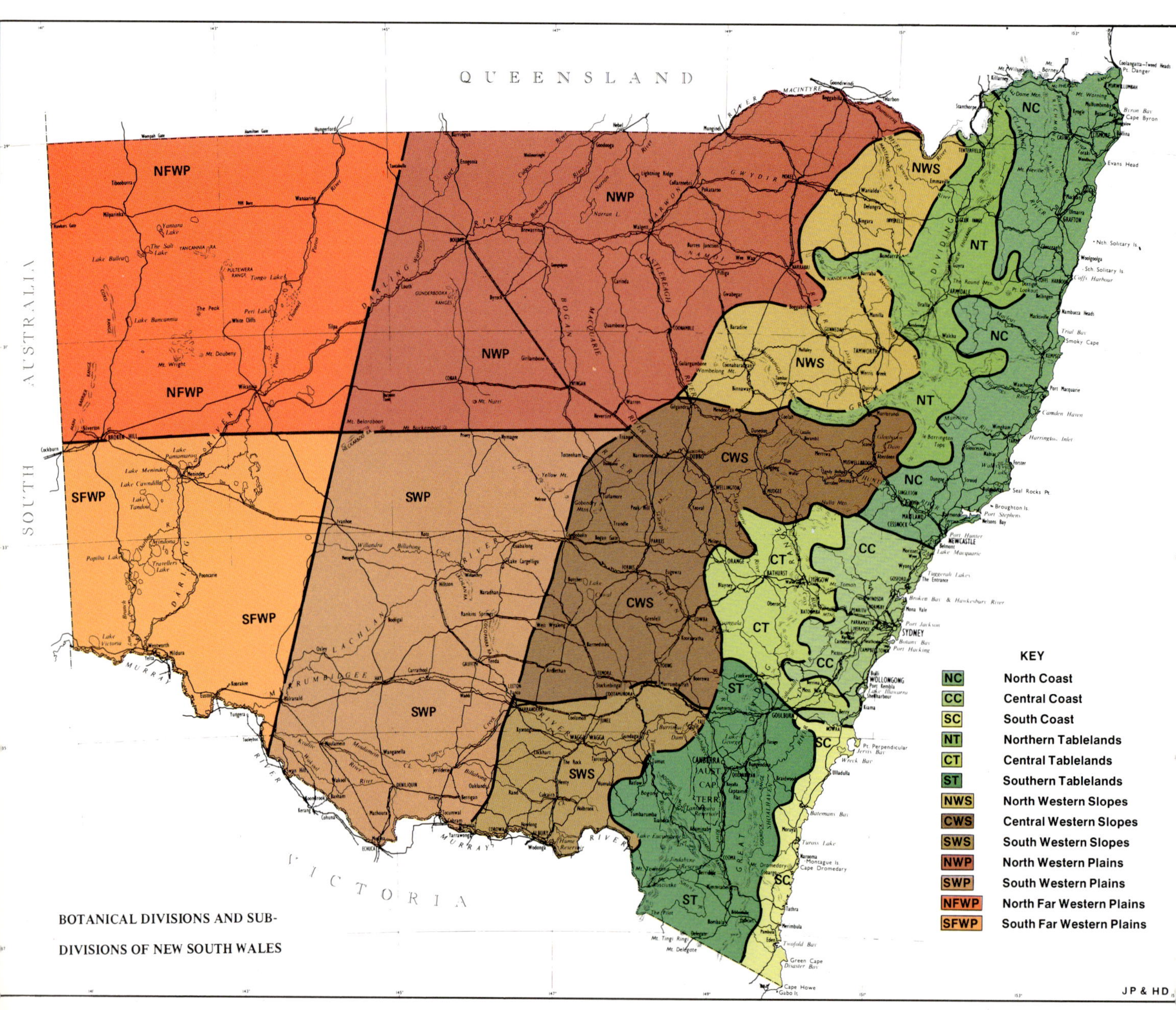

BOTANICAL DIVISIONS AND SUB-

DIVISIONS OF NEW SOUTH WALES

44

WATERPLANTS IDENTIFICATION GUIDE ALISMATACEAE to ZOSTERACEAE

The treatment here is designed to aid in the identification of waterplants. The photographs and descriptions are best used in conjunction with the keys but can be used by themselves to identify plants.

The following botanical texts supply more information on many of the plants.

References for Identification Guide

ASTON, H. I. (1973) "Aquatic Plants of Australia". (Melbourne University Press: Melbourne) 368 pp.

BEADLE, N. C. W., EVANS, O. D. and CAROLIN, R. C. (1972) "Handbook of the Vascular Plants of the Sydney District and Blue Mountains". (A. H. & A. W. Reed: Sydney) 724 pp.

COOK, C. D. K., GUT, B. J., RIX, E. M., SCHNELLER, J. and SEITZ, M. (1974) "Water Plants of the World". (Dr. W. Junk: The Hague) 561 pp.

HARTLEY, W. (1979) "A Checklist of Economic Plants in Australia". (CSIRO, Melbourne) 214 pp.

HARTOG, C. den (1970) "The Sea-grasses of the World". (North-Holland: Amsterdam) 275 pp. + 20 pl.

HOLM, L. G., PLUCKNETT, D. L., PANCHO, J.V., HERBERGER, J. P. (1977) "The World's Worst Weeds". (University Press of Hawaii), 609 pp.

SCULTHORPE, C. D., (1967) "The Biology of Aquatic Vascular Plants". (Edward Arnold: London) 610 pp.

WILLIS, J. H. (1972) "A Handbook to Plants in Victoria". 2nd edition. (Melbourne University Press: Melbourne) 2 Vols.

Water Plantain

Alisma plantago-aquatica

An erect emergent native perennial up to 1.5 metres tall. Leaves basal, blades 10-25 cm long and 7-10 cm wide, with usually 7 prominent parallel veins connected by numerous transverse veins. Petiole up to 80 cm long, flattened on one side and with small wings at the base. Inflorescence an open panicle up to 60 cm long by 40 cm wide on a leafless peduncle. Panicle branches whorled, bracteate. Flowers bisexual, approximately 1 cm in diameter with 3 free, green sepals to 2 mm long and 3 free, pale pink petals about 4 mm long. Stamens 6, carpels numerous (about 20), free, laterally compressed. On maturity the 1-seeded carpels, 2-2.5 mm long, fall singly. *Seed photograph page 437.*

Growth Biology

Flowering takes place in summer after a period of rapid vegetative growth in spring. Flowers open for 1 day, then shrivel. Seeds are buoyant for some time, and this aids in dispersal. Seeds germinate readily in moist soil.

Habitat

Beside creeks, lakes and in swamps in water to 50 cm deep. Also persists in drying mud.

Economic Significance

In rare situations growth is obstructive to water flow, but the plant is generally a desirable part of the aquatic environment. The fruits and leaves are eaten by animals and birds.

Distribution

Widespread but not common in New South Wales.
NC, CC, NT, CT, ST, SWS, Victoria.

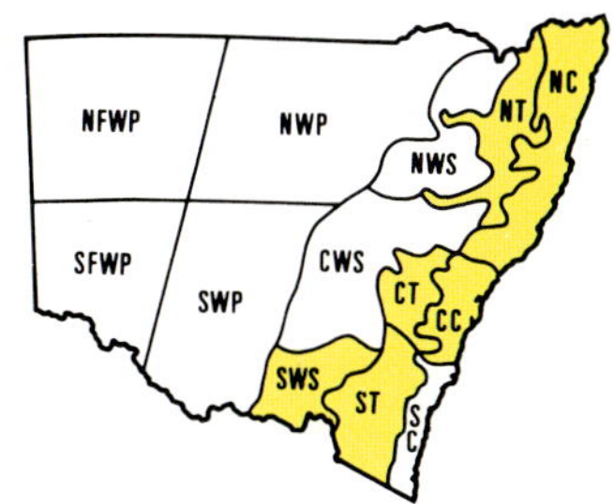

Left: Young plants of *Alisma plantago-aquatica,* one plant with an inflorescence of flowers, buds and young fruits

Right: Flowers and fruits of *Alisma plantago-aquatica*

Starfruit

Damasonium minus

A native erect, emergent annual or short-lived perennial up to 1 metre high with floating and/or emergent leaves. Leaves with a petiole to about 30 cm long, blades 5-10 cm long, 1.5-4 cm wide, cordate, with 3-5 parallel veins connected by numerous finer transverse veins. Inflorescence an open panicle up to 50 cm long with deciduous bracts and borne on a leafless peduncle. Flowers bisexual, about 6 mm in diameter. Sepals 3, green, about 1 mm long. Petals 3, about 3 mm long, white or pale pink. Stamens 6, carpels free, usually 9 (6-10). Fruiting carpels 5-6 mm long, flattened, in the shape of an isosceles triangle, all joined along the shortest side forming a star-shaped fruit 10-12 mm across. Fruiting carpels eventually fall singly. Appearance and habit resemble a smaller, less leafy, *Alisma plantago-aquatica. Seed photograph page 437.*

Growth Biology

Germinates in September-October and flowers in November-December in south-western New South Wales.

Habitat

Stationary or slowly-flowing, ephemeral or permanent fresh water up to 30 cm deep.

Economic Significance

This plant will grow in competition with rice and is a major problem in some aerial and sod-sown crops or where the stand is thin. It seldom grows densely in channel systems, where it is of minor importance.

Distribution

NC, CC, NT, CWS, SWS, NWP, Queensland, Victoria, Tasmania, Northern Territory, South Australia, Western Australia.

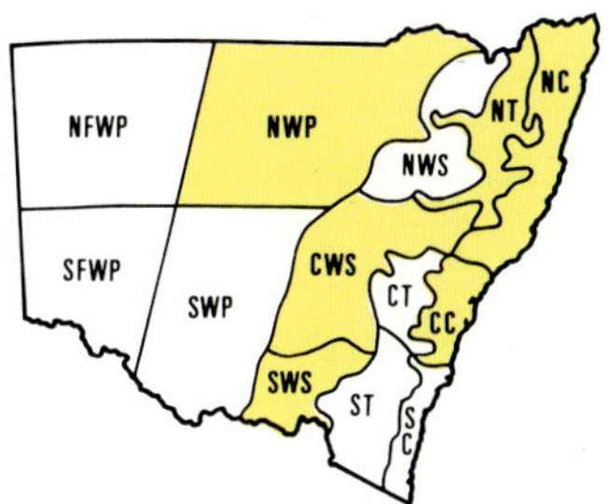

Left: *Damasonium minus,* a common weed of rice crops; (insert) flowers and characteristic star-shaped fruits of *Damasonium minus*

Sagittaria

Rooted, emergent aquatic perennial herbs. Most species are monoecious. Inflorescence mostly erect with 1-8, 3-flowered whorls. Flowers mostly unisexual, with female flowers in the lower whorls of the inflorescence and male flowers in the terminal whorls; occasionally the female flowers with a ring of functional or non-functional stamens. Sepals and petals 3. Sepals reflexed in male flowers. Petals usually white. Stamens whorled, numerous. Carpels numerous, single-seeded.

There are a number of species of Arrowhead; some are used as ornamentals in rock pools, e.g., *S. sanfordii*. One species *(S. sagittifolia)* produces underground tubers which are eaten by ducks, and, in some countries by man. The generic name is derived from the Latin, "sagitta" meaning arrow.

1. Mature emergent leaves linear or lanceolate   **S. graminea**
1a. Mature emergent leaves sagittate   **S. montevidensis**

Sagittaria graminea with broad-linear leaves and inflorescences shorter than the leaves

Sagittaria

Sagittaria graminea ssp. *platyphylla*

An introduced rhizomatous perennial with linear or ovate, rarely cordate leaves up to 25 cm long and 10 cm broad. Submerged leaves often very long (to 50 cm) with no obvious blade development. Female flowers with thickened, recurved pedicels in fruiting stage.

This species is very plastic in leaf size and shape and although most Australian material has been identified as ssp. *weatherbiana,* the fruiting characteristics indicate that it would be better included under ssp. *platyphylla.* Another species, *S. sanfordii,* is similar but has narrower leaves. *Seed photograph page 437.*

Growth Biology

Seeds presumably germinate in spring. Established plants produce leaves in early spring, flowering in late summer, with fruit being produced in autumn or early winter. Spread seems mainly by rhizomes. Capable of forming floating mats which will take root in suitable locations.

Habitat

Shallow water to 50 cm deep, banks of rivers or streams.

Economic Significance

Occasionally a problem, obstructing flows in drainage channels and creeks in some inland localities. Reported a significant weed in north central Victoria. Not generally regarded as troublesome and cultivated widely as an ornamental.

Distribution

CC, ST, Queensland, Victoria, America.

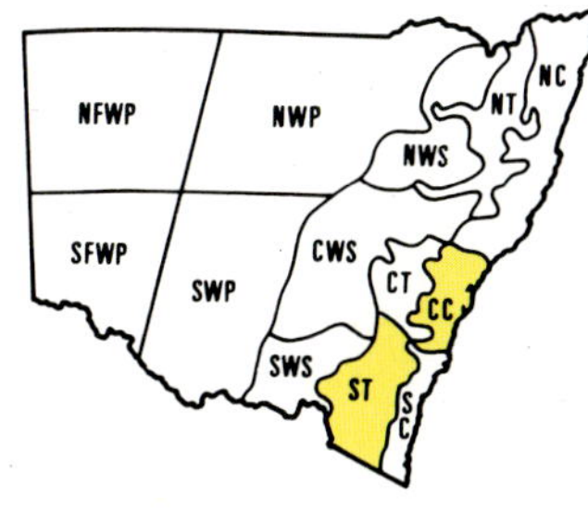

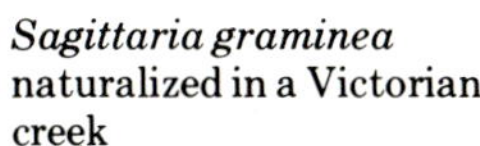
Sagittaria graminea naturalized in a Victorian creek

Arrowhead

Sagittaria montevidensis

An introduced caespitose perennial up to 1 metre high that may behave as an annual, with submerged, floating, but mainly emergent leaves. Emergent leaves strongly sagittate (floating leaves ± sagittate) up to 25 cm long and 20 cm broad, with the lobes up to 15 cm long and 10 cm across. Female flowers with (ssp. *calycina*) or without (ssp. *montevidensis*) a ring of functional stamens. Petals white (ssp. *calcyina*) or white with purple base (ssp. *montevidensis*). *Seed photograph page 437.*

Growth Biology

In areas where it causes most problems, this species usually behaves as an annual, germinating in spring, growing rapidly and flowering in summer.

Habitat

Both subspecies are cultivated and also have become naturalised in drainage channels, rice crops and in some of the now permanent swamp areas created by irrigation works.

Economic Significance

Occasionally a problem, obstructing drainage channels or competing with growing rice.

Distribution

SWP, North and South America and widely cultivated.

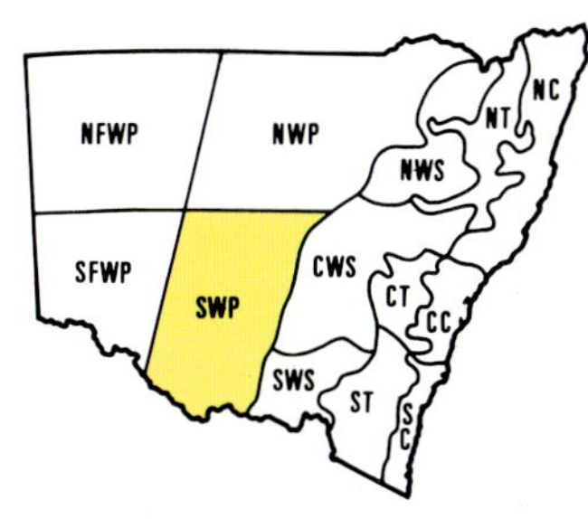

Left: *Sagittaria montevidensis* obstructing a supply channel

Right: *Sagittaria montevidensis* showing its characteristic sagittate leaf on a long petiole

                          Alternanthera philoxeroides

An introduced stoloniferous, glabrous perennial forming dense floating or rooted mats up to 1 metre deep. Stolons to about 10 metres long, rooting at the nodes. Leaves opposite, sessile, entire, 2-7 cm long, 1-2 cm wide. Inflorescence axillary, on a peduncle 1-4 cm long, usually shorter than the leaves, a subglobose head about 1 cm diameter of few to several flowers. Flowers bisexual with 5 white, papery perianth segments about 4 mm long. Stamens 5. Ovary superior, 1-locular.

Growth Biology

Flowering and maximum growth occurs in mid-summer. Viable seed has not been recorded in Australia. Propagates vegetatively, and each node can develop into a new plant. Forms a dense mat of hollow interwoven stems near the surface of the water and may extend 15 metres or more from the shore. A floating mass can support the weight of a man.

Alternanthera philoxeroides producing axillary inflorescences from the base of one of a pair of opposite leaves

Habitat

Static or flowing water, either attached to the bank or free-floating, or away from water in high rainfall areas. The plant responds to high levels of nutrients as occur in the Georges River at Liverpool, New South Wales. Grows primarily in fresh water but will withstand 10 per cent sea-strength salinity or up to 30 per cent in flowing sea water. Leaves are apparently killed by frost, but the species is known to thrive near Albury, New South Wales.

Economic Significance

A. philoxeroides was a major problem in southern United States of America, where an estimated 27 000 hectares are infested. Although the plant has not been eradicated, the problem in that country has been reduced in many localities following the introduction of the insects *Agasicles hygrophila* and *Vogtia malloi* that feed solely on this species. These insects were released in Australia in 1977. Their impact on this plant is outlined in the section on biological control, pages 484-7.

A declared noxious aquatic pest in New South Wales, it is extremely difficult to eradicate by chemical or mechanical means and its spread into other parts of Australia should be prevented. Mechanical removal to clear waterways provides temporary control, but also hastens the spread of the weed.

Distribution

Since 1944 it has been identified at East Maitland, Williamtown, Carrington, Duck Creek at Clyde, Woomargama near Albury, Georges River at Liverpool, and Clarence River at Grafton, all in New South Wales.

NC, CC, SWS.

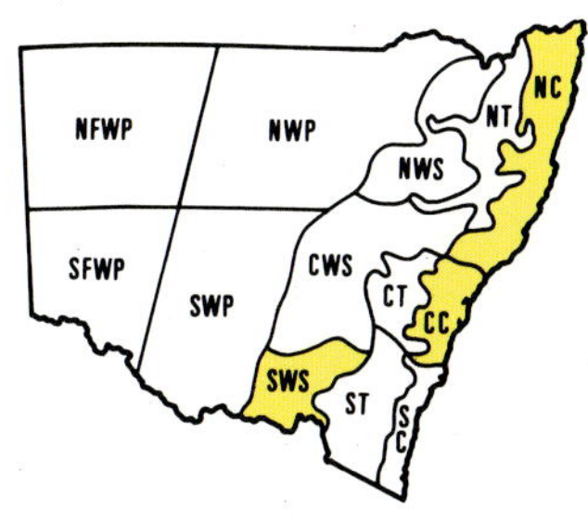

Alternanthera philoxeroides growing out from the banks of the Georges River

CM

Hydrocotyle bonariensis

An introduced creeping rhizomatous or stoloniferous perennial herb with erect, shiny leaves. Stems prostrate, rooting at the nodes and producing either a leaf or a short lateral stem bearing a terminal inflorescence. The lateral stem may or may not have nodes with roots. Leaves peltate, orbicular, up to 6 cm diameter, with crenulate margins, borne on a petiole up to 20 cm long, rarely much longer in thick vegetation. Inflorescence stalk approximately as long as the petiole. Umbels compound, 1-6 cm diameter, yellow.

Habitat

Sand or soils near the coast, but can grow into water and appears able to survive both underwater and floating on the surface while still connected to the shore.

Economic Significance

May be advantageous for stabilising sandy soils on the coast. Occasionally a weed of urban yards.

Distribution

NC, CC, SC, South Australia, America.

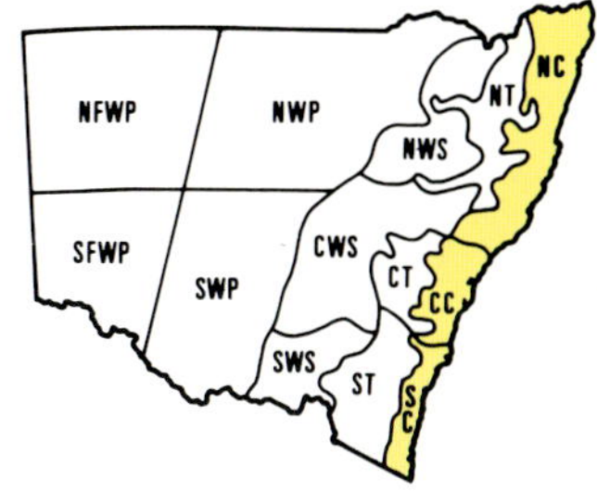

Left: *Hydrocotyle bonariensis* with peltate leaves and typical umbellate inflorescence

Right: *Hydrocotyle bonariensis* growing into a sandy creek near Kurnell

Lilaeopsis polyantha
showing the recurving
fruiting inflorescence and
septate leaves

Lilaeopsis

Lilaeopsis polyantha

Note: Up to 5 species of *Lilaeopsis* have been distinguished in Australia, but the distinction between these taxa appears far from clear. All New South Wales material is here treated as *L. polyantha.*

A native emergent, occasionally submerged, perennial. Stems creeping, rooting at the nodes. Leaves 2 or more per node, usually up to 15 cm long (occasionally longer) and 3 mm wide, cylindrical, hollow with transverse septa. The leaf outline frequently slightly contracted at the septa. Umbels simple, 1-8 flowered on slender peduncles arising at the nodes. Bracts few, minute. Flower pedicels filiform, 2-10 mm long. Sepals 5, minute. Petals 5, to 1.5 mm long, white or pink. Stamens 5. Ovary inferior with 2 locules. Fruit 2-4 mm long, ridged.

The cylindrical leaves could be confused with leaves of some monocots, but *Lilaeopsis* can be distinguished by the septate cylindrical leaves, constricted at the septa, arising from the nodes of a creeping stem. *Seed photograph page 437.*

Habitat

Fresh or brackish water up to 30 cm deep or on wet mud on the margins of water bodies.

Economic Significance

None known.

Distribution

NC, CC, NT, CT, ST, CWS, SWS, Victoria.

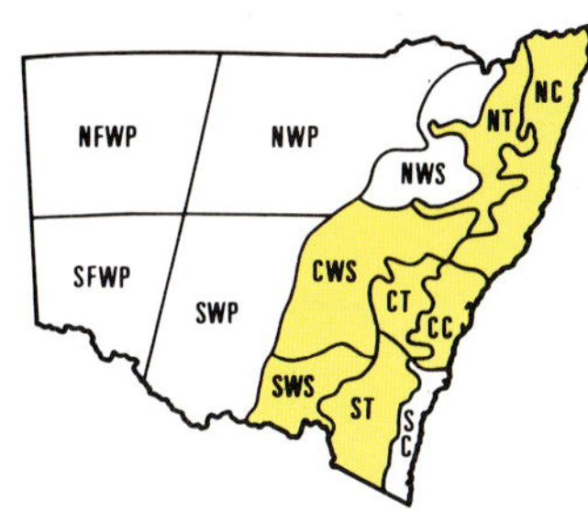

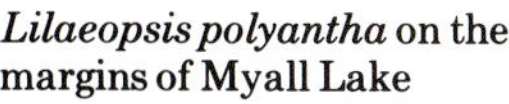
Lilaeopsis polyantha on the margins of Myall Lake

Flowers and fruits of *Sium latifolium*

Broadleaf Water Parsnip

Sium latifolium

Introduced decumbent perennial herb, rooting where nodes come into contact with soil or water. Up to 1 metre tall with hollow stems up to 3 cm in diameter. Submerged leaves rarely found in Australia, emergent leaves once-divided (pinnatisect), up to 20 cm long with 2-12 pairs of sessile, ovate leaflets, the basal pair up to 5 cm long and 3 cm wide with each subsequent pair being smaller than the pair below it. The terminal leaflet frequently lobed. Leaflets with regularly toothed margins. Umbels compound, both lateral and terminal and on long stalks. Bracts large, leaflike. Sepals 5, minute. Petals 5, white. Stamens 5. Ovary 2-celled, inferior. *Seed photograph page 437.*

Habitat

Along creek banks, around water-bodies and in drains.
Our material does not seem to be as aquatic as *S. latifolium* is in Europe. In Australia the plants may grow into the water from the bank but would rarely be regarded as truly aquatic, except in clear spring water such as occurs in the Mt Gambier district of South Australia.

Economic Significance

May grow in drains and retard water flow.

Distribution

NC, SC, NWS, CWS, Victoria, South Australia, Western Australia, Europe.

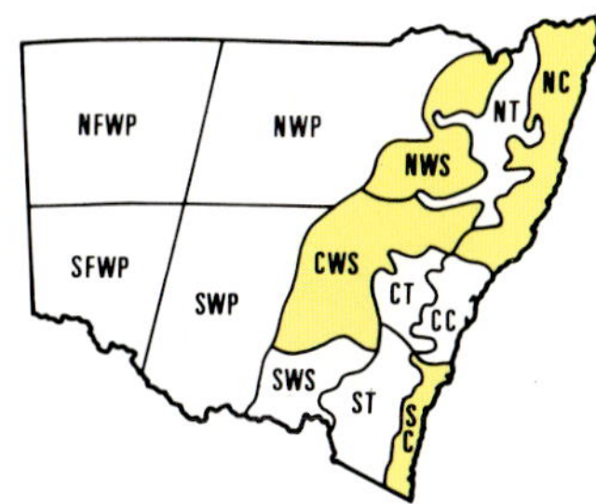

Vegetative growth of *Sium latifolium* produced prior to flowering

Aponogeton

Tuberous or rhizomatous perennials with linear submerged leaves and petiolate floating leaves. Flowers bisexual, borne in simple or forked emergent spikes. Perianth segments 1-6. Stamens 6 or more. Ovary superior, 3-6 free carpels.

1. Inflorescence yellow, usually a simple spike, rarely with smaller lateral branches below.
A. elongatus
1a. Inflorescence white, normally with 2 ± equal branches.
A. distachyos

Cape Pond Lily

Aponogeton distachyos

Introduced emergent perennial with tuberous basal stem. Leaves arising from near the apex of the tuber, petioles reportedly to 1 metre long, but normally about 20cm with floating elliptic leaves usually to about 15 cm long and 5 cm wide. Inflorescence emergent, showy, forked, white. Each fork to 10 cm long with 6-12 ovate white bractlike perianth segments, each subtending a smaller flower with 15-20 stamens, having dark-coloured anthers and 3-6 carpels. Flask-shaped fruiting carpels enlarge to 1 cm long, crowded along the inflorescence branch.

Growth Biology

A. distachyos is one of the few winter-growing aquatics. Seeds seem capable of germinating at almost any time of the year, but this is less common in the warmer months. The tuber is formed very quickly, and once mature is able to withstand a range of extreme environmental conditions. Juvenile leaves have poorly differentiated blades. After mature leaves are formed, the plant may flower. Flowers are produced almost all year, again less commonly in summer. Each plant will produce one to several inflorescences over a short period of time. Flowering and fruiting takes approximately two months. After a brief break, flowering starts again.

Habitat

Static and flowing fresh water to about 50 cm deep, but seems restricted by high temperature.

Economic Significance

Potentially a minor weed in cooler areas. Frequently cultivated in ponds.

Distribution

CC, Victoria, South Australia, South Africa.

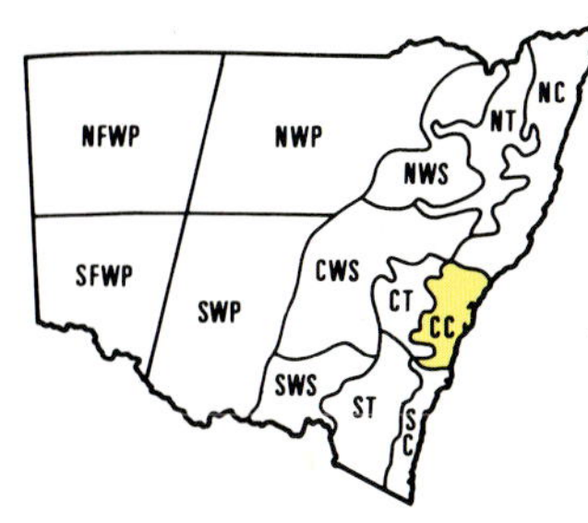

Left: Inflorescence of *Aponogeton distachyos,* the white bracts are clearly visible each side of the yellow or brown-tipped stamens

Right: *Aponogeton distachyos* growing at the Botanic Gardens, Sydney

Aponogeton elongatus

Tuberous native perennial with mainly submerged leaves to 30 cm long and 4 cm wide with a few floating leaves having blades to 15 cm long and 2-4 cm wide. Inflorescence simple, emergent, to 15 cm long. Flowers well spaced on inflorescence, yellow. Petals 2. Stamens approximately 10, apparently variable and apparently not all functional. Carpels 3.

Growth Biology

Little is known about seed germination in this species. It seems that leaf growth is only initiated from the tuber when it is close to the mud/water interface. If flooding buries the tuber deeper it can lie dormant for many years until further floods remove the burden. Growth occurs only in the summer months with flowering and fruiting in late summer, early autumn.

Habitat

Both shallow and deep pools of rivers with muddy bottoms. This species is capable of growing under shade and can be found in pools overtopped by tree branches.

Economic Significance

Harvested commercially as a local industry on the North Coast for use in the aquarium plant trade. The tubers were (in New South Wales—they still are in Queensland and Northern Territory) eaten by the aborigines. The distribution is too restricted in New South Wales for the species to be of economic significance.

Distribution

NC, Queensland, Northern Territory.

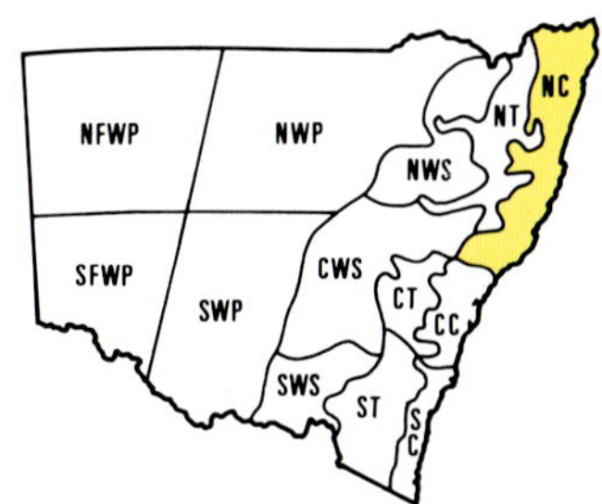

Left: A decumbent yellow inflorescence of *Aponogeton elongatus* resting on the glossy floating leaves

Right: Whole plant, including tuber, of *Aponogeton elongatus*

Water Lettuce

Pistia stratiotes

An introduced floating, tufted, stoloniferous, monoecious perennial up to 15 cm tall and 30 cm wide with fibrous roots up to 1 metre long. Stolons radial, terminating in a new plant. Leaves closely overlapping (like a lettuce) without a petiole, hairy, 6-15 cm long, 4-8 cm wide, a characteristic pale-green colour, with 7-15 ridges. Inflorescence 1-few per plant, axillary, solitary, about 2 cm long, with a whitish slit spathe larger than and encircling the spadix. Flowers unisexual, without a perianth. Female flower solitary, basal. Male flowers 2-8, terminal, separated from the female by a membranous cup. Male flowers each with a single stamen.

The fine fibrous root mass and lettuce-like leaves characteristic of *Pistia stratiotes*

Growth Biology

Under tropical conditions where plant nutrient levels are adequate, growth is rapid, and many new daughter plants are produced on stolons from the parent plants. Small inconspicuous flowers shaped like a miniature Arum Lily *(Zantedeschia* sp.) flower, to which family it belongs, are partially concealed in mature plants. In some countries seeds about 2 mm long, oblong in shape and slightly tapered at each end, are produced. Seeds float for a few days then sink. Seeds not recorded in Australia.

Pistia is frost sensitive and does not thrive in temperate zones. An attempt to cultivate this species in an outdoor aquarium at Griffith in south-western New South Wales failed. The plant withered after exposure for one month to an ambient temperature varying from 10°C to 18°C. Similarly plants cultivated outdoors in Sydney rarely survive the winter. Like other floating plants, it can survive for long periods when stranded on mudbanks or other damp situations.

Habitat

Nutrient-rich stationary or slowly-flowing water.

Economic Significance

Resembles *Eichhornia* and *Salvinia* in forming floating obstructions, but in Australia it is not considered as potentially troublesome as these plants. *Pistia* forms a suitable habitat for mosquitoes that carry parasites responsible for malaria, filariasis and encephalomyelitis.

Significant and troublesome infestations occur in coastal Queensland. Reportedly a serious weed in Africa, Asia and the Caribbean. Dense mats will adversely affect light penetration, oxygen levels, fish life, bank up at structures and increase water loss through transpiration.

This plant was recognised in ancient times as having medicinal value, including the treatment of skin diseases, ringworm, asthma and as a laxative. Used widely in aquaria.

Distribution

NC, Queensland, Cosmopolitan.

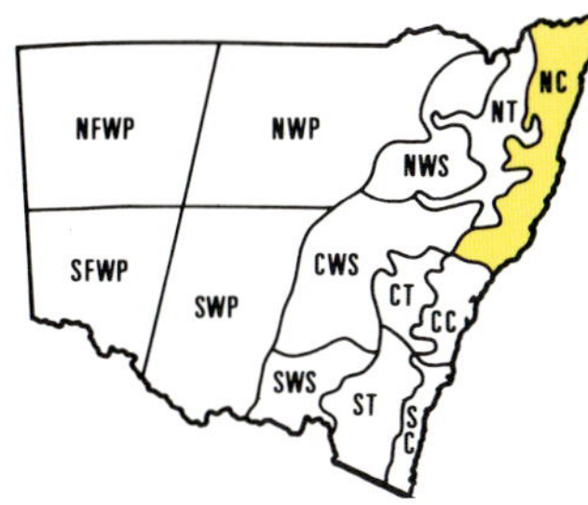

Young growth of *Pistia stratiotes* blocking a drain in Florida, U.S.A.

Arum Lily

Zantedeschia aethiopica

Introduced robust erect perennial herb to 1 metre high with a short, thick rhizomatous rootstock. Leaves petiolate, large, cordate or sagittate, bright green, up to 50 cm long and 25 cm wide with obtuse basal lobes. Spathe white, funnel-shaped, large up to 20 cm long. Central spadix bright yellow with female flowers on the lower quarter and the remainder covered with male flowers. Fruit yellow-green, a 1-5 celled berry.

Growth Biology

Little is known about germination in this species. Plants produce new leaves in late winter, flowering soon after. During the warmer months the plants grow and flower, with the leaves mostly dying back in winter.

Habitat

In wet or damp places, along stream or channel banks or in drains and floodways.

Economic Significance

Probably not of much significance in New South Wales, but does cause some problems in other southern States. Has the potential to cause problems in some circumstances. An attractive garden plant.

Distribution

CC, Victoria, South Australia, Western Australia.

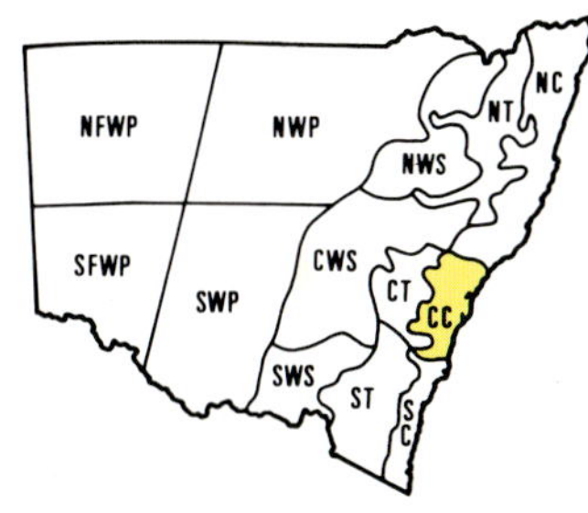

Left: Inflorescences of *Zantedeschia aethiopica*

Right: *Zantedeschia aethiopica* along a creek bank in Western Australia

Azolla

Azolla

Floating native perennials 1-3 cm long, about 1 cm wide and 3 mm deep. More or less pinnately branched. Branched or simple roots to 4 cm long form at some nodes. Leaves alternate, overlapping, 2-lobed. Upper lobe thick, photosynthetic with a nitrogen-fixing blue-green alga (*Anabaena* sp.) in the cavity; lower lobe membranous, lower surface only in contact with water. Sporangia frequently present but supposedly not fertile. Sporangia when present 2-4 in axils of lower lobes, both megasporocarps and microsporocarps formed on the one plant. Megasporocarp less than 1 mm long with a single megaspore. Microsporocarp 1.5-2 mm long, globular, containing many stalked microsporangia, each containing 4 or more groups of microspores. The stage normally seen is the sporophyte, which can be perpetuated vegetatively almost indefinitely. Like all ferns, *Azolla* possesses a gametophyte generation. The gametophyte is minute, free-floating.

The generic name "Azolla" is derived from the Greek *a*, without; *zoe*, life, because the plants are killed by drought.

There are two species of *Azolla* in New South Wales.

1. Mature roots branched with small fine side branches. Plants regularly pinnate.
Azolla pinnata
Ferny Azolla

1a. Mature roots simple, no hairlike branches. Plants irregularly pinnate.
Azolla filiculoides
var. **rubra**
Pacific Azolla

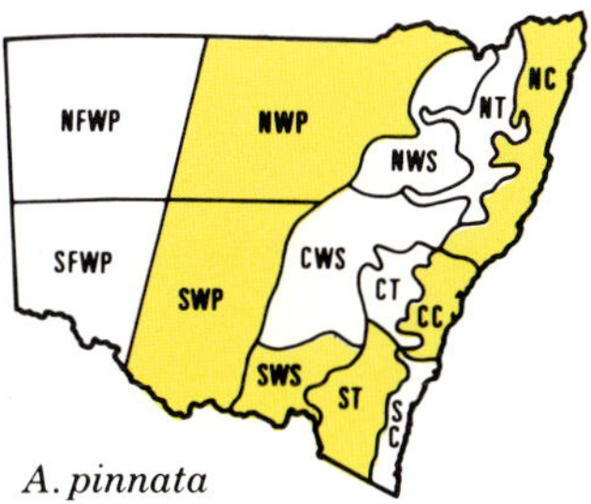

A. pinnata

Left: *Azolla filiculoides* on the left and *A. pinnata* on the right

Growth Biology

During the spring or in shaded situations the upper parts of the plant are green, whilst in summer and autumn and when exposed to full sunlight they usually become dark red. Plants can double leaf area in 7 days. Dense growth is usually an indication of high nutrient levels.

Habitat

Common on still or slowly-flowing water where there is little wave action, or banked up at regulators or other obstructions.

Economic Significance

Depending on the situation, *Azolla* can be an undesirable weed or a beneficial plant. In Vietnam, yields have been increased by cultivating *Azolla* in ricefields. The blue-green alga *Anabaena azollae* living in the leaves of *Azolla* fixes nitrogen from air. Farmers fertilise the paddy, encouraging *Azolla* growth. Later, as decay sets in, nitrogen is released.

Occasionally plants can cover the surface of farm dams, lagoons or billabongs in sufficient density to deter stock from watering and/or block pump inlets. It is also an attractive plant for use in garden pools.

Distribution

A. filiculoides is common in New South Wales, whereas *A. pinnata* is much less common. Cosmopolitan.

A. filiculoides: NC, CC, CT, ST, NWS, CWS, SWS, NWP, SWP, SFWP, Victoria, Tasmania, South Australia.

A. pinnata NC, CC, ST, SWS, NWP, SWP, Queensland, Victoria, Northern Territory, South Australia.

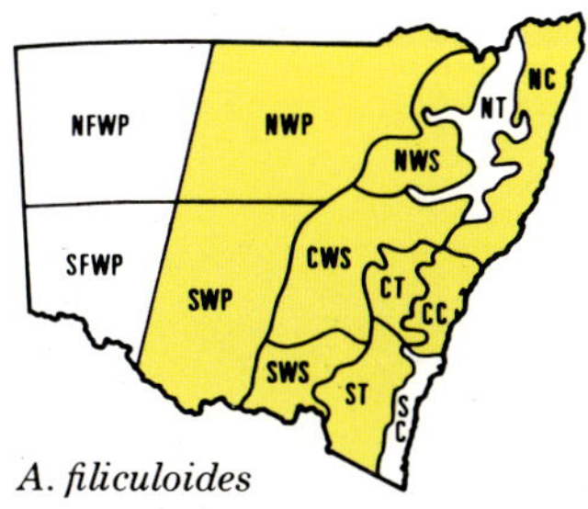

A. filiculoides

Left: *Azolla filiculoides* on a
lagoon in northern Victoria

Watercress

Rorippa nasturtium-aquaticum
[Synonym ***Nasturtium officinale***]

An introduced emergent, decumbent perennial up to 60 cm high. Stems hollow, usually 1-2 cm diameter, up to 1 metre long, rooting at the nodes. Leaves alternate, pinnate, up to 18 cm long, stipulate. Leaflets opposite, up to 1 cm long by 0.5 cm wide, 1-4 pairs, with one terminal. Inflorescence a many-flowered raceme. Flowers bisexual, 4-6 mm diameter. Sepals 4, about 3 mm long, green. Petals 4, white, 5-6 mm long. Stamens 6 (2 + 4). Ovary elongated. Fruit a siliqua, 10-18 mm long, 2-2.5 mm wide. *Seed photograph page 438.*

Rorippa nasturtium-aquaticum showing the alternate pinnate leaves and the small white flowers

Growth Biology

Summer flowering with fruit produced late summer-autumn.

Habitat

Found on the margins of or in still or slowy-flowing water of drains, creeks and lakes. May be emergent in water up to 50 cm deep or growing on mud. In the clear water of Piccaninnie Pond, Mt Gambier, South Australia, this plant grows submerged in water up to 1.5 metres deep and typically about 10°C in temperature.

Economic Significance

R. nasturtium-aquaticum is a member of the mustard family and widely cultivated for use as a fresh salad plant or cooked vegetable. It is a good source of iron, iodine and vitamins A, B and C. When cultivated, 3 or 4 harvests can be made each year. However, it should not be consumed from areas likely to be contaminated from disease organisms, such as urban drainage channels. Occasionally causes obstruction in small supply and drainage channels.

Distribution

NC, CC, NT, CT, ST, NWS, CWS, Queensland, Victoria, Tasmania, South Australia, Western Australia.

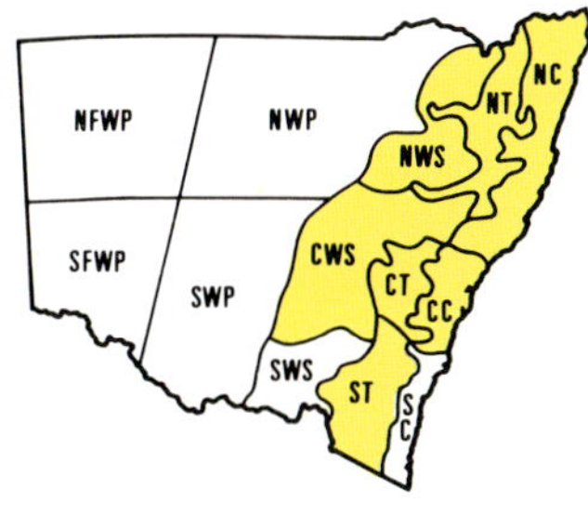

Rorippa nasturtium-aquaticum blocking a drain near Griffith

The characteristic
poppy-like flowers of
Hydrocleys nymphoides

Water Poppy   *Hydrocleys nymphoides*

An introduced stoloniferous perennial with floating and emergent leaves and emergent flowers. Stems, trailing, may be several metres long. Leaves may all arise from the base but are usually in groups along the stem. Leaf blades almost round, up to 12 cm across, often much less. Petiole attached at the margin, with several (usually 3-13) parallel veins arising all along the midrib with numerous interconnecting veins. Petioles terete, crisp, with externally obvious septa. Sepals 3, green, about 2 cm long, 0.8 cm wide. Petals 3, yellow, delicate, 3-4 cm long, 3-4 cm wide. Stamens numerous, filaments purple; carpels 6.

Growth Biology

Buds are produced in profusion and flowering occurs for an extended period in warmer months. Flowers short-lived, lasting 1-2 days. Towards the end of the growing season small plantlets are produced. These break off and rise to the surface. They can float for several days at least and will take root in mud.

Habitat

Stationary or slowly-flowing water up to 1 metre deep and with high levels of nutrients.

Economic Significance

Could be a potential weed in warm nutrient-rich situations. In Valencia Creek near Maffra, Victoria, this species has thrived for about 10 years. A cultivated ornamental.

Distribution

Naturalised in Queensland and Victoria, not recorded from New South Wales.

Hydrocleys nymphoides naturalized on a farm water storage in Gippsland, Victoria

Watershield

Brasenia schreberi

A native stoloniferous or rhizomatous perennial, rooting at the nodes. Leaves floating, oval, peltate, up to about 5 cm long and 8 cm wide. Petiole to about 2 metres in length, dependent on water depth. Stems, petioles and undersurface of leaves covered with clear mucilage. Leaf undersurface often red. Flowers solitary, axillary, floating. Perianth segments 6-8, dark maroon. Stamens numerous, variable (12-36). Carpels free, 4-18. The mucilage on the submerged parts is a useful identification characteristic.

Growth Biology

Flowers during summer and fruit formed in late summer-autumn. After flowering above the water surface, fruits develop and ripen under water. Little is known about seed germination in this species. Some leaves may overwinter and new leaf growth commences in late spring or early summer.

Habitat

Still, shallow fresh-water lagoons or backwaters on sandy or muddy bottoms. Prefers acid water conditions.

Economic Significance

A food source for waterfowl.

Distribution

NC, CC, SWP, Queensland, Victoria, Africa, Asia, America.

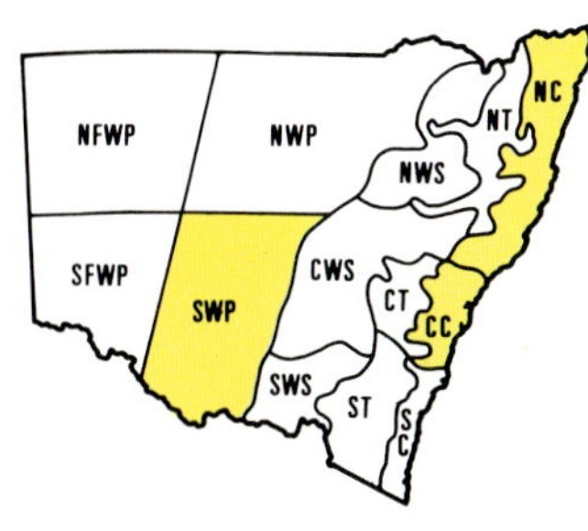

Left: *Brasenia schreberi*

Right: *Brasenia schreberi* on the margins of a small natural lake

Fanwort   *Cabomba caroliniana*

An exotic submerged perennial up to 2 metres long, with a very thin gelatinous coating, emergent flowers and a few small floating leaves. Submerged leaves opposite or whorled, upper leaves almost sessile, lower with a petiole to 3 cm long, initially trichotomously or more branched, subsequently mainly dichotomously branched to produce smooth ultimate segments after 3-7 branchings. The final divided leaf is roughly fan-shaped, up to 5 cm across. Floating leaves few or absent early, peltate, entire, narrow-elliptic, usually less than 2 cm long. Flowers bisexual, axillary, solitary, 1-2 cm across, perianth segments 6, free, in two whorls, 0.5-1 cm long, white or cream with yellow bases. Stamens 6. Carpels 2-4.

The submerged leaves could be confused with *Myriophyllum elatinoides* or *Ceratophyllum demersum,* but *C. demersum* has strictly dichotomously divided leaves with serrulate segments and *M. elatinoides* has pinnate leaves.

Growth Biology

This is basically a summer-growing and flowering plant.

Habitat

Cabomba caroliniana grows in ponds, lakes and quiet streams in water to 3 metres deep. It will also continue to grow free-floating.

Economic Significance

A species, not naturalised in Australia, that causes considerable obstruction to canals and drains in southern United States of America. In nutrient-rich waters, *C. caroliniana* has the potential to the be a serious weed, particularly because of its resistance to herbicides. The species has little value to wildlife. This and other species are widely used as aquarium plants.

Distribution

Not yet recorded from New South Wales. There are about 7 species of *Cabomba* in the world in warm and temperate climates.

Left: *Cabomba caroliniana* showing its initially trichotomously and subsequently dichotomously branching leaves

Callitriche

Native and introduced aquatic or semi-aquatic prostrate herbs. Leaves opposite, entire. Flowers minute, unisexual, both sexes on the same plant. Perianth absent. Male flower consists of 1 stamen. Female flower of one 2-carpellary, 4-locular ovary. Fruit small, laterally flattened, dividing into 4 articles, each frequently winged or keeled.

The introduced *C. stagnalis* is the most economically important species of *Callitriche*. The other species are only mentioned here.

Species of *Callitriche* have many forms and the taxonomy is confused.

1. Fruit longer than wide.   **2.**
1a. Fruit wider than long.   **3.**

2. Lower leaves 3-nerved, spathulate; mature fruit pale, wing conspicuous.

                                                 C. stagnalis
2a. Lower leaves 1-nerved, linear; mature fruit dark-brown, wing conspicuous.   *C. umbonata*

3. Leaves rhomboidal, sometimes with a lateral tooth; fruit wing one-third or more the width of the segment.   *C. muelleri*
3a. Leaves ovate; fruit slightly winged (less than one-third the width of the segment).   *C. sonderi*

Common Starwort

Callitriche stagnalis

Leaves variable, more or less spathulate, 3- , 5- or 7-nerved, about 1 cm long by 5 mm broad, but frequently larger or smaller. Fruit subsessile, green, approximately 1.5-2 mm wide, wing obvious and continuous. In irrigated crops this species can often be recognised by the series of very short internodes at the stem tip. This results in an almost rosette-like arrangement of the terminal leaves.

Growth Biology

Seems to grow, flower and fruit from spring to autumn.

Habitat

Usually in shallow fresh water or mud, but can grow in water to about 1 metre deep.

Economic Significance

Infrequently causes obstruction of water flow in drainage channels.

Distribution

CC, SC, CT, ST, NWP, SWP, Queensland, Victoria, Tasmania, South Australia, Western Australia, Europe, Asia, America.

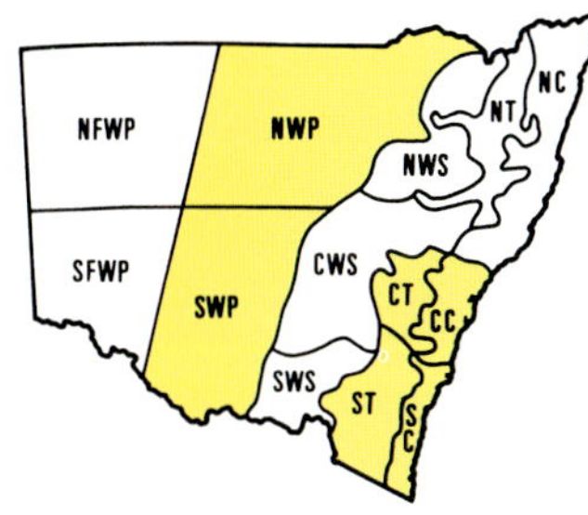

Left: The small terminal leaf rosettes often characteristic of *Callitriche stagnalis*

Right: *Callitriche stagnalis* in shallow running water in the Sydney region

Casuarina

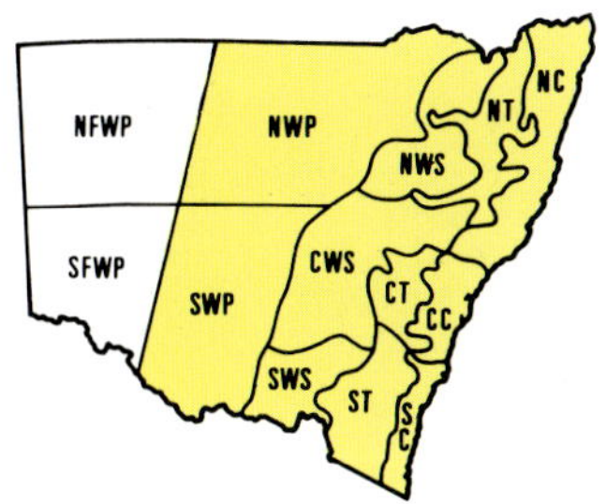

Native dioecious or monoecious shrubs or trees with slender, wiry, articulate branchlets ("leaves"). The true leaves are reduced to ribs which project at each node as whorls of teeth on the ultimate branchlets. These teeth are clearly visible after the branchlet is broken at a node. Flowers unisexual. Female in thickened woody spikes ("cones"); each flower with 2 lateral woody bracteoles, no perianth and a 1-locular ovary. Male flowers in whorls in non-woody spikes; each flower with two lateral bracteoles, 1-2 perianth segments and 1 stamen. Of the many species only two grow in wet habitats.

1. Large shrubs or trees, usually in or near saline areas on the coast; leaf-teeth usually 12-16 in each whorl; cones subglobose to cylindrical, usually about 12 mm diameter.   *C. glauca*
1a. Trees on or near banks of freshwater creeks or rivers; leaf-teeth usually 8-10 in each whorl; cones globular, mostly about 8 mm diameter.   *C. cunninghamiana*

River Sheoak   ***Casuarina cunninghamiana***

A large native dioecious tree to almost 30 metres tall with more or less drooping branchlets. Leaves reduced to 8-10 teeth in each whorl at the node. Cones globular, approximately 8 mm diameter. This species suckers readily.

Habitat

Found on the banks of freshwater streams in the eastern half of the State, especially the coast and tablelands.

Economic Significance

An extremley important species that stabilises river and stream banks, reducing erosion. This was the first protected tree species in New South Wales. The timber is useful and the wood is an excellent fuel. Used extensively in ornamental plantings in both wet and dry areas. The foliage is palatable to stock.

Distribution

NC, CC, SC, NT, CT, ST, NWS, CWS, SWS, NWP, SWP, Queensland, New Guinea.

Left: *Casuarina cunninghamiana* on the banks of the Cudgegong River; (insert) branches from male (left) and female (right) trees

Swamp Sheoak — *Casuarina glauca*

A native dioecious tree to 20 metres tall with ascending branchlets. Leaves reduced to 12-16 teeth in each whorl at the nodes. Cones cylindrical or almost globular, approximately 12 mm diameter. This tree suckers readily.

Habitat

Usually found in saline areas near estuaries.

Economic Significance

A valuable fuel and timber tree, but now valued more for its ability to colonise or stabilise saline areas and estuary banks. In places on the North Coast regrowth after clearing has caused some problems. Used as an ornamental.

Distribution

NC, CC, SC, CWS.

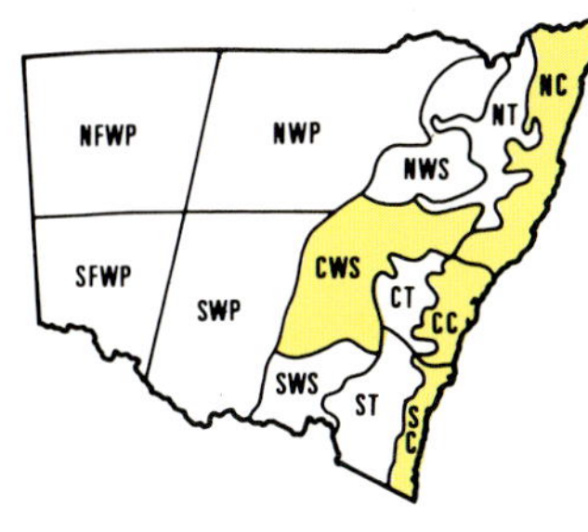

Left: Young trees of *Casuarina glauca* lining a tidal creek

Right: A branch from a female tree of *Casuarina glauca*

Ceratophyllum demersum
with its dichotomously
branched leaves

Hornwort

Ceratophyllum demersum

A native, submerged, unattached monoecious perennial up to 2 metres long. Leaves whorled, 1-4 cm long, dichotomously branched, usually 2-4 times. Segments ± cylindrical with small teeth. Flowers unisexual, axillary, solitary, surrounded by an involucre of several bracts each 1.5-2 mm long. Male flowers with numerous stamens on a common stalk. Female flower with 1 sessile ovary. Fruit black, 4.5 mm long (excluding spines) with 1 terminal and 2 laterally separated basal spines longer than the fruit (see photograph). Occasionally one or both of the basal spines may only partially develop.

It can easily be confused with other aquatic plants, resembling some species of *Myriophyllum* when observed in the natural habitat. In *Myriophyllum* the leaves are pinnate whereas in *Ceratophyllum* the leaves are forked and have jagged projections on the margins, with an appearance similar to deer antlers (see photograph). The type of leaf is described by the generic name *Ceratophyllum* from the Greek "keras", horn and "phyllon", leaf. Another species, *C. submersum,* is common in South-East Asia and may occur in Australia.

The fruit of *C. submersum* has 1 spine instead of 3, and the leaves are more forked.

Growth Biology

Flowers and fruits are produced at the leaf axils in the summer. Fertilisation takes place beneath the water surface. Stamens separate from the plant and float to the surface, where pollen is released. The pollen is heavier than water and sinks to the stigmas of the submerged female flowers. Flowers and fruits are fairly common in late summer in static water around 25°C. Fruits are denser than water and sink to the bottom.

The plant overwinters by thickened lateral tips which sink to the mud and grow again under favourable conditions. These dormant apices contain increased quantities of starch.

The 3-spined fruits of
Ceratophyllum demersum

Habitat

Found in temperate to tropical zones in water up to 10 metres deep. It is tolerant of low light intensities. Usually a rootless plant, either floating at varying levels in the water, anchored in the mud or caught on logs or other aquatic plants. Frequently found in lagoons and swamps where water is static or slow-moving. It generally favours slightly alkaline water and grows best when the water is rich in nitrogen.

Economic Significance

In New Zealand it is a serious weed in lakes supplying water to hydro-electric stations, where intake structures have been blocked. In Australia it creates problems from time to time, obstructing flows, blocking irrigation and drainage structures, or creating a nuisance in fishing and water sport areas.

Distribution

Native to Europe, Asia, Africa, North America, cosmopolitan. NC, CC, SWS, SWP, Queensland, Victoria, Northern Territory, Western Australia.

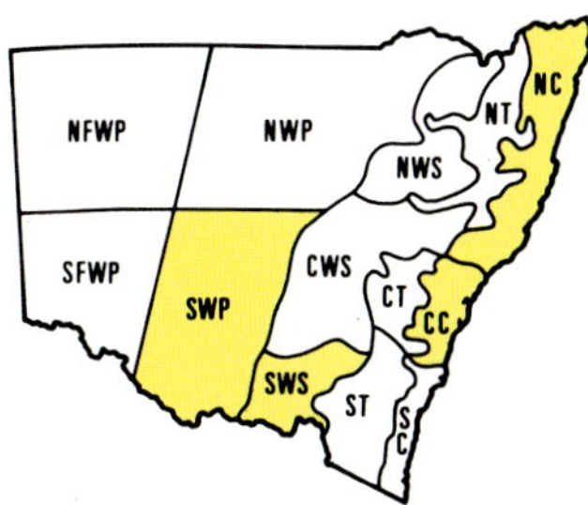

Left: Submerged growth of *Ceratophyllum demersum* in a channel

Bushy Starwort
Wild Aster

Aster subulatus

An erect introduced biennial or annual herb to 2 metres tall. Stems often much-branched, especially towards the top. Leaves sessile, narrow-triangular; lower leaves to 10 cm long, upper leaves much less. Inflorescences numerous, to 5 mm diameter in flower and 1 cm diameter in fruit. Involucral bracts linear-lanceolate, in 3-4 unequal rows; approximately 25 blue, white or pink ray florets, all female; approximately 10 tubular, bisexual disc florets. Seeds compressed, narrow, to 2 mm long, minutely hairy with a well-developed pappus of approximately 30 soft hairs. *(Seed photograph page 437.)*

Growth Biology

Seeds germinate in spring or summer. Plants grow during the warmer months, flowering in summer and autumn. Under suitable conditions the plant can survive during winter and may behave as a short-lived perennial.

Habitat

Damp places, including drains, dam overflows and margins of dams, tanks and lakes.

Economic Significance

Although widespread, *A. subulatus* usually does not cause problems, but in special circumstances it may warrant control.

Distribution

In all regions of the State, Queensland, Victoria, South Australia, Western Australia.

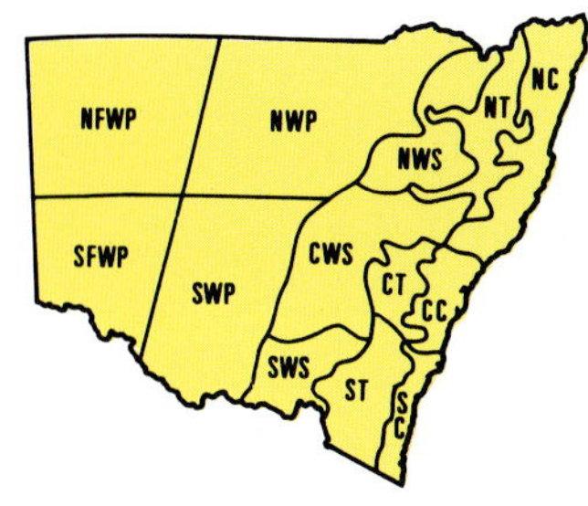

Left: *Aster subulatus* growing on the damp margins of a lake

Right: An inflorescence and stem base of *Aster subulatus*

Waterbuttons

Cotula coronopifolia

A native decumbent, emergent perennial to 60 cm high, rooting from the lower nodes. Leaves alternate, to 7 cm long and 1-2 cm wide, stem-clasping with a sheathing base. Margins entire to deeply lobed. Inflorescence a tight head 6-12 mm diameter, solitary, terminal on peduncles to 8 cm long. Involucre with 2 rows of few, nearly equal bracts 2-5 mm long. Outer single row of flowers female, corolla absent, producing a winged seed 1.5-2 mm long. Inner disc flowers bisexual, with a 4-lobed corolla, producing a non-winged seed less than 1.5 mm long.

Growth Biology

Flowers throughout year except in late winter. Fruits from spring to late winter.

Habitat

Still or slowly moving saline, brackish or fresh water to 50 cm deep. Plant will grow out over deeper water with roots reaching towards the bottom. Plants can grow on wet or drying ground.

Economic Significance

Infrequently a problem in drainage channels.

Distribution

NC, CC, SC, ST, SWS, SWP, SFWP, Queensland, Victoria, Tasmania, South Australia, Western Australia.

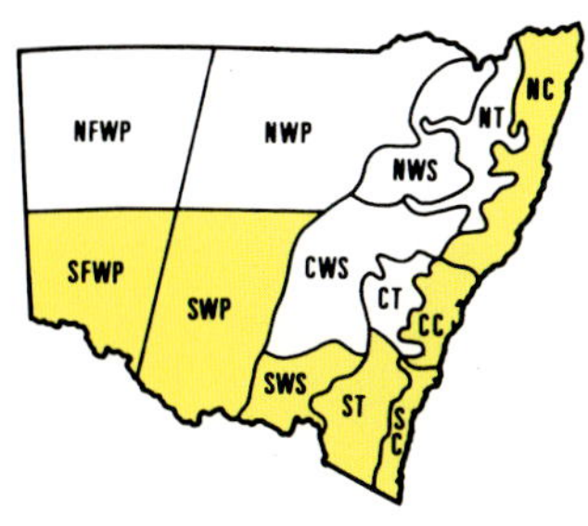

Left: *Cotula coronopifolia* in a drainage channel in Victoria

Right: The yellow flower head and deeply-lobed leaves of *Cotula coronopifolia*

Eclipta

Eclipta

Native prostrate, scrambling, occasionally erect, annual herbs. Leaves covered with short stiff hairs, lanceolate, to about 5 cm long. Flower heads approximately 1 cm diameter, axillary and terminal on variable length pedicels. Involucre of approximately 2 rows of chaffy bracts. Outer ligulate florets female, white or yellow; inner disc florets hermaphrodite. Stamens 5. Ovary single, with a single carpel. Pappus reduced to small teeth or absent. Ray fruit triangular, disc floret subterete.

The two species are quite similar. *E. prostrata* (White Eclipta) has larger inflorescences than *E. platyglossa*. *E. prostrata* has white inflorescences and *E. platyglossa* has yellow inflorescences. *Seed photograph page 437.*

Habitat

Both species may occur as a weed in damp places, e.g., river banks, drainage channels or even on drying ground after flooding.

Economic Significance

Either species may occasionally produce sufficient growth to be a problem in small drainage or supply channels in irrigation systems in the north-east of New South Wales.

Distribution

E. platyglossa NC, CC, SC, NWS, CWS, SWS, SWP, NWP, SFWP, Queensland, South Australia.
E. prostrata NC, Queensland, Northern Territory, Europe, Africa, Asia.

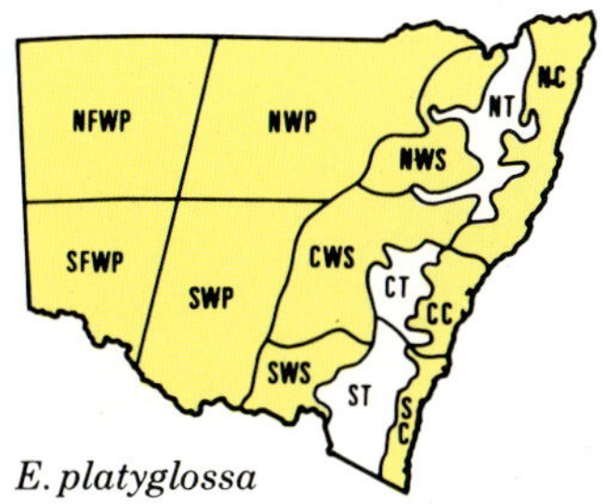

E. platyglossa

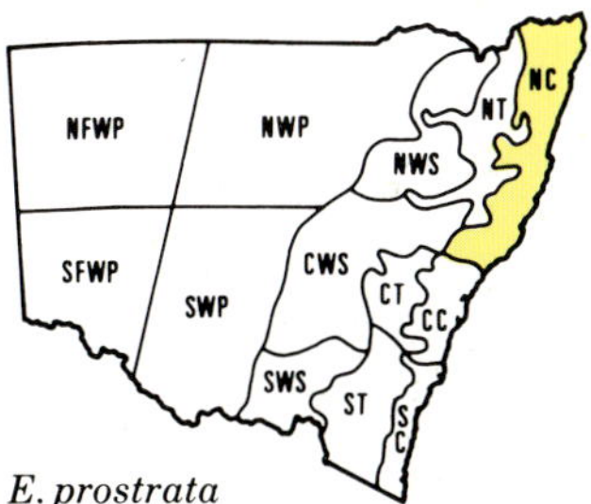

E. prostrata

Left: Tangled growth of *Eclipta platyglossa;*

Insert: *Eclipta platyglossa* showing the weak scrambling stems and the small yellow flower head

Ipomoea diamentinensis

A native stoloniferous perennial with stems to 4 metres long, rooting at the nodes, ± inflated when growing over water. Leaves alternate, erect on petioles to 10 cm long with blades to 15 (?) cm long and 5 cm wide (on New South Wales plants), usually hastate or sagittate. Inflorescence usually 1-flowered (ocasionally few-flowered), axillary, usually on a pedicel (or peduncle) subequal or shorter than the petiole. Flower bisexual, about 5 cm diameter. Sepals 5, persistent, 0.5-1 cm long. Corolla fused, white or pale pink, 5-lobed, 3-5 cm long. Stamens 5. Ovary superior, 2-celled.

Growth Biology

Plants grow in summer and flower in February-March. In New South Wales the plant is not common and flowers are rarely reported.

Habitat

Stationary or slowly-flowing fresh water, usually rooted on mud banks and growing out across deep water or climbing on bank vegetation.

Economic Significance

Nil in New South Wales.

Distribution

NWP, NWS, Queensland, Northern Territory.

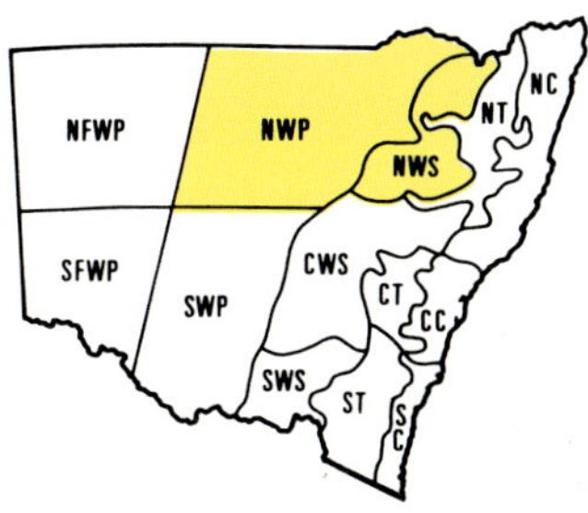

Left: *Ipomoea diamentinensis* growing from the bank of the Angledool River

Crassula
Swamp Crassula

Crassula helmsii

A native prostrate or decumbent perennial forming loose or tight mat-like growths. Stems much-branched, intertwined, to 50 cm long but usually only a few cm high unless supported by surrounding vegetation. Leaves opposite, joined at the base by a small sheath about 1 mm long. Leaves ± succulent, more or less elliptical in section up to 1.5 cm long and less than 2 mm wide. Flowers solitary, axillary, pedicellate on pedicels subequal to the leaves, bisexual. Calyx 4-lobed, green, about half the length of the petals. Petals 4, white, about 2 mm long. Stamens 4. Carpels 4, free, each usually with 2-6 seeds 0.5 mm long.

Growth Biology

Summer flowering and growing.

Habitat

In or on the margins of stationary or slowly-flowing water usually less than 30 cm deep (rarely in deeper water).

Economic Significance

Seldom obstructive in channels. Because of its low and dense growth *Crassula* has potential for planting or encouraging as a competitive species in the beds of intermittently-flowing drainage channels.

Control

Usually not necessary. Resistant to many foliage translocated herbicides.

Distribution

CC, NT, CT, ST, SWS, SWP, Victoria, Tasmania, South Australia, Western Australia.

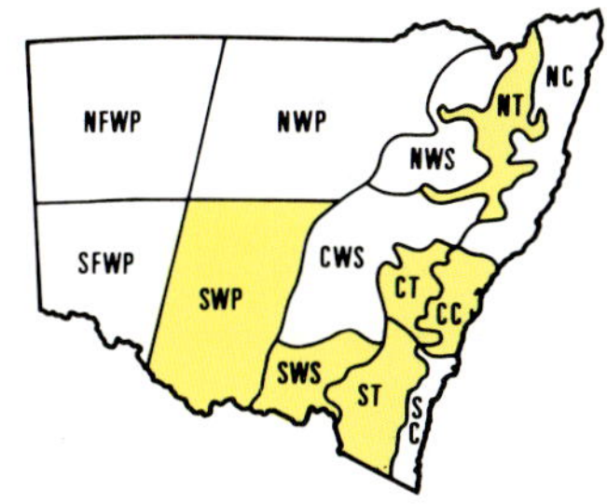

Left: The small, semi-succulent, opposite leaves and small white flowers of *Crassula helmsii*

Right: *Crassula helmsii* along a channel near Griffith

Most genera of the Cyperaceae have species which grow in either permanently or seasonally wet habitats. Individuals may be difficult to classify as aquatics or non-aquatics, because many species are capable of growing in either situation. The non-aquatic species frequently grow on swamp margins or creek banks where they may be periodically inundated for long periods.

Identification can be difficult and is often best left to appropriate experts. A key is provided here to the genera most likely to be classified as aquatics in the broad sense. Only selected species are treated in more detail. For many practical purposes a generic identification may be adequate.

1. Flowering glumes arranged in two opposite rows (distichous), spikelets flattened. Spikelets 1-many-flowered. **2.**

1a. Flowering glumes arranged in spirals. Spikelets not flattened (some *Scirpus* and *Fimbristylis* spp. slightly flattened); perianth bristles present or absent. **3.**

2. Lowest 1-2 glumes sterile, perianth bristles or scales absent, upper rhachis or rhachilla straight, rhachilla internodes ± equal in length; stamens 3-1; style 3-2-fid, rarely entire; nut equally 3-angular or lenticular. Culms without nodes. ***Cyperus***

2a. Lowest 2-several glumes sterile; perianth of 6 or fewer bristles or absent. Upper rhachis or rhachilla distinctly zig-zag, upper internodes much longer than lower internodes. Stamens 6-1 (usually 3); style 3-fid. Nut cylindrical, more or less 3-angular. Culms often 1-2 noded. ***Schoenus***

3. Flowers unisexual, perianth bristles absent. Nut enclosed in utricle. Spikelets single-flowered. Inflorescence a paniculate, spicate, umbellate or solitary arrangement of spikes. Caespitose perennials with flattened grass-like leaves and 3-angular stem. ***Carex***

3a. Flowers bisexual; perianth bristles present *or* absent. Nut not enclosed in utricle. **4.**

4. Inflorescence a single spike, spikelet or compact head (with individual spikelets not obvious). **5.**

4a. Inflorescence with more than one distinct spikelet, either a few large, discrete clustered spikelets or inflorescence branches present and inflorescence axis clearly visible. **12.**

5. Inflorescence terminal, not subtended by an obvious bract. Any bracts less than half the length of the inflorescence. **6.**

5a. Inflorescence apparently lateral, subtended by an obvious bract, as long as or longer than the inflorescence. Perianth bristles or scales present. **10.**

Left: Large plants of *Lepironia articulata* growing behind Waterlilies in a lagoon off the Colo River

6. One flower (-2, but then on separate branches) per inflorescence; alpine.
Oreobolus
6a. Two or more flowers (but possibly only one spikelet) per inflorescence. Mostly not alpine. **7.**

7. One bisexual flower per spikelet. Inflorescence globular; large caespitose perennial to 1 metre high. ***Gymnoschoenus***
7a. Two or more bisexual flowers per spikelet. If inflorescence nearly globular then several flowers per spikelet and not large caespitose plants. **8.**

8. Inflorescence a single, many-flowered, cylindrical or conical spikelet more or less the same diameter as the stem. Stems ("leaves") $\pm$ cylindrical, usually septate in the more robust species. Base of style dilated and persistent; perianth bristles or scales present.
Eleocharis
8a. Inflorescence obviously of greater diameter than the stem, 2-many flowered.

9. Style base dilated and deciduous; perianth bristles absent, plants not large, caespitose with numerous basal leaves. ***Fimbristylis***
9a. Style base not dilated; perianth bristles present *or* absent. Plants either with few true leaves or, if leaves present, these mostly scattered along the stems. ***Scirpus***

10. Stems triquetrous, not septate. Perianth bristles present.
Scirpus
10a. Stems terete, septate. Perianth scales present. **11.**

11. Nut smooth. Spikelet solitary, elongated, almost lanceolate in outline, not partially enclosed by leafy base of bract. Glumes closely overlapping and perianth scales not exserted. Large caespitose perennials to 2 metres or more high. Leaves reduced to basal sheathing scales. ***Lepironia***
11a. Nut with 8 prominent ridges. Inflorescence globular and/or partially enclosed by the leafy base of the bract, with many spikelets. Individual spikelets mostly obvious, glumes loose and perianth scales $\pm$ exserted. Caespitose perennials up to 1.5 metres high, a few basal leaves.
Chorizandra

12. Spikelets with more than 2 perfect flowers. Tufted plants with mostly basal leaves. **13.**
12a. Spikelets with one or two perfect flowers. Plants tufted or rhizomatous; leaves basal or cauline. **17.**

13. Base of style dilated. **14.**
13a. Base of style not dilated. **15.**

14. Style base deciduous.   *Fimbristylis*
14a. Style base persistent.   *Bulbostylis*

15. Perianth scales 2, membranous, enclosing the narrow-cylindrical nut (approx. 1 mm long). Small plants with 1-flowered spikelets arranged in spikelet-like spikes.   *Lipocarpha*
15a. Perianth bristles and/or scales 6 or 0.   **16.**

16. Perianth scales of two types in one flower; 3 bristles with enlarged tips alternating with 3 simple bristles. Glumes conspicuously awned.
                                                             Fuirena
16a. Perianth scales or bristles the same in each flower, 6 or 0; when present always much narrower than the broader compressed (at least on one side) ± trigonous or lenticular nut.   *Scirpus*

17. Perianth scales or bristles present.   **18.**
17a. Perianth absent.   **19.**

18. Style with 2 branches; sterile glumes 3-4; stamens (1-) 3; perianth bristles (0-) 1-6-7 antrorsely scabrous. Style base dilated, persistent, clearly distinct from nut. Ligule absent.   *Rhynchospora*
18a. Style with 3 branches; often several lower sterile glumes; stamens usually 3; perianth scales 6 or less, somewhat inflated at least at base. Style base not persistent. Ligule present.   *Lepidosperma*

19. Leaves with a ligule. Spikelets several on each branch, mostly with 2 flowers but only upper flower perfect (male and female) and fertile. Occasionally only 1 flower present. Lower flower sterile, male or absent. Caespitose perennials up to 2 metres high with terete stem. Leaves long, linear. Inflorescence paniculate, each branch subtended by a more or less leafy bract.   *Gahnia*
19a. Leaves, if present, without ligule. Lowest flower perfect if 2-flowered, middle or middle and upper flower(s) fertile if 3-flowered (rarely 1-flowered in *Baumea teretifolia,* a rhizomatous perennial).

20. Leaves flat, 3-ranked, scabrous on margin and keel. Style 2-3-fid; stamens 2. Stoloniferous or rhizomatous perennials with more or less terete hollow stems 1-2 metres high. Spikelets in dense corymbose clusters.   *Cladium*
20a. Leaves laterally compressed or terete, 2-ranked, not scabrous (reduced in *B. juncea*). Style 3-fid; stamens usually 3. Rhizomatous perennials with terete, or more or less compressed, pithy or septate stems. Spikelets in more or less erect panicles.   *Baumea*

Baumea

Native rhizomatous perennials with stems terete or ± compressed, septate or non-septate. Leaves terete or laterally compressed. Spikelets arranged in open or contracted panicles. Each spikelet with (1-) 2 or 3 flowers with the lowest or middle one perfect. Stamens usually 3. Style 3-fid.

A genus with approximately 12 spp. in New South Wales. *B. articulata* is probably the only truly aquatic species, although the other species all grow in wet or damp places. *B. rubiginosa* is treated here as an example of the other less truly aquatic species.

Baumea articulata showing the long septate leaves and open inflorescence

Jointed Twigrush

Baumea articulata

A native rhizomatous perennial to 2.5 metres tall. Leaves terete, septate with large air spaces, about 1 cm diameter. Panicle open, drooping, brownish, to 50 cm long. Lower or middle floret perfect. *Seed photograph page 438.*

Habitat

Coastal lagoons or swamps, usually in standing water less than 1 metre deep, often on deep mud.

Economic Significance

Suitable habitat for water birds and valuable component of marsh and lake flora.

Distribution

NC, CC, SC, NT, Victoria, Tasmania, South Australia, Western Australia.

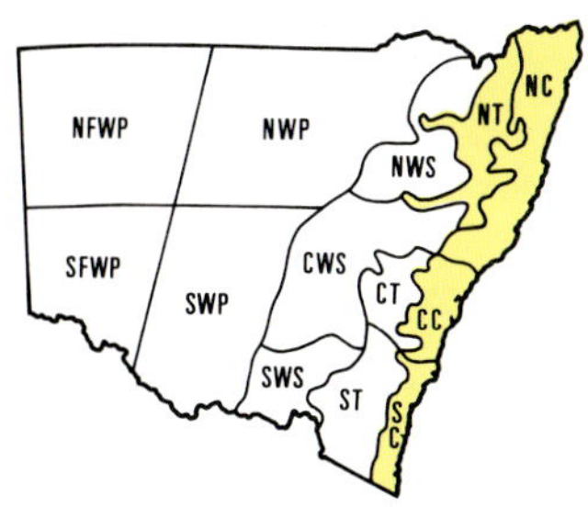

Baumea articulata in an old water storage near Sydney

Baumea rubiginosa

A native rhizomatous perennial to 1 metre high. Leaves present, compressed, with distinct laminae, sometimes subseptate. Inflorescence reduced, an interrupted arrangement of brown or reddish-brown dense subglobular clusters of spikelets. Flowers usually 3 but often 2, glumes usually ciliate.

Habitat

Damp places, ephemeral swamps, lagoon and creek banks.

Economic Significance

Nil.

Distribution

NC, CC, SC, NT, CT, CWS, Queensland, Northern Territory, South Australia, Western Australia.

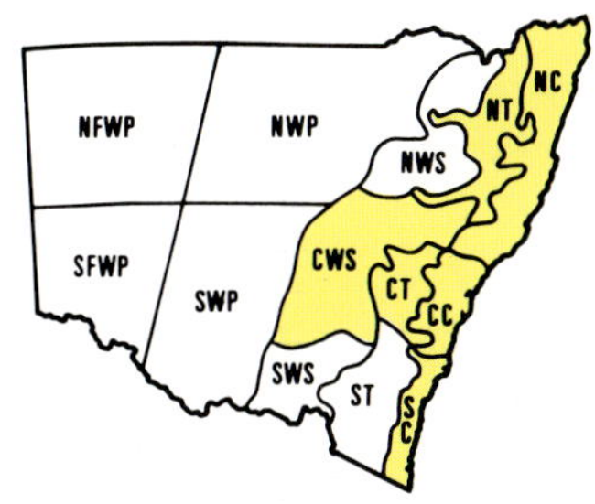

Left: *Baumea rubiginosa* normally grows in swampy conditions rather than in standing water

Right: The compact inflorescences of *Baumea rubiginosa*

Carex

Native caespitose or rhizomatous perennials. Stems often 3-angular. Leaves flattened and grasslike. Inflorescence a paniculate or racemose arrangement of spikes. Spikelets unisexual, often in separate spikes. Stamens usually 3. Female flower enclosed in a flask-shaped utricle subtended by a bristle. Style 2- or 3-fid. Nut compressed or 3-angular.

About half of the approximately 30 spp. in New South Wales grow in damp places, though none is truly aquatic. *C. fascicularis* and *C. appressa* are probably the species most frequently encountered in damp areas.

Tussock Sedge

Carex appressa

A native caespitose perennial to about 2 metres tall, usually less than 1 metre. Leaves about 5 mm wide, to 60 cm long (rarely up to 1 metre long) with scabrid margins. Inflorescence a contracted spikelike panicle to 25 cm long. Spikelets numerous, male and female intermixed, about 5 mm long. Style branches 2. *Seed photograph page 438.*

Habitat

Damp areas, lake and creek banks, ephemeral swamps; can survive periodic inundation.

Economic Significance

May cause a problem as a pasture weed in some damp areas. Useful for erosion stabilisation.

Distribution

NC, CC, SC, NT, CT, ST, NWS, CWS, SWS, NWP, SWP, Queensland, Victoria, Tasmania, South Australia, Western Australia.

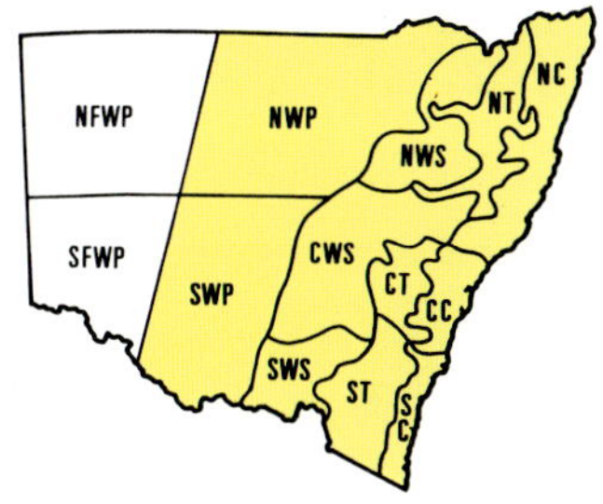

Left: *Carex appressa* on a dam bank near Rankins Springs

Right: *Carex appressa* showing the stiff appressed inflorescence

Carex fascicularis growing on a creek bank

Tassel Sedge *Carex fascicularis*

A native caespitose perennial less than 1 metre tall. Stems triquetrous. Leaves flattened, up to 7 mm wide and 40 cm long. Inflorescence a racemose arrangement of spikes subtended by a leafy bract much longer than the inflorescence. Spikes to 6 cm long, the terminal spike male, the lateral spikes female. Glumes with scabrid awns. Style branches 3. *Seed photograph page 438.*

Habitat

Swamps and creek banks; can withstand periodic inundation.

Economic Significance

Nil.

Distribution

NC, CC, NT, CT, ST, Queensland, Victoria, Tasmania, South Australia, Western Australia.

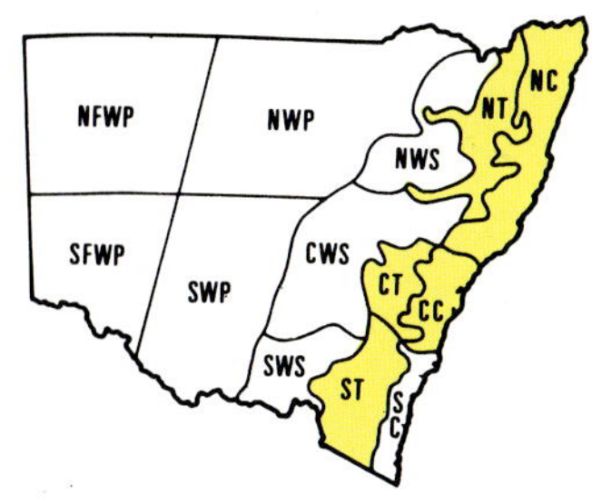

Inflorescence and base of
Carex fascicularis

Inflorescence and base of *Cyperus difformis*

CYPERACEAE

Cyperus

Annuals or perennials; caespitose, rhizomatous or tuberous. Plants usually leafy only at the base but often with leaflike inflorescence bracts. Stems usually trigonous but occasionally terete, to 2.5 metres high in a few species, to 1 metre high in most other species. Leaves mostly with flat blades, rarely reduced to scalelike sheaths. Inflorescence of 1—many heads in a simple or compound umbel. Spikelets 2—many flowered, mostly flattened. Glumes in 2 opposite rows (distichous). Flowers bisexual with 1-3 stamens, 2-3 stigmas and the style not dilated at the base, deciduous. Perianth bristles absent. Nut trigonous or biconvex.

Dirty Dora
Rice Sedge

Cyperus difformis

Erect caespitose native annual with reddish fibrous roots. Culms up to 50 cm high, strongly triquetrous, smooth, compressible. Leaves few, basal, soft, about the same length as the culms, 2-4 mm wide, sometimes reduced to loose sheaths; sheaths red-brown. Inflorescence bracts 1-3, spreading, the lowest longer than the inflorescence and nearly erect, pushing the inflorescence to one side. Inflorescence branches 3-11, usually simple, spreading, up to 5 cm long, terminating in dense globular heads 6-12 mm diameter. Sometimes the inflorescence is contracted into a single dense head. Spikelets linear, compressed, obtuse, 4-8 mm long, approximately 1 mm wide, 10-40 flowered. Glumes approximately 0.75 mm long, obtuse, scarcely keeled, faintly 3-nerved, dark red to brown, becoming variegated with age, with pale margins. Stamens 1, rarely 2; style 3-lobed. Nut almost as long as the glume, acutely triquetrous. *Seed photograph page 438.*

Cyperus difformis,
a common weed
of rice crops

Growth Biology

Flowers in summer in New South Wales. Spreads and grows with great rapidity and under ideal conditions completes a cycle of growth from seedling to flowering plants to seedling in less than two months. This is a growth cycle faster than the rice crop with which it is competing. In tropical countries *C. difformis* flowers and fruits throughout the year. Up to 50000 seeds per plant have been recorded in Italy with a germination percentage of 60 per cent.

Habitat

In drying mud or in eutrophic mud covered by water, ditches and channels, rice crops.

Economic Significance

A serious weed of rice crops. Generally not a significant weed of waterways. Together with *Echinochloa crus-galli,* provides formidable competition for young rice.

Distribution

NC, CC, SC, NT, ST, NWS, CWS, SWS, NWP, SWP, NFWP, Queensland, Victoria, Northern Territory, South Australia, Western Australia, pantropical. Reported a weed in 46 countries and grows to altitudes of 1400 metres in Indonesia.

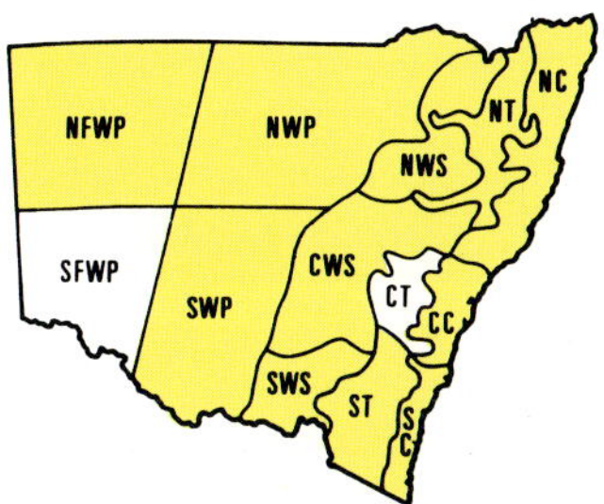

Umbrella Sedge

Cyperus eragrostis

Tufted perennial to 1 metre high with short, thick woody rootstock. Leaves 4-8 mm wide with scabrous margins and, sometimes, midrib; sheathing bases red-brown or purple. Inflorescence bracts 5-9, unequal, longest up to 30 cm and usually much exceeding the inflorescence, similar to the leaves. Inflorescence simple or compound, branches 0-12 cm long, each branch terminating in a subglobose cluster of spikelets 1-5 cm in diameter. Spikelets flattened, many flowered, 5-15 mm long, approximately 3 mm broad. Glumes dull-greenish when fresh, yellowish or reddish when mature or dry, 2-2.5 mm long, 3 evenly spaced nerves, falling off singly from the base of the spikelet, enclosing the nut. Stamen 1. Style 3-fid. Nut dark grey, half the length of the glume, triquetrous, distinctly apiculate with the persistent style base. *Seed photograph page 438*. The leaves of the young plants may be confused with *Cyperus rotundus*.

Habitat

Roadside gutters and pools, rocky creeks and in rice fields, disturbed sandy sites and irrigation channels.

Economic Significance

Occasionally a weed of small supply and drainage channels and in thin rice stands. In many situations this attractive sedge provides competition against less desirable species.

Distribution

NC, CC, NT, CT, ST, CWS, SWS, NWP, SWP, Queensland, Victoria, Tasmania, America, New Zealand.

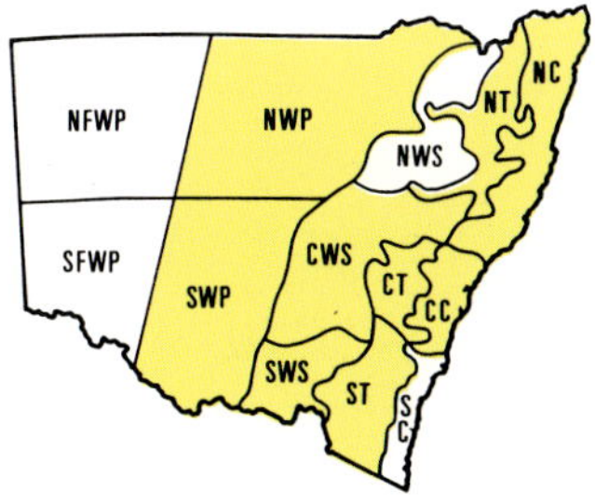

Left: *Cyperus eragrostis* in a supply channel near Leeton
Insert: Young and old inflorescences of *Cyperus eragrostis*

CYPERACEAE

Cyperus exaltatus

Robust caespitose native perennial to 2 metres high, with a short woody rhizome. Culms triquetrous, 2-5 mm wide in upper part. Leaves flat, approximately as long as the culm, 3-10 mm wide, with 2 obvious lateral veins on upper surface, margins scabrous. Sheaths long and wide, purplish brown. Inflorescence bracts 3-6, similar to the leaves, the lower ones much exceeding the inflorescence, to 90 cm long and approximately 2 cm wide, scabrous. Bracts on secondary umbels narrow and scarcely exceeding the inflorescence. Inflorescence compound with 5-10 branches to 18 cm long, each branch ending in a cluster of elongated brown spikes 2-9 cm long, 0.3-1.0 cm wide. Spikelets almost distichous, brown, 2-10 mm long, 1-2 mm broad, 6-20 flowered, glumes falling away singly from the base of the spikelet. Glumes 1-2 mm long, overlapping, mucronate, keeled, with 3-5 indistinct nerves towards the centre and broad scarious nerveless margins. Stamens 3. Style 3-fid. Nut less than half the length of the glume, triquetrous, pale. *Seed photograph page 438.*

Habitat

On river and creek banks or frequently-inundated alluvial floodplains. Infrequently occurs in drainage channels.

Economic Significance

Generally a valuable plant providing cover for waterfowl and stabilising banks against erosion. Occasionally cultivated in gardens.

Distribution

NC, CC, NWS, CWS, SWS, NWP, SWP, Queensland, Victoria, South Australia, Western Australia, Africa, Asia.

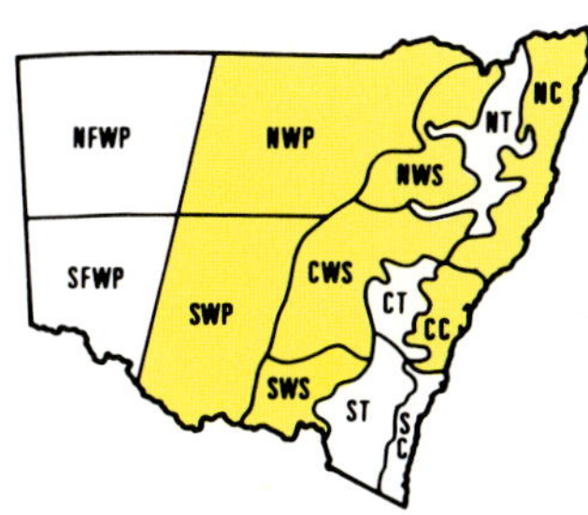

Left: *Cyperus exaltatus* in a supply channel near Yanco

Right: Inflorescence stem and base of *Cyperus exaltatus*

Cyperus involucratus
[Synonym ***C. alternifolius***]

An introduced caespitose perennial to 1 metre high. Stem triangular. Leaves reduced to reddish-brown sheaths. Inflorescence bracts 12-20, 10-30 cm long, approximately 1 cm wide. Inflorescence with as many branches as bracts, each branch to 10 cm long, much shorter than the subtending bract. Each primary branch terminating in a cluster of spikelets or a cluster of spikelets and 1 or more secondary branches. Each secondary branch terminating in a cluster of spikelets. Spikelets 6-40 flowered, to 1 cm long and 3 mm broad. Glumes to 2 mm long, 3-5 nerved, green or yellowish. Stamens 3. Style 3-fid. Nut trigonous. *Seed photograph page 438.*

Habitat

Mainly cultivated, but may persist almost indefinitely once planted, and capable of establishing vegetatively from garden refuse. Will grow on creek banks and in water to 60 cm deep.

Economic Significance

Grown commercially. Has become naturalised in other States, but probably of little weed potential.

Distribution

No naturalised records from New South Wales. Queensland, Africa, widely cultivated.

Left: The decorative spreading inflorescence of *Cyperus involucratus*

Right: *Cyperus involucratus* in a garden near Hanwood

Papyrus

Cyperus papyrus

Erect caespitose introduced perennial to 2.5 metres tall with a short, thick, woody rhizome. Stems triangular. Leaves reduced to brown sheaths. Inflorescence bracts 4-10, 7-15 cm long, approximately 1 cm wide, brown. Inflorescence compound, often with more than 100 branches 10-25 cm long, each with a sheathlike bract 3-5 cm long at the base, and 3-5 bracts each approximately 20 cm long and approximately 1 mm wide at the tip. The plant as commonly seen consists of large numbers of mainly sterile inflorescences. When fertile, the long, thin, ultimate bracts subtend 3-5 spikes, each with 20-30 spikelets. Spikelets 3-20 flowered, brown, to 1 cm long, 1 mm wide, ± cylindrical. Glumes to 2 mm long with a single nerve. Stamens 3. Style 2-fid. Nut trigonous, brown, approximately two-thirds length of glume.

Habitat

Mainly cultivated. Rarely survives as individuals after being discarded as garden refuse. Will grow in slowly-flowing water to 1 metre deep. Can grow and spread vegetatively while floating.

Economic Significance

Grown for the nursery trade. To date has not caused any weed problem but requires monitoring.

Distribution

CC, Africa, widely cultivated.

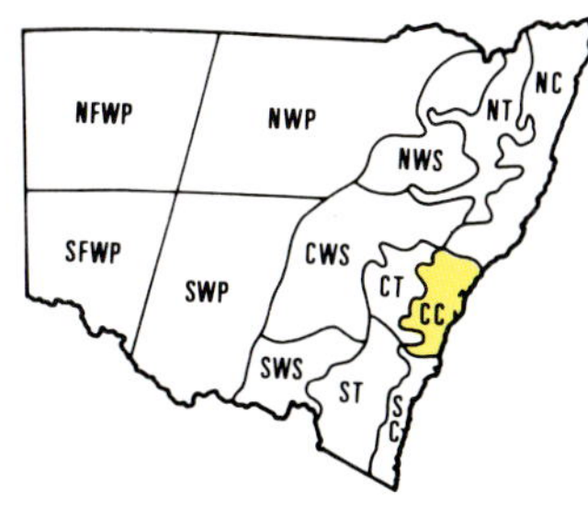

Left: Sterile inflorescences of *Cyperus papyrus*

Right: *Cyperus papyrus* in Centennial Park, Sydney

Spikerush

Eleocharis

Leafless, usually rhizomatous, annuals or perennials. Culms usually produced along a linear rhizome but occasionally tufted, may be terete or compressed, sometimes with obvious transverse septa. Leaf sheaths 1 or more, membranous or scarious ± mucronate. Inflorescence a single terminal spikelet with no bracts. Spikelet (few-) many flowered, globular to linear-cylindrical, ± similar in diameter to culm. Glumes spirally arranged, overlapping, usually deciduous. Lowest 1-2 glumes often sterile, more persistent than the fertile glumes. Flowers bisexual. Perianth bristles mostly 6, rarely up to 10, occasionally reduced or even absent, especially in lower florets. Stamens 1-3. Style 2-3-fid. Nut roughly orbicular with the persistent enlarged base of the style on the apex.

Common Spikerush

Eleocharis acuta

A native rhizomatous perennial less than 1 metre tall. Culms erect, 1-3 mm wide, terete or slightly flattened but trigonous below the spikelet, longitudinally striate. Sheaths more or less purplish at the base, almost truncate, prominently mucronate. Spikelets cylindrical 1-3 cm long, 3-7 mm wide. Glumes obtuse to acute, 3-5 mm long, brown, distinct midvein. Perianth bristles usually 7 (6-9). Stamens 3. Style 3-fid, deciduous. Nut 1.5-2 mm long, 1-1.2 mm wide. Style base whitish or discoloured, 0.4-0.6 mm high, 0.5-0.6 mm wide. *Seed photograph page 439.*

Habitat

In or alongside perennial wetlands, including channels.

Economic Significance

Infrequently obstructive in supply and drainage channels.

Distribution

All regions of New South Wales, Queensland, Victoria, Tasmania, South Australia, Western Australia.

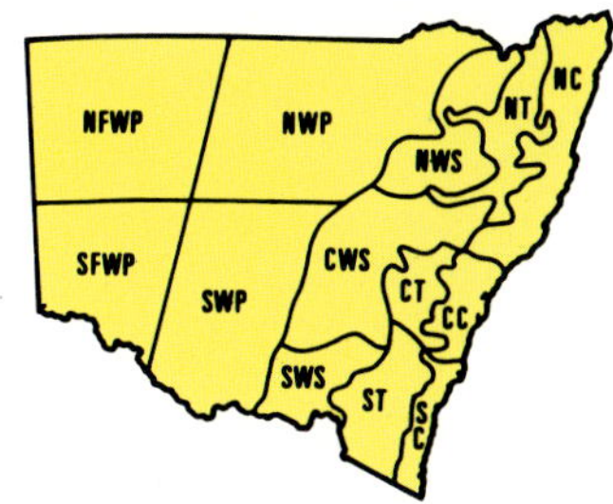

Left: *Eleocharis acuta* in a channel.
Insert: Inflorescences with exserted stamens

Sag

Eleocharis equisetina

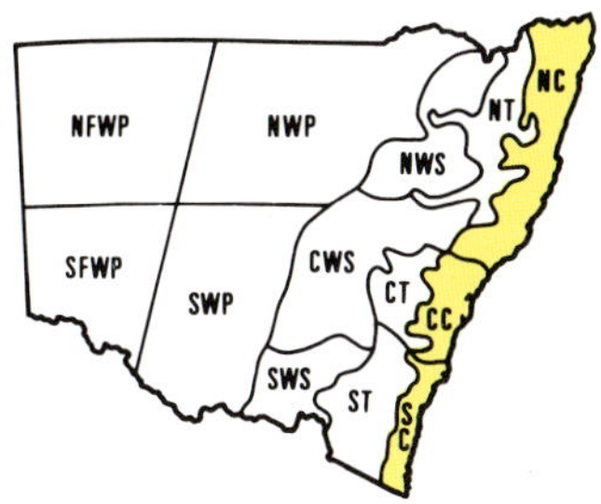

A native stoloniferous perennial to 1 metre tall. Culms erect, terete, obviously transversely septate, longitudinally striate, up to 4 mm diameter. Sheaths brown or purplish without a mucro. Spikelet cylindrical, 2-4 cm long. Glumes obtuse, 4-5 mm long, midvein distinct. Perianth bristles 6-7. Stamens 3. Style 3-fid. Nut approximately 2 mm long. Style base triangular half to two-thirds as wide as the nut, remainder of style persistent.

Habitat

More or less permanent water up to about 0.5 metre deep in swamps or lagoons.

Economic Significance

Part of habitat for waterfowl.

Distribution

NC, CC, SC, Queensland.

Pale Spikerush

Eleocharis pallens

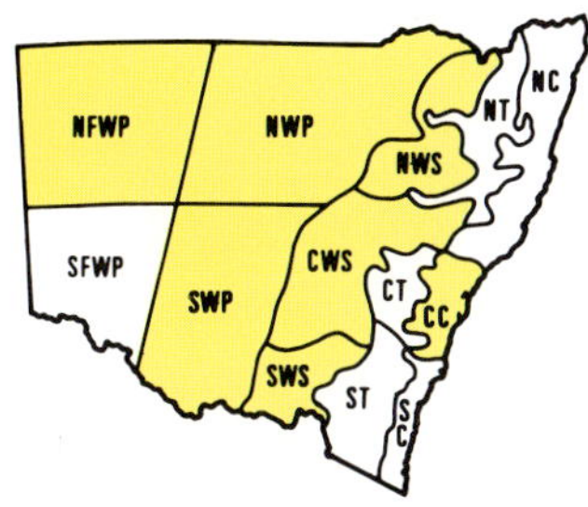

Tufted rhizomatous perennial to 50 cm high. Rhizomes short. Culms subterete, smooth, 0.5-1 mm wide. Sheaths appressed and upper sheath with a mucro up to 2 mm long. Spikelet linear-cylindrical, acute or subacute, frequently dark when young but brown or tawny later. Mostly 1-2 cm long and 2 mm wide. Glumes numerous, readily deciduous, ovate, acute, red-brown, 3-3.5 mm long. Perianth bristles 7-10. Stamens 3. Style 3-fid. Nut 1-1.4 mm long, approximately 1 mm wide, brown. Style base variable, one-third to three-quarters as wide as the nut.

Habitat

Ephemeral pools, swamps and periodically inundated floodplains on heavy soils.

Economic Significance

Occasionally obstructive in small channels.

Distribution

CC, NT, SWS, CWS, SWS, NWP, SWP, NFWP, Queensland, Victoria, Northern Territory, South Australia, Western Australia.

Left: *Eleocharis equisetina* in a natural swampy depression near Wyong
Insert: The narrow inflorescences characteristic of *Eleocharis*

Eleocharis plana

Eleocharis plana

Tufted rhizomatous native perennial to 80 cm high. Culms flat or slightly convex, 2-4 mm wide, longitudinally striate. Sheaths appressed, purplish, the uppermost with a rigid mucro 1.5-3 mm long. Spikelet linear, cylindrical, subacute, mostly 10-20 mm long (rarely longer) and 2-3 mm wide. Glumes more or less acute, faintly keeled, 3.5-4.5 mm long. Perianth bristles 6-8. Stamens 3. Style 3-fid. Nut brown, 1-2 mm long, approximately 1 mm wide. Style base about half as wide as the nut.

Habitat

Ephemeral pools, swamps and periodically inundated floodplains on heavy soil. Drainage and supply channels.

Economic Significance

Occasionally obstructive in intermittently-flowing supply and drainage channels.

Distribution

ST, NWS, CWS, SWS, NWP, SWP, NFWP, SFWP, Queensland, South Australia.

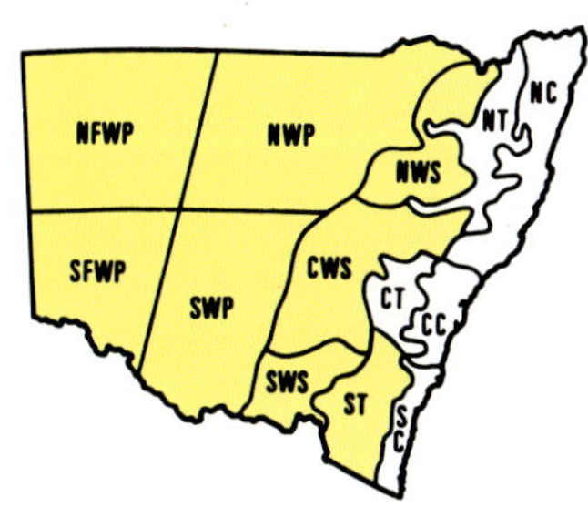

Eleocharis plana in a channel near Coleambally

Eleocharis pusilla

Slender rhizomatous native perennial to 25 cm high. Culms ± erect, to 0.5 mm wide. Spikelet ovate to lanceolate, 2-7 mm long, to 2 mm wide. Glumes few, pallid or deep red-brown, to 2.5 mm long. Perianth bristles few, small, occasionally absent. Stamens 3. Style 3-fid. Nut pale, obscurely trigonous, each face with 3-4 ribs. Style base small, one-third to one-half as wide as nut. *Spikelet photograph page 439.*

Habitat

A wide range of habitats from permanently wet to infrequently inundated sandy to heavy soils. Usually only grows in shallow water when aquatic.

Economic Significance

Has potential for cultivating as competitive plant above and below the waterline.

Distribution

NC, CC, NT, CT, ST, NWS, CWS, SWS, NWP, SWP, Queensland, Victoria, Tasmania, Northern Territory, South Australia, New Zealand.

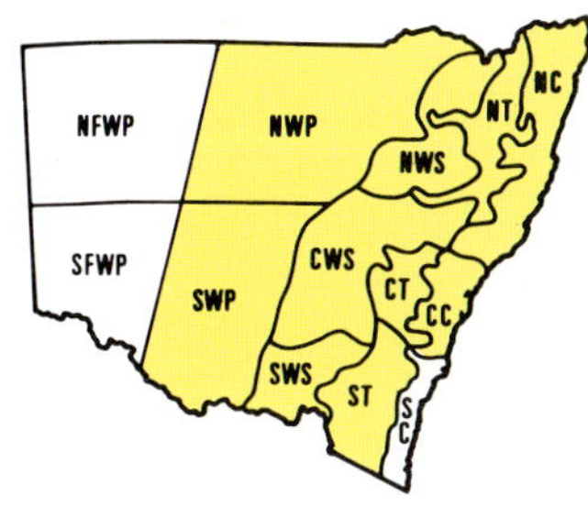

Left: Small flowering plants of *Eleocharis pusilla*

Right: Submerged growth of *Eleocharis pusilla* on the margins of a pond near Narrandera

Tall Spikerush

Eleocharis sphacelata

Aquatic rhizomatous native perennial. Rhizome 5-6 mm diameter. Culms to 2 metres (above water surface) about 1.5 cm diameter, rising closely along the linear rhizome, terete, transversely septate, longitudinally striate. Spikelet cylindrical, acute, 3-5 cm long, 8-9 mm wide, many flowered. Glumes densely packed, almost elliptic, obtuse, mostly 8-9 mm long with fairly prominent midrib. Perianth bristles 8-10. Stamens 3. Style 3-fid. Nut pale, 2.2-2.5 mm long, approximately 2 mm wide. Style base about two-thirds as wide as the nut.

Habitat

Fresh water in lakes, lagoons, dams, swamps, etc. Will grow in quite shallow water and in water to 2 metres deep.

Economic Significance

Readily grows in deep and slowly flowing water, e.g. dams, channels, etc., and can drastically reduce water flow and/or almost grow right across small farm dams. Provides excellent cover for waterfowl.

Distribution

NC, CC, SC, NT, CT, SWS, NWP, SWP, Queensland, Tasmania, South Australia, New Guinea, New Zealand.

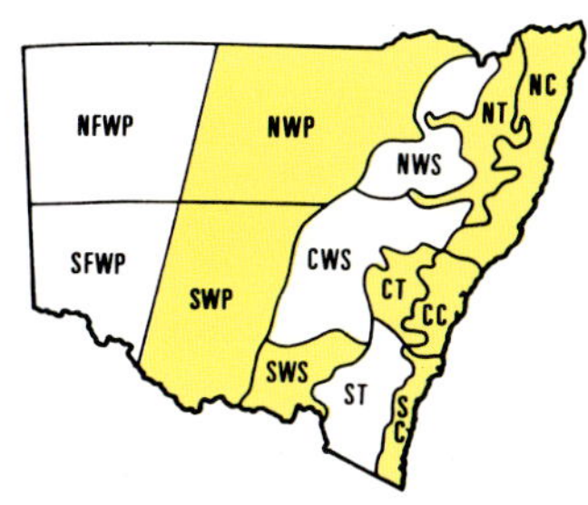

Left: *Eleocharis sphacelata* on the margins of an artificial lake in Victoria

Right: From left to right, (i) a fruiting inflorescence, (ii) an inflorescence with exserted stamens, (iii) an inflorescence with exserted stigmas and (iv) an immature inflorescence of *Eleocharis sphacelata*

Eleocharis sp.

An introduced rhizomatous perennial up to about 50 cm tall. Culms 3-5 mm diameter, terete, flattening readily when dry, longitudinally striate. Sheaths brown, darker at base, acute to obtuse, not mucronate. Spikelets cylindrical 1-5 cm long, usually less than 0.5 cm wide. Glumes 2 mm long, 1 mm wide, acute to apiculate, pallid with brown margins. Perianth bristles 6-7. Stamens 1-2. Style 3-fid, deciduous. Nut approximately 1 mm long and 0.5 mm wide, pallid, subtrigonous. Style base about one-quarter to one-third the width of the nut, annular with a central projection. *Seed photograph page 439.*

Habitat

Only known from in and around rice crops.

Economic Significance

Requires monitoring; may be a potential problem in rice crops.

Control

Report sighting to Department of Agriculture.

Distribution

SWP.

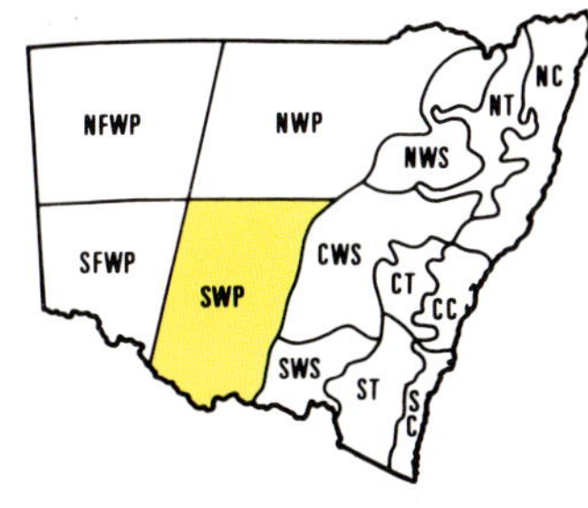

Left: An unidentified species of *Eleocharis* near Griffith

Right: Inflorescences of an as yet unidentified species of *Eleocharis*

Sawsedge

Gahnia

Native caespitose perennials to 2.5 metres high. Leaves flattened though often rolled, frequently scabrous. Inflorescence a large exserted panicle with each branch subtended by an obvious bract. Spikelets numerous, usually 2-flowered with only the upper perfect. Glumes dark brown to black, whorled around the rhachis. Stamens 3-6.

Of the 16 species occurring in New South Wales, none are truly aquatic, but *G. sieberana* probably occurs in wetter places than most of the other species.

Sawsedge
Swordgrass

Gahnia sieberana

A native caespitose perennial to 2.5 metres high. Leaves to 2 cm wide, 1.5 metres long, scabrous. Panicles to 1.5 metres long, open, stems hollow. Spikelets pedicellate, to 8 mm long, 2-flowered, with a terminal bisexual floret and a male or sterile floret below. Glumes dark-brown or black. Fruit frequently exserted at maturity, yellow to orange. *Seed photograph page 439.*

Habitat

Creek banks or beds of ephemeral creeks and swamps.

Economic Significance

Helps reduce erosion.

Distribution

NC, CC, SC, NT, CT, ST, Queensland, Victoria, Tasmania, South Australia.

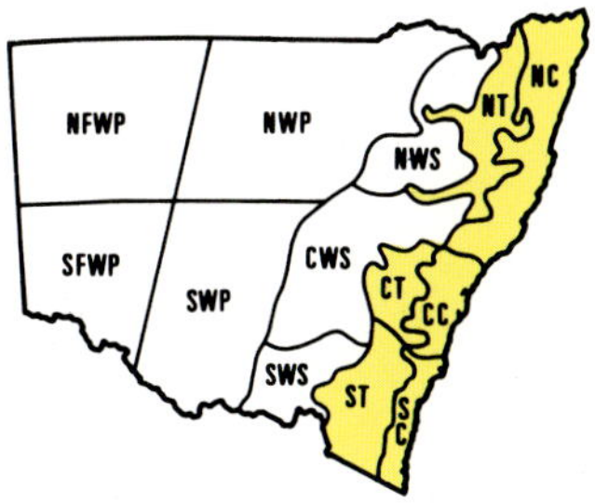

Left: A characteristic large tussock of *Gahnia sieberana* showing the large open inflorescences

Insert: Portion of an inflorescence of *Gahnia sieberana*

Lepironia articulata

A native rhizomatous perennial to 3 metres tall. Leaves reduced to basal sheathing scales. Stems terete, hollow, septate. Inflorescence a lateral ovoid dense spike to 3 cm long and 1 cm wide. Glumes overlapping, obtuse. Spikelets solitary. Perianth scales numerous. Stamens 8. Style 2-fid. Nut compressed, smooth.

Habitat

Swamps and margins of lakes and lagoons.

Economic significance

Waterbird habitat.

Distribution

NC, CC, Queensland, Northern Territory.

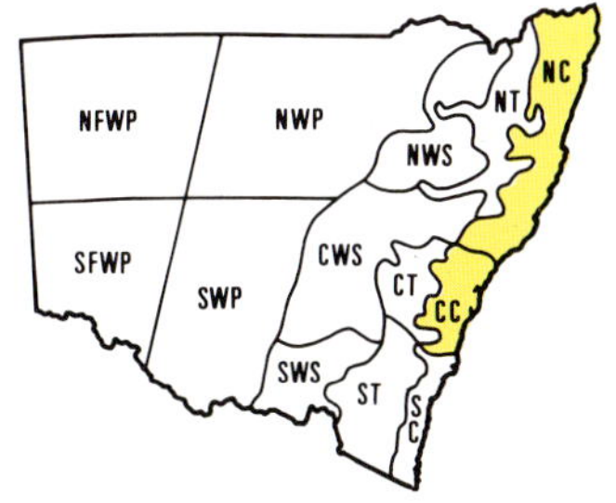

Left: The long cylindrical stems from a large clump of *Lepironia articulata*

Right: *Lepironia articulata* inflorescences with exserted stamens and styles and in fruit

Caespitose or stoloniferous perennials or annuals. Leaves present or reduced to sheaths. Inflorescence terminal but often apparently lateral, large and open or reduced, often subtended by 1 bract appearing as a continuation of the stem or by several smaller bracts. Glumes overlapping, spirally arranged, the lowest 1 or 2 often sterile. Flowers bisexual. Stamens 1-3. Style deciduous, 2-3 branched.

Of the approximately 30 species in New South Wales, only *S. validus*, *S. fluviatilis* and *S. fluitans* would normally be regarded as truly aquatic. *S. prolifer* and *S. mucronatus* are probably more frequently noticed in wet areas than most of the other species. There are several low-growing species which also occur in these habitats, e.g., *S. inundatus* and *S. crassiusculus,* but they are not truly aquatic and are rarely of concern, or rarely even noticed.

Submerged growth of *Scirpus fluitans*

Floating Clubrush

Scirpus fluitans
(including *S. productus*)

A native elongated submerged or terrestrial perennial with emergent inflorescences. In submerged plants the stem is usually about 1 mm diameter and up to 3 metres long, often branched. Leaves alternate, sheaths originally closed but splitting, to 3 cm long, without auricles or ligule. Blade to 2 mm wide and 15 cm long. Inflorescence a solitary terminal spikelet, usually emergent, about 5 mm long, with 5-10 bisexual flowers. Stamens 2 or 3. Style branches 2. Nut compressed.

Vegetative growth may be confused with *Potamogeton pectinatus*. The prominent stem-clasping ligule of *P. pectinatus* is absent in *S. fluitans*.

Habitat

Submerged plants usually in water less than 1 metre deep. Favours fast-flowing clear streams. Found up to altitudes of 900 metres. Terrestrial forms and intermediate forms often found.

Economic Significance

Nil.

Distribution

CC, SC, NT, ST, Victoria, Tasmania, South Australia, Western Australia.

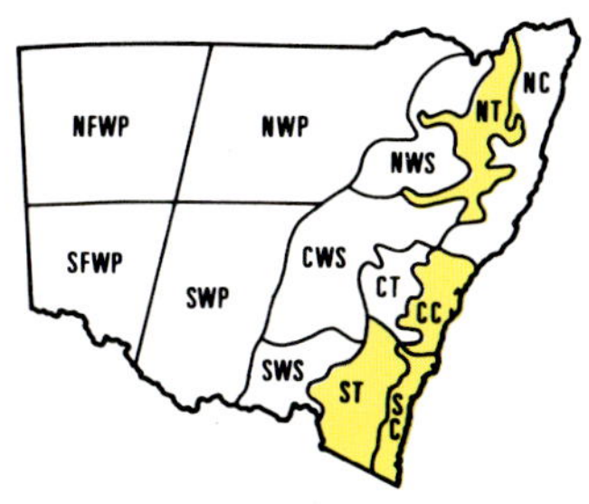

Scirpus fluitans in a creek flowing into the Colo River

Scirpus fluviatilis, (A)
rhizome and stem base,
(B) trans-section of stem,
(C) a group of spikelets and
(D) an inflorescence

Marsh Clubrush

Scirpus fluviatilis

A native rhizomatous perennial to 2 metres high, frequently forming tubers. Stems triquetrous. Leaves well developed, to 12 mm wide and 50 cm long. Inflorescence open, subtended by several leaflike bracts longer than the inflorescence. Spikelets many flowered, ovoid, to 2.5 cm long. Flowers bisexual. Stamens 3. Style branches 3. Perianth bristles 6, shorter than the nut. Nut ± trigonous, about 4 mm long. *Seed photograph page 439.*

The specific name "fluviatilis" is derived from the Latin "fluvius", river.

Growth Biology

Main growth in summer, with flowering from October to January.

Habitat

In shallow water along creeks, in shallow swamps, and sometimes in brackish water. May grow away from water in high rainfall areas.

Economic Significance

Shelter for wildlife and useful for stabilising banks, as it generally will not spread into deeper water. May become a weed of agricultural land in higher rainfall areas. Has potential to be a weed of rice.

Distribution

NC, CC, SC, SWP, Queensland.

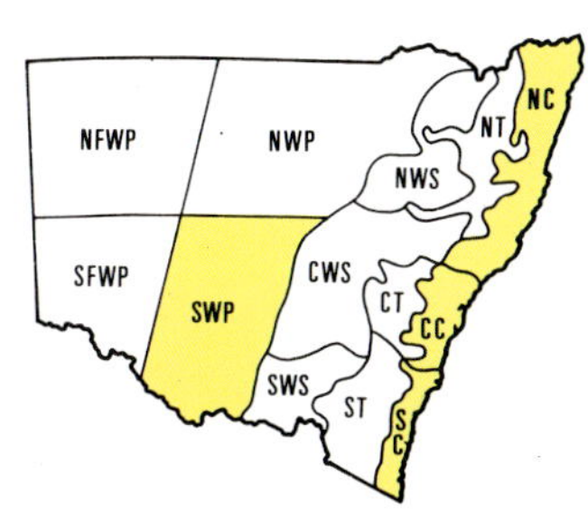

Scirpus fluviatilis on the margins of a large channel near Narrandera

CM

Scirpus mucronatus

A caespitose native perennial up to 1 metre high. Leaves reduced to sheathing bases. Stems acutely 3-angular. Inflorescence apparently lateral, a spherical cluster of up to about 12 sessile spikelets, subtended by a bract which appears to be a continuation of the stem. Spikelets many-flowered, up to 1.5 cm long and 0.5 cm wide. Perianth bristles about as long as the flattened nut, or longer. *Seed photograph page 439.*

Habitat

Creek and river banks, swamps and periodically inundated floodplains or billabongs.

Economic Significance

Component of habitat for water birds.

Distribution

NC, CC, NT, NWS, Queensland, Northern Territory, Western Australia.

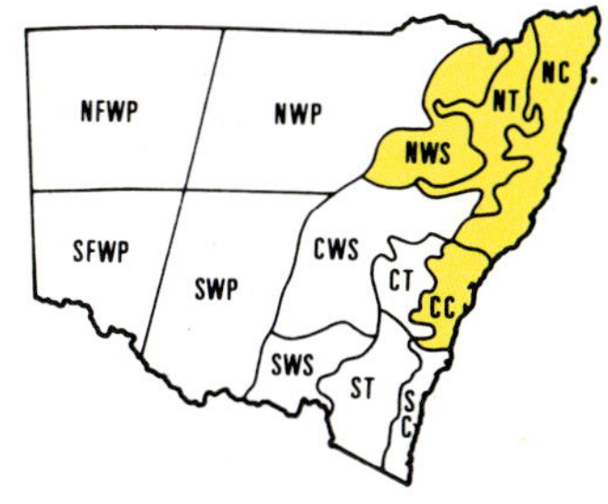

Left: *Scirpus mucronatus* on the banks of the Coldstream River

Right: Inflorescences of *Scirpus mucronatus*

CM

Scirpus prolifer

An introduced stoloniferous perennial to 50 cm high. Leaves reduced to a sheathing base. Stems soft, ± terete. Inflorescence terminal, a spherical cluster about 1.5 cm diameter of about 25 spikelets, but often with one or more branches longer than the spikelets bearing similar but smaller clusters of spikelets. The inflorescence is subtended by a few chaffy bracts shorter than the spikelets. Spikelets up to 1 cm long and about 2 mm wide, occasionally proliferating (producing vegetative shoots or plantlets). Perianth bristles absent. Stamens 3. Nut 3-angular. *Seed photograph page 439*.

Habitat

Small creeks and swamps, specially in disturbed areas.

Economic Significance

Not yet recorded from areas with flood irrigation, but could become a weed of supply channels.

Distribution

CC, SC, CT, Victoria, Western Australia, South Africa.

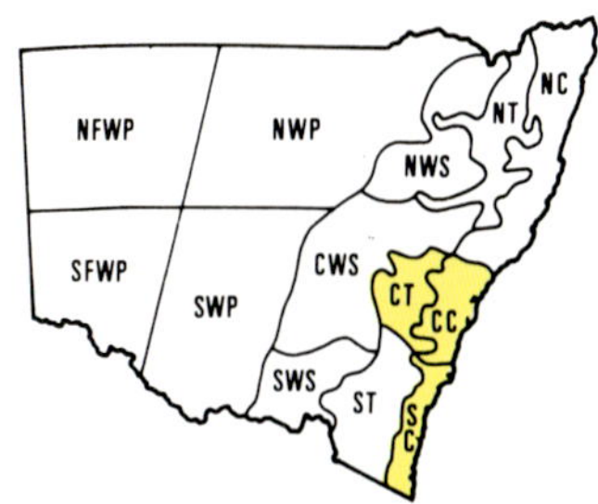

Left: *Scirpus prolifer* on the margins of a dam near Elanora

Right: Inflorescences of *Scirpus prolifer*

**River Clubrush
Great Bulrush**

Scirpus validus

A native erect rhizomatous perennial to 3 metres tall. Leaves reduced to sheathing bases or the occasional small blade developed. Stems terete, about 12 mm diameter with spongy pith and numerous incomplete septa. Inflorescence terminal, open (may be compact when young), consisting of numerous spikelets up to 1 cm long and 0.5 cm wide, subtended by chaffy bracts shorter than the inflorescence. Perianth bristles present (5-6). Stamens 3. Style branches 2. Nuts flattened, about 2 mm long. *Seed photograph page 439.*

Growth Biology

Summer flowering and fruiting with main growing period from October to April.

Habitat

Margins of slow-moving creeks and rivers, lakes and bore drains, and in swamps.

Economic Significance

Sporadically obstructive in drainage channels and small creeks. In areas of similar climatic conditions, such as California, U.S.A., this species of *Scirpus* is a problem. Where obstruction to water flow is not a consideration this is a useful species, preventing erosion and providing cover for wildlife.

Distribution

NC, CC, SC, NT, ST, NWS, NWP, SWP, SFWP, Queensland, Victoria, Tasmania, South Australia, Western Australia.

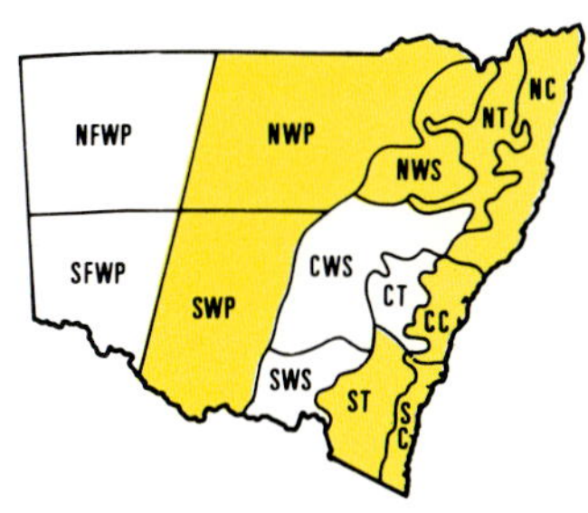

Left: *Scirpus validus* on the margins of a lake near Griffith

Right: Inflorescences of *Scirpus validus*

Submerged plant of
Aldrovanda vesiculosa

DROSERACEAE

Aldrovanda vesiculosa

A native submerged unattached perennial, stems to about 20 cm long, 2-3 cm across. Leaves in whorls, about 8 per whorl. Leaves modified to form insect "traps" on petioles 3-6 mm long. Traps ± circular, 3-5 mm diameter. Flowers solitary, axillary, bisexual, 6-8 mm diameter. Sepals 5, 3-4 mm long. Petals 5, white 5-6 mm long. Stamens 5. Ovary solitary. Style with 5 stigmatic branches.

Growth Biology

A tropical species which relies on trapping water fauna to supplement nutrient supply.

Habitat

Stationary or slow-moving water in lagoons or swamps, generally found within 1 metre of the surface.

Economic Significance

Nil in New South Wales.

Distribution

NC, Queensland, Northern Territory, Western Australia.

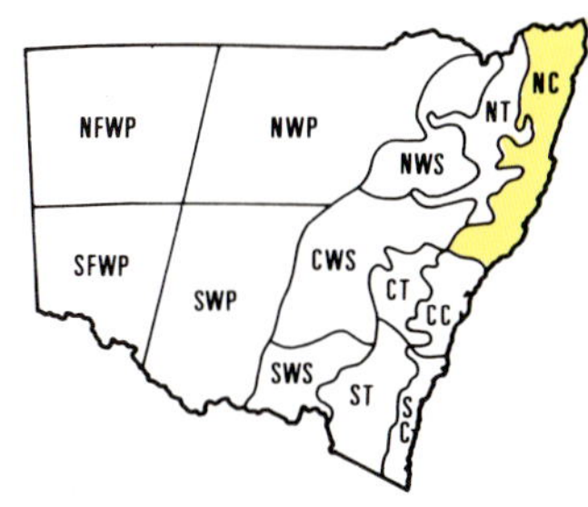

Aldrovanda vesiculosa clearly showing the insect traps

Elatine gratioloides
showing the weak stems
with widely-spaced pairs
of leaves common in
submerged plants

154

Waterwort

Elatine gratioloides

A native decumbent annual usually less than 20 cm tall. Stems rooting at the lower nodes. Leaves opposite, to 15 mm long and 5 mm wide, shortly petiolate with 2 small deciduous stipules. Leaf margins with minute glands at the ends of the lateral veins. Flowers bisexual, solitary, axillary with only one flower per leaf pair. Sepals 3, minute, membranous. Petals 3, about 1 mm long, pink. Stamens 3. Ovary 3-locular. Styles 3. Fruit 1-3 mm diameter with a terminal depression. Seeds 0.5-0.7 mm long, characteristically curved-cylindrical with ± hexagonal pits.

Habitat

In or on the margins of stationary or slowly-flowing water to about 40 cm deep. In rice crops it may appear as a green carpet submerged in shallow, clear water.

Economic Significance

Found frequently submerged on the unsown headlands and borrow-pits of rice crops in south-western New South Wales, but is not considered a problem.

Distribution

NC, CC, SC, NT, CT, ST, CWS, SWS, NWP, SWP, Queensland, Victoria, Tasmania, South Australia.

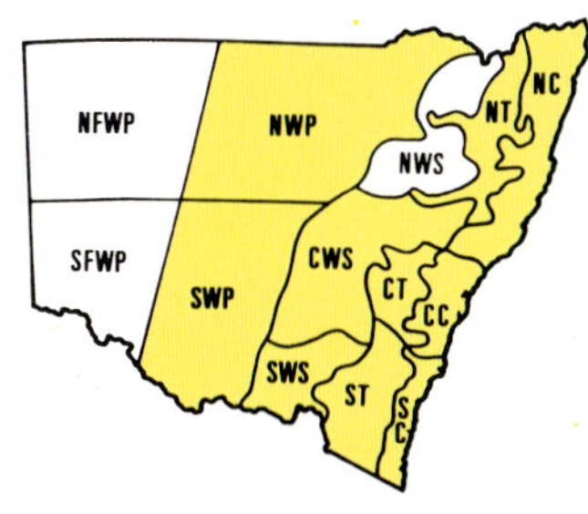

Submerged growth of *Elatine gratioloides* on the margins of a rice crop

A large family with many aquatic and semi-aquatic species. It is impossible to draw up a critical yet exhaustive list of grasses likely to be found in perennial or ephemeral wetlands in New South Wales. A key is provided to those genera which, under varying circumstances, could be considered to be aquatic or semi-aquatic or may be consistently associated with the margins of various water bodies. The species that are of economic significance on, in or around waterways are treated individually.

1. Fertile spikelets usually with 2 heteromorphous florets, the upper hermaphrodite or female, the lower male or neuter and then sometimes reduced to the lemma (rarely both hermaphrodite, as in *Isachne* spp.). Rhachilla not produced beyond the upper floret. Lower glume occasionally suppressed. Mature spikelets usually falling entire from their pedicels (disarticulate below the glumes), with the exception of *Arundinella*. Rarely subpersistent on a flat, indistinct and tardily disarticulating rhachis—as in *Stenotaphrum*.   **2.**
1a. Spikelets 1- to many-flowered, breaking up at maturity above the more or less persistent glumes, or if falling entire then not 2-flowered with the lower floret male or barren and the upper hermaphrodite, usually more or less laterally compressed or terete.   **27.**

2. Spikelets usually in pairs or triplets with one sessile and the other(s) pedicellate, or more rarely both pedicellate with one shorter and the other longer pedicellate, or rarely the pedicellate spikelet reduced to the pedicel or absent, those of each pair or trio alike in sex (homogamous) or different (heterogamous), falling entire at maturity from the usually articulate axes of variously arranged (solitary, digitate or paniculate), often spike-like racemes. Glumes often more or less rigid or at least firmer and longer than the lemmas (excluding awns). Lemmas membranous or hyaline, in the hermaphrodite or female spikelets the upper lemma often awned with a usually geniculate awn or reduced to an awn or muticous.   **3.**

2a. Spikelets solitary, or if in pairs or triplets then all virtually similar. Glumes usually herbaceous or membranous, more rarely indurated, the lower glume usually smaller than the upper or sometimes completely suppressed. Lower lemma usually resembling the upper glume in texture and often in size. Upper lemma from papery to very tough and rigid, usually awnless (occasionally mucronate).   **12.**

3. Internodes of the raceme-axis and also the pedicels stout, more or less thickened upwards, 3-angled or rounded, often more or less hollowed out on one side.   **4.**
3a. Internodes of the raceme-axis and pedicels slender, cylindrical or flattened. Lower floret of sessile spikelet neuter.   **7.**

Left: Tussocks of *Potamophila parviflora* and *Saccharum officianarum* (Sugarcane) growing in the Richmond River. Sugarcane and *Pennisetum pupureum* (Elephant Grass) are mixed with other introduced and native species on the banks.

4. Internodes and pedicels fused to form roughly semi-cylindric internodes. **5.**
4a. Internodes and pedicels not fused together. Upper lemma of sessile spikelet awned, hermaphrodite, the lower floret male. **6.**

5. Sessile spikelets more or less sunk in the hollow of the rhachis, upper fertile lemma awnless, the lower floret neuter. Spikelets of each pair more or less similar. Spreading decumbent grass with leaves less than 5 mm wide. *Hemarthria*
5a. Spikelets of each pair dissimilar, robust spreading decumbent grass with leaves more than 5 mm wide. *Rottboellia*

6. Racemes paired or digitate. Lower glume more or less winged on the keels, the upper not awned. *Ischaemum*
6a. Racemes solitary. Lower glume not winged, the upper awned with a setose awn. *Sehima*

7. Spikelets of each pair alike in sex and shape. **8.**
7a. Pairs (or trios) of spikelets dissimilar in sex and more or less in shape (or if those of some pairs of a raceme are alike then both male or barren). Fertile spikelets usually awned. **9.**
8. Caespitose perennial with narrow linear leaves. Racemes digitate or subdigitate, silky-villous with fulvous hairs. *Eulalia*
8a. Slender decumbent annual with lanceolate leaf-blades. Racemes 3-6, somewhat distant on a short axis, not manifestly villous.
Microstegium

9. Racemes in more or less compound espatheate panicles with the more or less elongated branches (peduncles) usually opposite or whorled on a more or less elongated axis. Joints and pedicels without a translucent middle line. Upper lemma of the sessile spikelet awned from the tip or 2-dentate or 2-lobed and awned from the sinus, rarely awnless. Caespitose grasses. **10.**
9a. Racemes spatheate or gathered into spatheate panicles. Racemes breaking up at maturity, the spathe and sometimes the lower (and then sterile) part of the raceme persistent. **11.**

10. Racemes 1- to 2- (rarely 3-) jointed, if regularly more than 1-jointed then the awn geniculate and twisted with a prominent column. Leaf-blades involute or convolute. *Chrysopogon*
10a. Racemes with 3 to several joints (rarely less on the same panicle). Awn bristle-like without a distinctly differentiated column, or absent. Leaf-blades flat or folded, the sheaths keeled. *Vetiveria*

11. Fertile spikelet either dorsally or laterally compressed, the lower glume more or less 2-keeled; callus short and obtuse; fertile lemma 2-fid and awned from the sinus, the awn usually glabrous. Racemes 2, at first approximate but soon sharply deflexed, the rhachis less markedly slender and fragile. Pedicellate spikelets developed.   ***Cymbopogon***

11a. Fertile spikelet more or less terete usually with a more or less pungent callus. Awn, or at least its column, hairy. Lower 1 to few pairs of spikelets usually sterile. Basal sterile spikelets forming a false involucre around the fertile spikelet on the 1-jointed racemes solitary in the ultimate spathes.   ***Themeda***

12. Main axis of the more or less spike-like panicle readily or tardily or partially disarticulating into segments, the spikelets not subtended by bristles or an involucre. Spikelets falling entire (sometimes tardily) from the rudimentary pedicels, solitary or 2-5 (rarely more) on short spike-like racemes that are more or less sunk in hollows on one side of a flattened, herbaceous or spongy, continuous or jointed common axis (at least the upper portion usually jointed and tardily disarticulating), or the longer racemes at least closely appressed to it. Stoloniferous perennials with rather broad, short, obtuse blades, short flowering culms and terminal and axillary inflorescences.   ***Stenotaphrum***

12a. Main axis of the panicle or racemes not disarticulating and the racemes persistent on the main axis, the spikelets not falling attached to the segments or, if so, then the spikelets or groups of spikelets variously subtended by bristles or by a spiny or bristly involucre.   **13.**

13. Spikelets (or some of them) or groups of spikelets subtended by 1 to several bristles which may be simple or connate into an involucre, or by branches bearing groups of bristles. Bristles or involucre falling with the spikelets at maturity. Bristles not united at the base, slender, smooth, antrorsely scabrous, ciliate or plumose, the spikelets solitary or in fascicles of 2-5.   ***Pennisetum***

13a. Spikelets not subtended by bristles or by an involucre (rarely subtended by a corona of very short, fine hairs).   **14.**

14. Both florets fertile, or if the lower floret male then its lemma very similar to the upper and indurated, chartaceous or thinly coriaceous. Glumes more or less equal and similar. Spikelets globular, small, awnless.   ***Isachne***

14a. Lower floret male or neuter, its lemma usually differing in texture or size from the lemma of the upper floret. Glumes usually unequal or dissimilar (rarely subequal and similar as in *Ottochloa* and then the spikelets not distinctly globular) or the lower (very rarely both) suppressed.   **15.**

15. Rhachilla disarticulating above the glumes and below the upper floret and not produced beyond it. Spikelets with 2 florets, the lower male or barren, usually with a palea. Glumes more or less persistent.

Arundinella

15a. Rhachilla disarticulating below the glumes.   **16.**

16. Fertile lemma thinly cartilaginous to indurate with more or less flat, thin or hyaline margins enveloping the margins of the palea; the minute, scale-like palea of the sterile floret often remaining attached to the base of the fertile floret. Spikelets borne on spike-like racemes that are digitately, subdigitately or racemosely arranged on a common axis.   **17.**

16a. Either the fertile lemma with more or less inrolled crustaceous margins and the palea of the sterile floret (if developed) not attached to the base of the fertile floret, or the spikelets not borne on slender racemes on a common axis, or both.   **18.**

17. Fertile lemma shortly awned, the margins firm but thinner than the back. Sterile lemma thinly cartilaginous like the fertile. Upper glume densely ciliate along the outer nerves, the rigid cilia at first adpressed then later spreading.

Alloteropsis

17a. Spikelets unawned, occasionally the fertile lemma terminating in a very minute, inconspicuous mucro. Fertile lemma with thin, flat, hyaline margins. Indumentum various but the upper glume not and the sterile lemma very rarely ciliate.

Digitaria

18. Spikelets awnless. Fertile lemma rugulose or smooth.   **19.**

18a. Either the glume(s) and/or the sterile lemma awned or cuspidate, or the partial rhachis produced into a long, awn-like point beyond the uppermost spikelet and then the upper glume and fertile lemma long acuminate. Fertile lemma smooth.   **26.**

19. Inflorescence of variously arranged (rarely solitary), simple or compound, usually more or less secund, often spike-like, dense or loose racemes, not an open, or contracted and cylindrical, panicle. Spikelets often in unequally pedicelled pairs or trios or solitary, alternately to the left and right of the median line of a usually triquetrous or compressed, dorsiventral rhachis.   **20.**

19a. Inflorescence either an open panicle or else contracted, cylindrical and spike-like.   **25.**

20. Back of the fertile lemma abaxial, the lower glume (if present) adaxial.   **21.**

20a. Back of the fertile lemma adaxial, the lower glume (if present) abaxial.   **22.**

21. Lower glume present and well developed. Racemes arranged racemosely along the main axis. ***Brachiaria***
21a. Lower glume absent or obsolescent. Racemes digitate or subdigitate. ***Axonopus***

22. Rhachis of the racemes produced into a distinct short bristle beyond the uppermost spikelet. Spikelets often very convex on the back and more or less depressed on the face, secund on the short, slender, sessile, spike-like racemes. ***Paspalidium***
22a. Rhachis of the racemes not produced into a bristle beyond the uppermost spikelet. **23.**

23. Lower glume little shorter than the upper and similar in appearance, both glumes shorter than and exposing the fertile lemma. Margins of the fertile lemma very narrowly subhyaline and minutely ciliolate upwards, entirely covering the tip of the palea. Delicate decumbent or scrambling perennial. ***Ottochloa***
23a. Lower glume very much shorter than the upper or absent. Spikelets shortly to very shortly pedicellate and usually more or less crowded on secund, spike-like racemes. **24.**

24. Lower glume generally absent. ***Paspalum***
24a. Lower glume present. Fertile lemma smooth and shining, acute or acuminately pointed, the margins inrolled below but more or less flat above and not enclosing the tip of the palea. Spikelets turgid, often stiffly hispid. ***Echinochloa***

25. Inflorescence a large or small, more or less open panicle. Spikelets not gibbous. ***Panicum***
25a. Inflorescence a contracted, cylindrical, spike-like panicle. Spikelets gibbous. ***Sacciolepis***

26. Rhachis not conspicuously produced beyond the uppermost spikelet. Lower glume one-third to one-half as long as the upper. Upper glume and sterile lemma both usually hispid with more or less stiff or bristly hairs. Upper floret hermaphrodite, the fertile lemma subcoriaceous or crustaceous, shining, the margins involute to near the tip then flat and not embracing the tip of the palea. Spikelets crowded in rows or irregularly along mostly simple, secund, spike-like branches of a panicle. ***Echinochloa***

26a. Partial rhachis conspicuously produced into a long awn-like point beyond the uppermost spikelet. Lower glume very much shorter than the upper. Upper floret usually female, much shorter than the lower staminate floret. Fertile lemma chartaceous but rather thin to subhyaline, embracing the tip of the palea. Panicles either loosely spike-like or spreading. ***Pseudoraphis***

27. Spikelets borne in open or contracted or spike-like panicles, less often in racemes or spikes and then with the lower or both glumes reduced or suppressed if on opposite sides of a continuous rhachis, or with 2 or more fertile florets if on one side of the rhachis. **28.**

27a. Spikelets sessile or shortly pedicellate along one side of the rhachis of digitate (rarely solitary or scattered) spikes or spike-like racemes and then with 1 fertile floret and 1- to 3-nerved lemmas, or on opposite sides of the rhachïs of solitary spikes or racemes. **53.**

28. Spikelets with 1 fertile floret, with or without 1 or 2 male or barren florets below it. **29.**

28a. Spikelets usually with 2 or more fertile florets, or if with 1 fertile floret then with sterile florets above it. **40.**

29. Stamens 3, 4 or 2. Palea 2-nerved. Glumes (or at least the upper glume) well developed or occasionally minute. **30.**

29a. Stamens usually 6. Palea 3- to 9-nerved. Glumes very small and hyaline or reduced to an obscure rim at the apex of the pedicel or suppressed. Aquatic grasses. **39.**

30. Spikelets with 3 florets, the terminal floret hermaphrodite, the lower 2 florets male or barren. All florets articulate above the glumes, the florets falling together. Both glumes longer than and concealing the florets. Sterile lemmas reduced to small scales adhering closely to the base of the fertile florets. Fertile floret chartaceous and tough, smooth and more or less shining though often thinly hairy, lanceolate to ovate. Spikelets strongly laterally compressed, the glumes strongly keeled and often winged on the keels. Panicle usually spike-like, occasionally somewhat interrupted and branching. ***Phalaris***

30a. Spikelets with 1 floret. Glumes (or at least the upper glume) well developed. **31.**

31. Spikelets falling entire at maturity, singly from the axis of slender, spike-like panicles or racemes. Lemma delicate 1- to 3-nerved. Spikelets borne singly and sessile or very shortly pedicellate on the continuous axis of a solitary, terminal spike or spike-like raceme. Lower glume absent. Upper glume awnless or mucronate or with a very slender awn up to 3 mm long. Low rhizomatous or stoloniferous perennials. ***Zoysia***

31a. Spikelets usually breaking up at maturity, the rhachilla disarticulating above the more or less persistent glumes, or if falling entire then with firmly membranous-awned or 5-nerved lemmas. **32.**

32. Lemmas more or less indurated and rigid at maturity, more or less terete or slightly dorsally compressed, with convolute or involute margins, tightly enveloping the grain, with a terminal awn or awns or cleft into usually awned segments, rarely awnless. **33.**

32a. Lemmas hyaline or membranous at maturity and awnless or dorsally awned, or more rarely terminally awned from the entire and obtuse or bifid or minutely toothed apex, or if the lemma indurated then the awn clearly dorsal or the glumes ciliate on the keels or the spikelets and lemmas more or less laterally compressed. **34.**

33. Lemma terminating in a single undivided awn. Lodicules 2 or 3. Callus short and blunt. Lemma elliptical, lanceolate or ovate, thinly crustaceous, smooth and shining, slightly dorsally compressed, with a slender, straight, often early deciduous awn, or awnless. ***Oryzopsis***

33a. Lemma 3-awned. Awns very unequal, the central much longer, very robust and geniculate and twisted below the bend, the laterals slender, straight or curved and not or scarcely twisted. Lemma convolute at the apex not concealing the palea which extends beyond the point of insertion of the awns. Spikelets very large, the glumes 3-5.5 cm long and more than 5-nerved. Rhachilla produced beyond the fertile floret, sometimes bearing an aborted floret. ***Anisopogon***

34. Lemmas 1- to 3-nerved, awnless. Glumes and lemma very similar in texture, thinly to very firmly membranous or hyaline, often somewhat shining. Grain free in a delicate pericarp, falling free from the lemma and and palea. ***Sporobolus***

34a. Lemmas usually 3- to 5-nerved, frequently awned. Glumes differing in texture from the lemma, usually longer and firmer than the hyaline lemma, but if shorter than the lemma or if the lemma is indurated then the glumes herbaceous membranous and dull. Grain usually with an adhering pericarp. **35.**

35. Rhachilla disarticulating below the glumes, the spikelet falling entire at maturity. Panicles dense and spike-like. **36.**

35a. Rhachilla disarticulating above the glumes. **37.**

36. Glumes both awned from the entire apex or from the sinus between the 2 obtuse, short lobes of the apex, the awns very delicate, 4-7 mm long. Lemma very thin, with a broad, blunt, minutely toothed tip, awnless or with an awn up to 2 mm long from the tip. ***Polypogon***

36a. Glumes awnless, usually more or less ciliate on the keel. Lemma membranous with a slender dorsal awn inserted in the lower third. ***Alopecurus***

37. Lemmas membranous or at least very thin, obtuse or truncate, smooth or softly hairy. Panicles spreading or contracted but not or scarcely spike-like. Spikelets (at least in the New South Wales species) less than 8 mm long. Palea as long as, to much shorter than the lemma or absent.
Agrostis

37a. Lemmas thinly chartaceous to indurated, more or less scaberulous or scabrous at least upwards, at least never completely smooth and shining. Palea usually well developed. **38.**

38. Spikelets less than 9 mm long. Lemma lanceolate, or broadly oblong (and then very small), usually more or less indurated at maturity and usually minutely (often 4-) toothed at the apex, minutely or shortly awned or rarely awnless, the awn not twice as long as the lemma.
Deyeuxia

38a. Spikelets 9-16 mm long. Lemma lanceolate to narrowly lanceolate-oblong, obtuse, very compressed and keeled, thinly chartaceous, scabrous or scaberulous upwards, awnless or with a minute dorsal awn behind the tip. Glumes not awned. Panicles spike-like, narrowly oblong to cylindrical, tapering upwards, dense with closely overlapping spikelets, pale. Coarse perennial of coastal sand-dunes, forming compact tufts and with stout, extensively branched rhizomes and very long, inrolled, rather rigid leaves. **_Ammophila_**

39. Glumes reduced to a hyaline rim. Floret strongly compressed, both lemma and palea 1-keeled and ciliate on the keels, the palea 3-nerved, the lemma 5-nerved, chartaceous. Slender, rhizomatous perennial.
Leersia

39a. Glumes small and hyaline. Floret not compressed or keeled. Palea 5-nerved, the lemma 7-nerved, both thinly membranous. Coarse tufted perennials found chiefly at the edges of or on rocky islets in moderately fast-flowing rivers. **_Potamophila_**

40. Lemmas or rhachilla joints bearing long silky hairs which envelop the lemma (at least in fertile florets); lemmas awnless or with a straight awn from the tip, often thin. Tall grasses with large plume-like panicles. **41.**

40a. Lemmas and rhachilla joints either glabrous or hairy, but in the latter case the hairs not enveloping the lemma or if so the lemma bearing a geniculate awn. Low to moderately tall grasses. **43.**

41. Leaves very long and crowded at the base of the culms. Spikelets unisexual. Lemmas of pistillate spikelets clothed with long hairs.
Cortaderia

41a. Leaves distributed along the culms. Tall stout "Reeds". **42.**

42. Lemmas hairy. Rhachilla joints naked.   ***Arundo***
42a. Lemmas naked. Rhachilla joints hairy.   ***Phragmites***

43. Lemmas with more than 3 nerves (usually 5). Lemma awned from the back or from the sinus of a 2-lobed tip with a geniculate or straight awn or mucro or awnless. Glumes and/or lemmas often with thin shining margins.   **44.**
43a. Lemmas 1- to 3-nerved.   **49.**

44. Lemma awned from the back, or, lower lemma awnless, upper lemma dorsally awned, apex entire or notched. Ligule membranous.   **45.**
44a. Lemmas entire or shortly 2- to 5-toothed at the apex, awnless or awned from the tip, the awn straight or curved but never geniculate.   **46.**

45. Spikelets mostly 4- to 10-flowered. Panicles loose or loosely contracted. Spikelets about 1 cm or more long, scarcely compressed. Lemma with a rather long geniculate awn from the back. Aquatic grasses.   ***Amphibromus***
45a. Florets 2, lowest floret hermaphrodite and awnless, the upper staminate with a short awn near the tip, the spikelets falling entire at maturity, compressed. Swampy ground, drains.   ***Holcus***

46. Spikelets dimorphous, the fertile and sterile intermixed on the same inflorescence. Fertile spikelets 2- to 3-flowered. Sterile spikelets with numerous, rigid, awn-tipped lemmas. Spikelets borne on one side of the main axis of a spike-like panicle. Annuals or perennials.   ***Cynosurus***
46a. Spikelets all alike on the same inflorescence.   **47.**

47. Lemmas laterally compressed and keeled (only slightly so in some species of *Poa* but then often with web-like hairs on the callus). Panicles usually loose and spreading at maturity, more rarely somewhat contracted. Spikelets small (less than 10 mm) awnless, borne in loose or contracted panicle.   ***Poa***
47a. Lemmas rounded on the back at least in the lower part, sometimes slightly keeled upwards. Nerves of the lemma parallel, not or scarcely converging at the summit, awnless. Spikelets in open or contracted panicles.   **48.**

48. Nerves prominent. Plants usually rather tall, inhabiting fresh-water marshes or forests. Margins of sheaths fused at least below. Lodicules firm, short, truncate, often fused. Hilum linear, as long as the grain.   ***Glyceria***
48a. Nerves indistinct. Plants rather low, inhabiting saline or alkaline soils or marshes.   ***Puccinellia***

49. Spikelets more or less densely arranged (or at least some or most of the spikelets overlapping) and very shortly pedicellate or sessile on a solitary terminal spike or digitate or subdigitate secund branches of the inflorescence. Spikelets not or only weakly compressed. Glumes slightly keeled. Lemmas rounded on the back, obtuse or obscurely notched, the lateral nerves obscure. Perennial, with rigid, rather tall culms. Creek banks.

Leptochloa

49a. Spikelets distinctly or shortly pedicellate or sessile in open or contracted or spike-like or interrupted panicles or spikes, or racemose along the branches of a panicle but not in digitate or simple spikes.

50.

50. Lemmas entire, awnless and not produced into awn-like points. Lemmas 3-nerved, various in texture. Spikelets in open or contracted panicles, rarely in racemes or few together in spike-like clusters on short branches.　　　　*Eragrostis*

50a. Lemmas distinctly or obscurely notched and then often with a small mucro from the sinus, or awned, or tapering into awn-like points, or at least truncate and more or less erose and then the spikelets racemose on elongate panicle branches.　　　　**51.**

51. Spikelets densely crowded on the almost globular or oblong, approximate or remote, spicate branches on the simple or sparsely branched main axis. Glumes and lemmas tapering into awn-like points.

Elytrophorus

51a. Spikelets racemose on elongate branches of the panicle, the lemmas truncate and more or less erose or notched and sometimes mucronate but not or scarcely awned.　　　　**52.**

52. Lemmas truncate and/or erose or minutely notched, not or scarcely mucronate mostly glabrous, rarely fringed with hairs.　　　*Leptochloa*
52a. Lemmas rather distinctly notched and mucronate, fringed with hairs in the lower part.　　　　*Diplachne*

53. Spikelets in 1 or 2 rows on one side of the usually continuous rhachis of digitate spikes or spike-like racemes. Lemma 1- to 3-nerved. Spikelets 1- to 2-flowered with only 1 fertile. Fertile floret without an imperfect floret above it though often the rhachilla produced and rarely bearing a minute abortive floret.　　　　**54.**
53a. Spikelets on opposite sides of the rhachis of solitary spikes or spike-like racemes.　　　　**55.**

54. Glumes shorter than the floret.　　　　*Cynodon*
54a. Glumes longer than the floret.　　　　*Brachyachne*

55. Lemmas 5- to 9-nerved, at length indurated. Spikelets 1- to many-flowered, solitary, not sunken in hollows in the rhachis, the latter continuous or articulated. Spikelets placed flat-wise to the rhachis. Ligule auriculate. Perennials. *Agropyron*

55a. Lemmas usually 1- to 3-nerved, hyaline or membranous. Spikelets 1- to 2-flowered, borne singly and more or less sunken in hollows or depressions in the articulate rhachis. **56.**

56. Glume 1, thin, very short, very much shorter than the floret. Lemma thinly indurated, shortly awned, 3-nerved. Spikelets distant on the very slender rhachis. *Psilurus*

56a. Glumes 1 or 2, coriaceous, longer than the floret. Lemma hyaline, awnless. Spikelets rather closely spaced on the rhachis. **57.**

57. Lower glume very small or suppressed in the lateral spikelets, the upper glume broad and facing the rhachis. Lemma of the first (and usually only) floret with its back adjacent to the rhachis of the spike. *Monerma*

57a. Both glumes well developed in the lateral spikelets, placed side by side in front of the cavity in the rhachis. Lemma with one side adjacent to the rhachis of the spike. *Parapholis*

Para Grass

Brachiaria mutica

An introduced strongly stoloniferous or ± decumbent perennial. Stolons up to 5 metres long, rooting at the nodes. Culms decumbent, rooting from the lower nodes, up to 2 metres high, nodes densely hairy. Sheaths hirsute, ligule a rim of hairs, blades up to 30 cm long, 6-20 mm wide. Panicle up to 20 cm long with few to several racemes along an elongated common axis. Racemes 2-8 cm long, becoming progressively shorter towards the top of the panicle. Spikelets 3-3.5 long, glabrous, with one bisexual flower and a lower male flower. Lower glume one-third to one-half as long as the spikelet. Upper glume the same size as the spikelet. Lower floret male, lemma similar to upper glume. Fertile floret approximately 3 mm long, finely rugose to almost smooth. *Spikelet photograph page 440.*

Growth Biology

Rarely sets fertile seeds. In Australia appears to be established vegetatively. Spread readily by the floating stolons becoming detached then re-establishing when lodged on a bank. From here it grows rapidly into the water and/or any adjacent cultivated land. Mainly summer-growing, and is frost susceptible.

Habitat

In shallow brackish or fresh water, swamps, seasonally inundated floodplains or on dry ground in high rainfall areas. The long stolons will float across 4-5 metres of slowly moving water.

Economic Significance

An important tropical fodder species, but in New South Wales important only as a serious weed of supply and drainage channels and as a weed of sugarcane. A rapid summer grower, it can successfully compete with the fastest-growing crops.

Distribution

NC, Queensland, Northern Territory, Tropics.

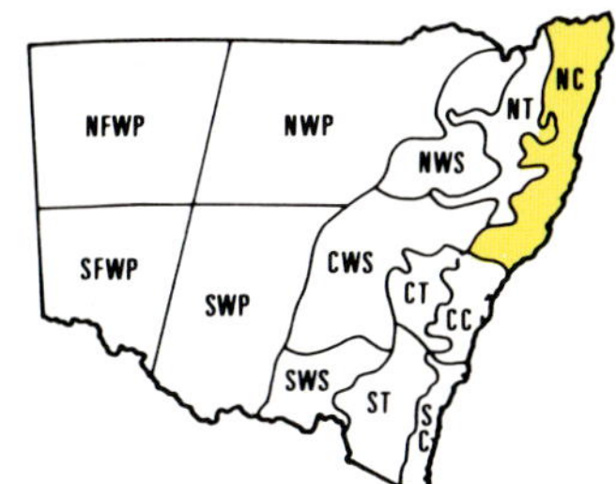

Left: *Brachiaria mutica* lining a flood mitigation channel on the North Coast.

Insert: Inflorescences of *Brachiaria mutica*

CM

Brown Beetle Grass

Diplachne fusca
(including ***D. muelleri***)

A native erect annual or short-lived perennial to 1 metre high. Ligule membranous, laciniate, approximately 5 mm long, blades up to 30 cm long, 2-6 mm wide, flat or inrolled. Panicle erect, open, 10-50 cm long. Spikelets with 4-many bisexual flowers, a characteristic leaden-green colour before maturity. Glumes 1-nerved, unequal, 3-5 mm long. Lemmas ± dorsally flattened, 4-5 mm long, 3-nerved with hairs on the nerves below the top, with a small mucro in an apical notch. *Seed photograph page 440.*

Habitat

Permanent or ephemeral water up to 80 cm deep. In drains, rice crops.

Economic Significance

Regarded as a pasture species with great potential overseas. In New South Wales a weed of rice crops. Reduces water flow through small, intermittently flowing channels.

Distribution

NC, CC, CT, NWS, CWS, SWS, NWP, SWP, NFWP, Queensland, Northern Territory, South Australia, Western Australia, Africa, Asia.

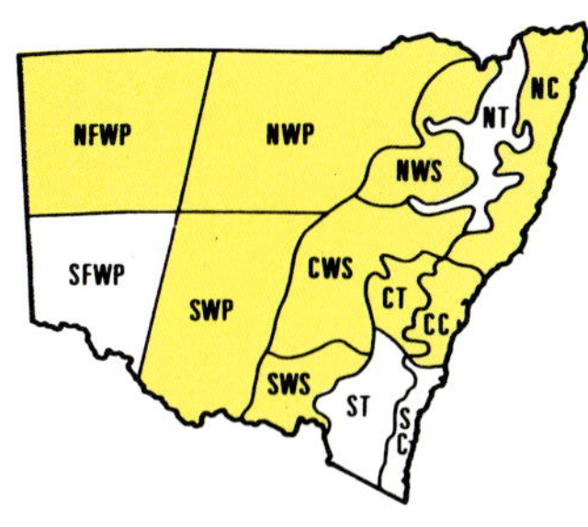

Left: *Diplachne fusca,* a common weed of rice

Right: *Diplachne fusca* in ponds at the Inland Fisheries Research Station at Narrandera

Echinochloa

Mostly annual caespitose or decumbent grasses frequently associated with ephemeral waterways and rice crops. Inflorescence a panicle of secund spike-like branches. Spikelets with an upper hermaphrodite and lower sterile floret, usually ovate to elliptical, frequently awned, usually hispid on the nerves of the glumes and the lower lemma. Fertile lemma ovate to elliptical, very convex on the back, hard, smooth and shining, the margins inrolled holding in all but the tip of the palea.

Echinochloa colona showing the characteristic open inflorescence

Awnless Barnyard Grass

Echinochloa colona

A probably native decumbent, occasionally ± erect annual to 60 cm high. Ligule mostly absent, blades 5-30 cm long, 3-8 mm wide, sometimes with transverse purple or black bands. Panicle erect, 4-15 cm long, 6-20 mm wide with 2-many branches, each 1-2.5 cm long. Spikelets crowded, acute, 2-3 mm long. Lower glume up to half the length of the spikelet, acute. Upper glume as long as the spikelet, 5-7 nerved. Sterile lemma similar to upper glume, awnless. Fertile lemma 1.5-2.5 mm long, obscurely 5-nerved. Palea slightly shorter than lemma. *Spikelets and florets photograph page 440.* The absence of awns on the spikelets differentiates this from some forms of *E. crus-galli.*

Habitat

Creek and river banks, drains, channel banks, rice crops. Tends to grow in damp soil rather than truly aquatic.

Economic Significance

Can be a serious weed of cultivation in cash crops. May also cause problems in other summer crops. Does not thrive in as wide a climatic zone as *E. crus-galli;* nevertheless, it is a major weed of the world.

Distribution

NC, CC, NWS, CWS, NWP, SWP, Queensland, Victoria, Northern Territory, Western Australia, Tropics.

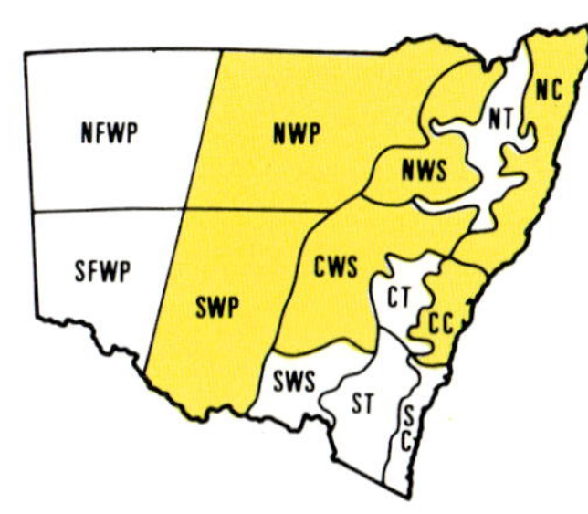

Echinochloa colona on the margins of a rice crop

Barnyard Grass   *Echinochloa crus-galli*

A very variable, more or less erect annual, usually less than 1.5 metres tall. Ligule absent, leaf blades flat, up to 35 cm long and 5-25 mm wide. Leaves hairless, except for occasional hairs at base, with faintly serrated margins and may carry variable reddish markings. Panicle erect, up to 25 cm long, 8 cm wide with few to many spike-like racemes, the axis 3-angled, scabrous. Lower racemes up to 10 cm long and decreasing in length towards the top. Spikelets 2-4.5 mm long, acute, awn small to very obvious up to 5 cm long. Lemma 2-3.2 mm long, palea almost as long. *Spikelets and florets photograph page 440*. Ligules absent, a characteristic that enables distinction of *Echinochloa crus-galli* from rice seedlings.

Growth Biology

In summer seeds are produced in profusion and a single plant has yielded up to 7000 seeds. Seeds germinate readily with an optimum germination temperature of 20°C.

Habitat

Ideally in cultivated land inundated with water such as rice crops and beds of flood mitigation and drainage channels.

Economic Significance

E. crus-galli is a major weed of rice, germinating with the crop and providing vigorous competition from an early stage. It grows faster than rice and tests have shown dense stands in the growing crop may remove over half the soil nitrogen, and this, plus crowding, may reduce yields up to 30 per cent. In temperate and tropical crops it is ranked as one of the three worst weeds of the world.

Distribution

NC, CC, SC, NT, CT, ST, NWS, CWS, SWS, NWP, SWP, NFWP, Queensland, Victoria, South Australia, Western Australia.

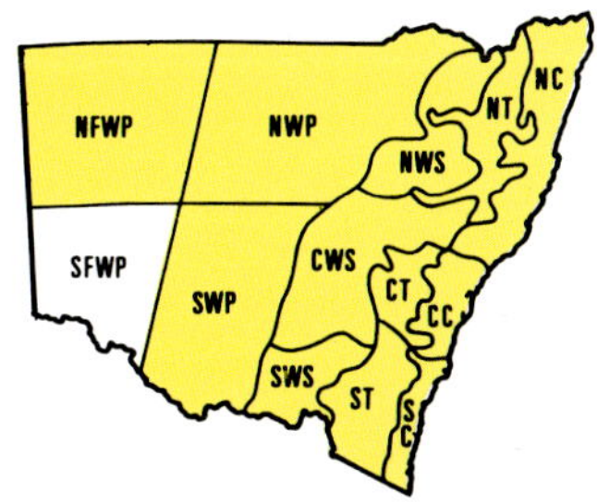

Left: *Echinochloa crus-galli* in a rice crop near Griffith

Right: Inflorescences of different forms of *Echinochloa crus-galli* (A, B, C) and *E. oryzoides* (D), both weeds of rice

Hairy Millet

Echinochloa oryzoides

An introduced erect annual to 1.5 metres high. Sheaths loose around the culms. Ligule absent. Leaves to 20 cm long, 9-12 mm wide. Panicle 10-25 cm long, initially erect, later drooping. The panicle branches are up to 4 cm long, shorter towards the top. Spikelets 4-7 mm long, 2-2.5 mm wide, in 2s or 3s. Lower glume one-third to one-half the length of the spikelets, acute, 3-5 nerved. Upper glume 5-7 nerved, scabrous-pubescent, the length of the spikelets, acuminate. Sterile lemma the shape and size of the spikelet with a scabrous awn up to 5 cm long. Fertile lemma 3.5-4.5 mm long including the short, 0.5 mm tip. Palea almost as long as the lemma. *Spikelets and florets photograph page 440.*

Habitat

Rice crops, drains, channels and other irrigation works.

Economic Significance

A weed of rice crops, but not nearly as significant as *E. crus-galli.*

Distribution

CWS, SWS, SWP, Europe

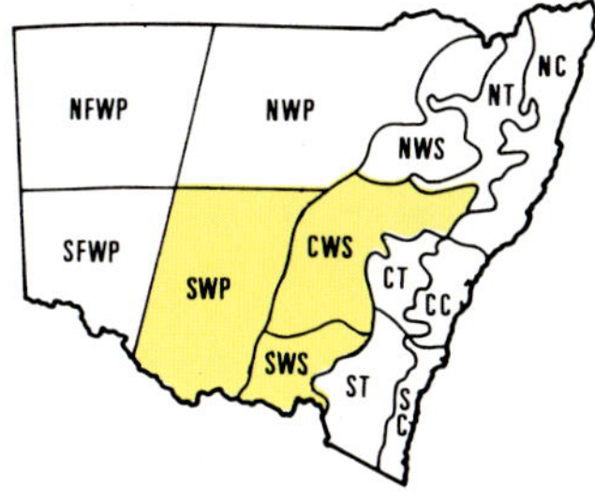

Left: The characteristically drooping inflorescences on a plant of *Echinochloa oryzoides*

Swamp Barnyard Grass

Echinochloa telmatophila

A native robust erect or slightly decumbent annual to 2 metres high. Culms often stout and up to 1 cm diameter. Ligule absent, blades flat, to 35 cm long and 18 mm wide. Panicle dense, 20-35 cm long, erect or slightly drooping. Panicle branches 2-10 cm long. Spikelets 3-4.2 mm long, in 2s or 3s. Lower glume one-third to one-half the length of the spikelet, 3-5 nerved, acute. Upper glume as long as the spikelet with an awn up to 7 mm long, 7-nerved. Sterile lemma similar to the upper glume, 7-nerved, spinulose on the lateral nerves with an awn 1-4 cm long. Fertile lemma 3-4.1 mm long, 1.25-1.5 mm wide, cuspidate. Palea almost as long as lemma. *Spikelets and florets photograph page 440.*

Habitat

Creek and river banks, creek beds, drainage channels.

Economic Significance

Can reduce flow in drainage or supply channels.

Distribution

NC, CC, Queensland, Western Australia.

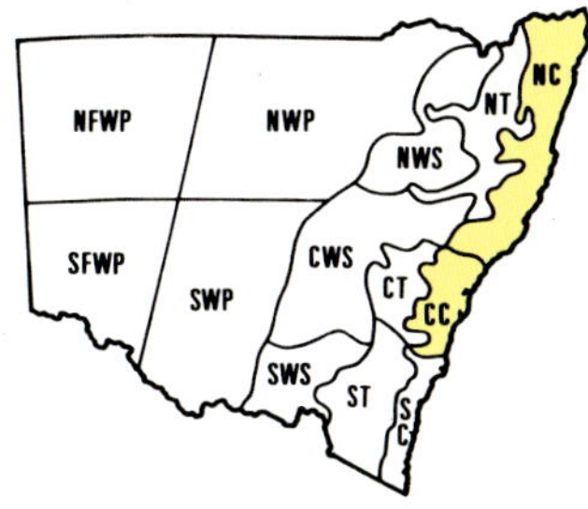

Left: The gently curving inflorescences and broad leaves of *Echinochloa telmatophila*

Canegrass *Eragrostis australasica*

An erect, loosely divaricate native perennial to 3 metres high. Culms hard, much-branched, approximately 5 mm in diameter at base, glaucous, virtually leafless. Inflorescence open, loose, drooping, up to 30 cm long, terminal or axillary. Spikelets 5-many-flowered, 5-12 mm long. Lemma obtuse, 2-3 mm long, 3-nerved with the lateral nerves closer to the midrib than the margins.

Habitat

Claypans and other seasonally inundated slowly flowing waterways and lakes in the west of the State.

Economic Significance

Cover for native fauna and wild pigs. Rarely eaten by stock.

Distribution

ST, NWS, SWS, NWP, SWP, NFWP, SFWP, Queensland, South Australia.

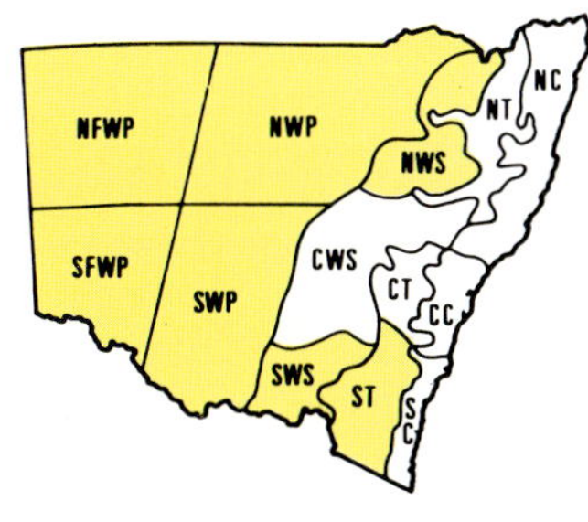

Left: *Eragrostis australasica* on the margins of a claypan near the Bulloo Overflow

Right: Inflorescences of *Eragrostis australasica*

Glyceria

Erect rhizomatous or stoloniferous perennials up to 2 metres high. Ligules membranous, usually prominent, blades flat or folded, up to 2 cm wide. Inflorescence a narrow or open panicle up to 50 cm long. Spikelets with 3-many bisexual flowers disarticulating above the glumes and between the florets, 5-12 mm long. Glumes 1-5 nerved usually shorter than the lowest lemma. Lemmas obtuse, entire or toothed, awnless.

There are four species of *Glyceria* in New South Wales. They can be distinguished by the following key.

1. Lower glume 3-5 nerved, 7-8 mm long.   *G. latispicea*
1a. Lower glume 1-nerved, 1.5-6 mm long.   **2.**

2. Lower lemmas 7-10 mm long.   *G. australis*
2a. Lower lemmas 3-5 mm long.   **3.**

3. Lemmas 3-toothed, spikelets 12-25 mm long.   *G. declinata*
3a. Lemmas entire, spikelets 5-12 mm long.   *G. maxima*

Glaucous Sweetgrass

Glyceria declinata

An erect introduced perennial to 1.5 metres tall. Ligule to 5 mm long, blades flat, to 30 cm long and 1 cm wide. Panicle erect, more or less reduced, with only the lower branches more than 2 cm long, sparse. Spikelets 4-15 flowered, 13-25 mm long. Glumes 1-nerved, 2-3 mm long. Lemmas 4-5 mm long, obtuse, 3-toothed. *Seed photograph page 440.*

Habitat

Creek or river banks, swampy ground, drains or channels.

Economic Significance

May cause some problems in reducing water flow in small drainage channels, but not a common species.

Distribution

SC, ST, SWP, Europe, America.

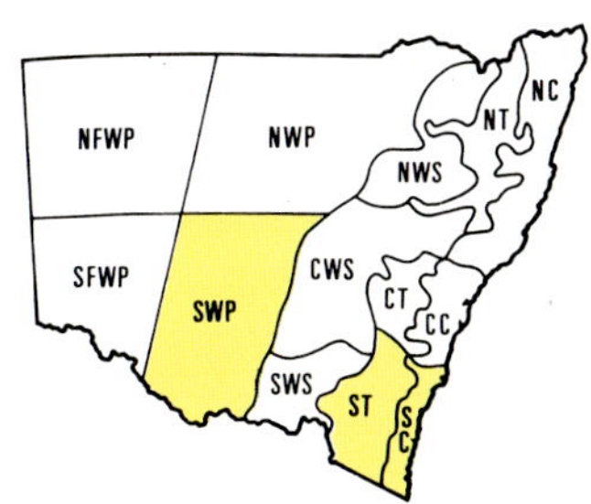

Left: *Glyceria declinata* in a drainage channel

Right: *Glyceria declinata*

Glyceria maxima on the
banks of a channel near
Rochester, Victoria

Reed Sweetgrass Water Meadow Grass

Glyceria maxima

Robust introduced perennial to 2.5 metres high. Ligule membranous, approximately 5 mm long, blades up to 40 cm long, 2 cm wide. Panicle open to 35 cm long. Spikelets numerous, with 4-8 bisexual flowers, 5-12 mm long. Glumes 1-nerved, 2-3 mm long. Lemmas obtuse, entire, 2-4 mm long. *Spikelets and florets photograph page 440.*

Habitat

Creek and river banks, in water on lagoon and dam margins up to 1 metre deep, swampy ground, drains and channels.

Economic Significance

Planted as a pasture species on swampy ground, but can impede flow in creeks, channels or drains.

Distribution

NT, CT, ST, SWS, SWP, Queensland, Victoria, Europe, Asia.

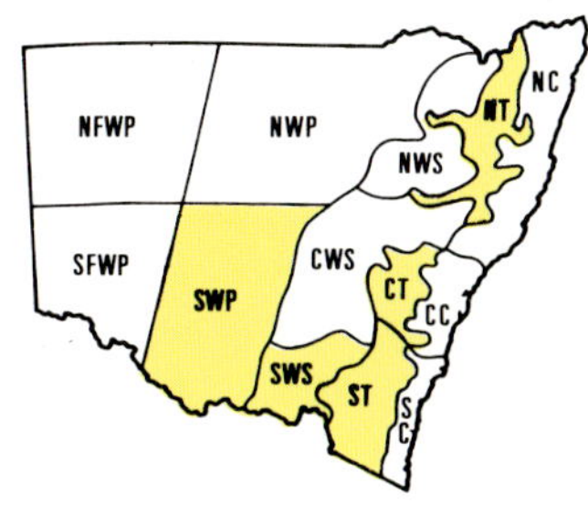

Inflorescences of *Glyceria maxima*

Inflorescences of
Hemarthria uncinata

Matgrass

Hemarthria uncinata

Decumbent, creeping native perennial to 80 cm tall. Ligule extremely short. Blades flat, or folded when dry, to 15 cm long, 2-5 mm wide, with a few scattered hairs near the junction of sheath and blade. Racemes solitary, terminal, 6-14 cm long, 0.5 cm wide, with spikelets closely appressed to the rhachis. Sessile spikelets 6-10 mm long, pedicel fused to the axis. Pedicellate spikelet slightly longer and narrower, pedicel free. Lower glume tapering into a variable point, as long as the spikelet, 7-8 mm long. Upper glume similar in length. Sterile lemma 2-nerved, shorter than the glumes. Fertile lemma nerveless, shorter than the glumes. Palea small, nerveless.

Habitat

Swamps or damp ground, creek banks.

Economic Significance

Perhaps of some value as a pasture species on damp ground.

Distribution

NC, CC, SC, NT, CT, ST, NWS, CWS, SWS, Queensland, Victoria, Tasmania, South Australia, Western Australia.

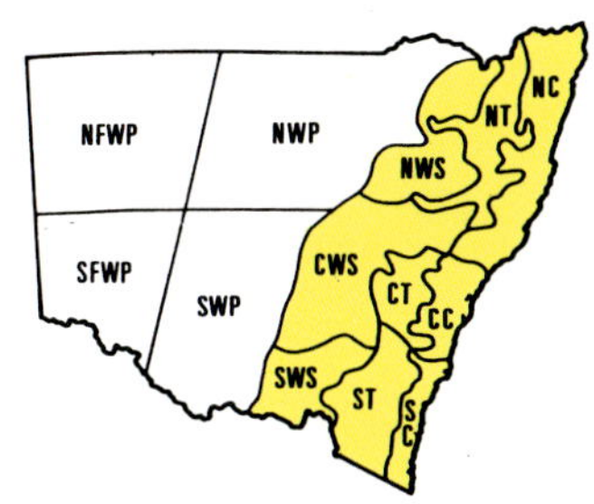

Hemarthria uncinata beneath *Melaleuca quinquenervia* in Centennial Park, Sydney

CM

Swamp Millet

Isachne globosa

A native decumbent or semi-erect perennial to 70 cm high. Ligule a rim of cilia 1.5-3 mm long. Blades to 12 cm long, 3-8 mm wide. Panicle finally open, more or less pyramidal, 2-10 cm long with spikelets mostly produced towards the ends of the branches. Spikelets solitary, 2-2.5 mm long. Glumes subequal to the spikelet. Lower floret staminate, 2-2.5 mm long, about 1 mm wide. Upper floret female or bisexual, approximately 1.75 mm long on a short stipe. Palea the same length as the lemma. *Spikelets and florets photograph page 441.*

Habitat

On river and creek banks, in lagoons, swamps or on seasonally inundated floodplains.

Economic Significance

Nil.

Distribution

NC, CC, SC, NT, CT, SWS, Queensland, Victoria, South Australia, Asia.

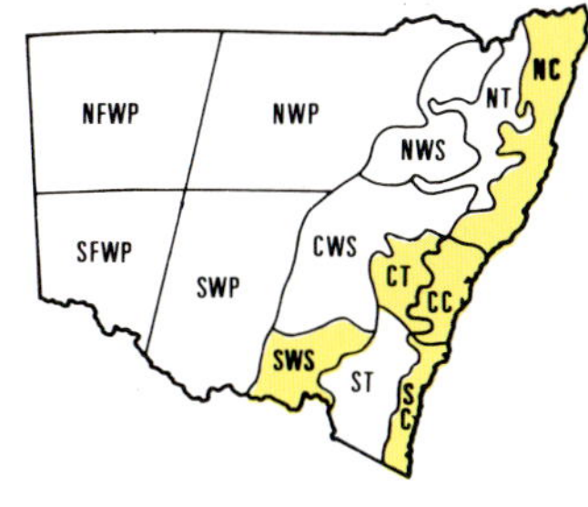

Left: Plants of *Isachne globosa* with the characteristic open inflorescence with stiffly spreading branches

Right: *Isachne globosa* on the banks of a creek running into the Colo River

Leersia

Perennial (or annual) rhizomatous and/or stoloniferous species, frequently with trailing decumbent stems. Inflorescence a stiff to lax ± open panicle. Spikelets flattened, bisexual, 1-flowered. Glumes reduced to a rim on the top of the pedicel. Lemmas 5-nerved, usually with a row of stiff hairs along the keel enclosing a subequal palea also with a row of stiff hairs along the keel. Stamens 6, 2 or 3 in Australia. Lodicules 2.

Swamp Rice Grass

Leersia hexandra

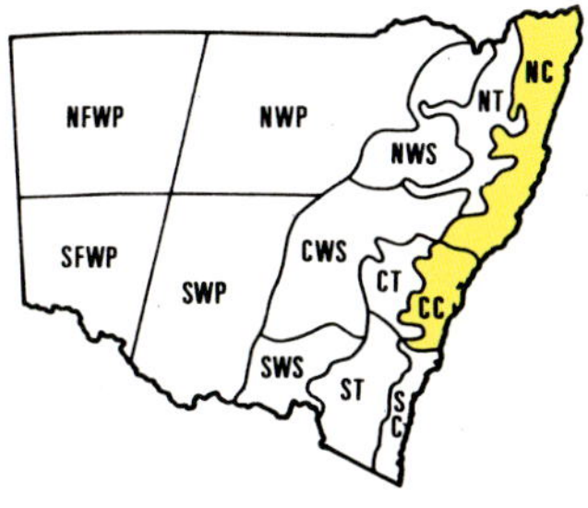

A native rhizomatous decumbent or erect perennial to approximately 1 metre high. Culms may root at lower nodes and frequently float on the surface of the water. Ligule membranous, 1-3 mm long, blades usually 5-20 mm long, 3-10 mm wide, usually lightly scabrous on both surfaces. Panicle 5-15 cm long, open or closed depending on stage of development. Branches of panicle usually alternate, 3-10 cm long with 1-5 racemes of spikelets on each branch. Each raceme with 3-10 shortly pedicellate spikelets. Spikelets with 1 bisexual floret, glumes reduced to an obscure minute persistent rim on the pedicel. Lemma 5-nerved, 3.4-5 mm long, 1.2-1.5 mm wide, keeled, with the keel and margins ciliate. Palea slightly shorter than the lemma. Stamens 6, grain falling within lemma and palea. *Spikelets photograph page 441.*

Growth Biology

Little is known of the biology of this plant. Seeds few or absent and spread is vegetative.

Habitat

River and creek banks, swamps, drains and channels. Will grow in water less than 1 metre deep, but usually grows out from the bank into water.

Economic Significance

In New South Wales it is common on north coast wetlands. Reportedly a weed of rice in Queensland and occasionally a weed in irrigation canals and flood mitigation drains. Also a valuable fodder with a crude protein content around 12 per cent and low in fibre content. In Venezuela, India, Burma and Angola it is a significant grazing plant.

Distribution

NC, CC, Queensland, Northern Territory, Tropics.

Left: *Leersia hexandra* on the banks of the Coldstream River
Insert: Inflorescences of *Leersia hexandra*

Open and partially
enclosed inflorescences of
Leersia oryzoides

Leersia oryzoides

An introduced rhizomatous scabrous perennial up to 1.5 metres tall. Culms decumbent to erect. Leaves up to 40 cm long and less than 1.5 cm wide; ligule membranous, 1-2 mm long. Inflorescence normally an open panicle up to 25 mm long, sometimes remaining enclosed within the upper sheath. Spikelets numerous, bisexual, flattened, 4-7 mm long; glumes reduced to a narrow rim at the top of the pedicel; florets 1; lemma with stiff hairs along the keel, 5-nerved, enclosing a sub-equal palea also with a line of stiff hairs along the keel. Stamens 2 or 3.

Growth Biology

The plants spread rapidly vegetatively in or on the margins of wet areas. The flowers are mostly chasmagamous (opening to exsert anthers and stigmas) but the inflorescence may remain enclosed in the upper sheath and then the spikelets are cleistogamous (remaining closed—usually leading to self-pollination). Cleistogamy apparently occurs during colder weather. Establishment by seed does not seem to have been very common over the years, but this species now seems to be spreading.

Habitat

In or near margins of slowly flowing water or on the banks of swiftly-flowing waterways.

Economic Significance

Potentially a serious weed in rice crops, particularly in sod-sown crops and rice land cropped on a short rotation.

Distribution

SWP, Victoria.

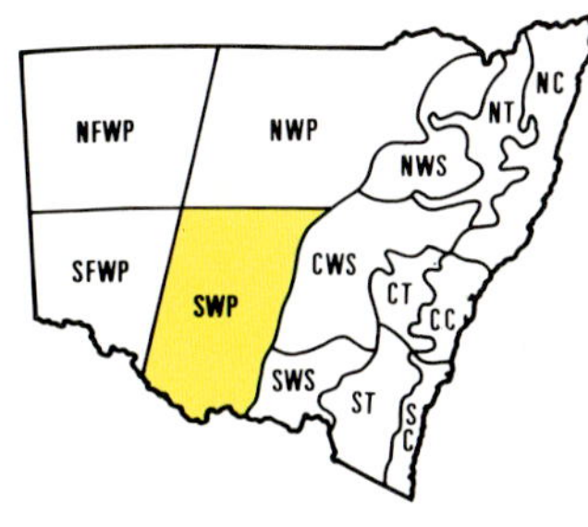

Leersia oryzoides in a supply channel near Leeton

Umbrella Canegrass

Leptochloa digitata

A native robust, shortly-rhizomatous perennial often over 2 metres tall. Stems rigid, often branched, to 5 mm diameter. Leaves to 6 mm wide, usually less than 12 cm long. Inflorescence of 6-20 subdigitately arranged racemes, each to 10 cm long. Spikelets arranged in 2 rows along one side of the raceme, crowded, 2.5-4 mm long, usually with 3 or more bisexual spikelets. Glumes unequal, 1-2 mm long. Lower lemmas about 2mm long, 3-nerved. Paleas shorter than the lemmas. *Seed photograph page 441.*

Habitat

Banks or islands of ephemeral watercourses in semi-arid summer dominated rainfall areas.

Economic significance

In its natural habitat *L. digitata* is an important erosion-limiting species. It may become a problem in some irrigation schemes in semi-arid summer rainfall areas.

Distribution

NWS, CWS, NWP, SWP, NFWP, SFWP, Queensland, Northern Territory, South Australia, Western Australia.

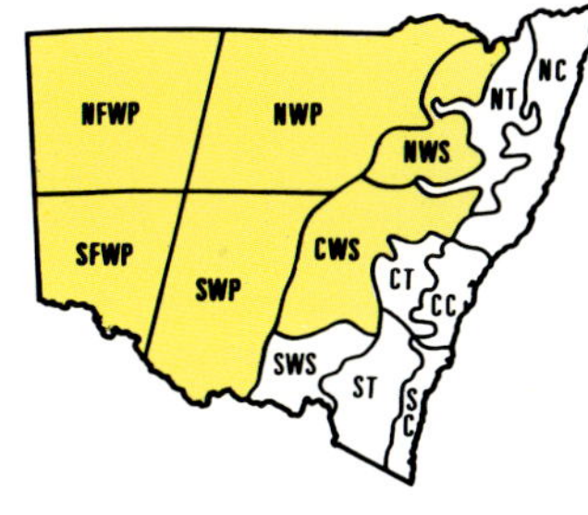

Left: *Leptochloa digitata* lining a supply channel near Warren

Right: A clump of *Leptochloa digitata* showing the stiff cane-like stems and digitate inflorescence

Panicum

Stoloniferous, rhizomatous or caespitose perennials or annuals. Ligules
ciliate or membranous. Inflorescence an open or contracted panicle.
Spikelets disarticulating below the glumes with the upper floret bisexual,
the lower sterile or male. Lower glume usually shorter than the upper.
Upper glume usually subequal to the spikelet. Sterile lemma often similar
to upper glume. Fertile lemma firm, usually smooth, with inrolled margins
holding a slightly shorter palea.

Blackseed Panic

Panicum bisulcatum

A native decumbent short-lived perennial, rooting at the nodes. Culms to 70 cm high. Ligule membranous, lightly hairy, 0.5-0.75 mm long. Blades to 15 cm long, 4-12 mm broad. Panicle open, to 40 cm long and nearly as broad. Spikelets with 1 lower sterile and 1 upper fertile floret, 2.4-3 mm long. Lower glume about half the length of the spikelet. Upper glume similar in size to spikelet. Sterile lemma subequal to the upper glume. Fertile lemma approximately 2 mm long. Palea subequal to the lemma. *Spikelets photograph page 441.*

Habitat

Creek and river banks and swamps along creeks and rivers. May grow into the water.

Economic Significance

Helps reduce erosion of creek and river banks.

Distribution

NC, CC, NWS, Queensland, Victoria.

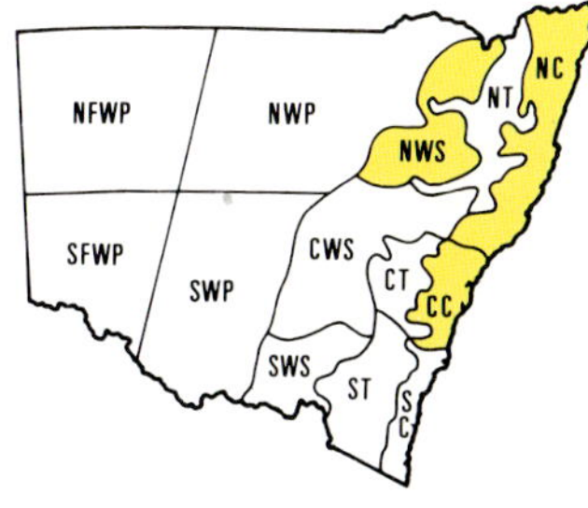

Left: *Panicum bisulcatum* on the banks of the Coldstream River

Right: Inflorescences of *Panicum bisulcatum*

Swamp Panic   *Panicum paludosum*

A native robust erect annual or short-lived perennial to 1.5 metres high. Culms erect, semi-floating or floating. Ligule membranous, hairy above, approximately 3 mm long. Blades to 30 cm long, 8-20 mm wide, flat. Panicle erect, open, to 45 cm long. Spikelets with 1 sterile and 1 bisexual floret 3.5-4.5 mm long, approximately 1 mm wide. Lower glume one-quarter to one-third as long as the spikelet. Upper glume the same size as the spikelet. Sterile lemma similar to the upper glume. Fertile lemma 2.5-2.75 mm long. *Florets photograph page 441.*

Habitat

In stationary or slowly-flowing streams, or lagoons in water to 1 metre deep.

Economic Significance

May reduce flow in some drains or channels.

Distribution

NC, Queensland, Northern Territory.

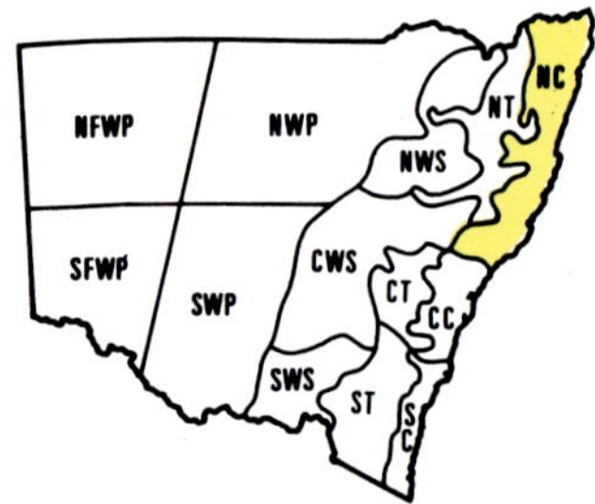

Torpedo Grass

Panicum repens

An introduced strongly rhizomatous perennial. Rhizomes wiry, to 7 metres long, apices sharp, covered with hard, thickened scales. Culms ascending, up to 1 metre high, clothed below with cataphylls (bladeless sheaths). Nodes and sheaths glabrous. Ligule truncate, ciliate above, 0.5-1.0 mm long. Blades to 20 cm long, 2-8 mm wide, flat or folded. Panicle ± exserted, 7-15 cm long, ± open. Spikelets 2.2-2.6 mm long, approximately 1 mm wide. Lower glume one-fifth to one-third as long as the spikelet. Upper glume same size as spikelet. Lower floret staminate, lemma similar to upper glume. Upper floret bisexual, approximately 2 mm long and 1 mm wide.

Growth Biology

Seeds are rarely produced in New South Wales and little is known about germination in this species. The plant spreads vigorously by its branching rhizome system, which can survive reasonably long dry periods, frost, and intermittent flooding. The plant is summer growing, and the tops are susceptible to frost.

Habitat

Mainly coastal sandy soils, but can become established in heavier soils and in drains and intermittently used channels.

Economic Significance

A serious weed in crops and pastures and difficult to eradicate. While not significant as a weed of drains and channels, it can spread easily along such systems, invading agricultural land. Because of its hardiness, *P. repens* could be used for erosion control in areas remote from agricultural land. In other areas its use is not recommended.

Distribution

NC, CC, CWS, NWP, Mediterranean, tropical and sub-tropical.

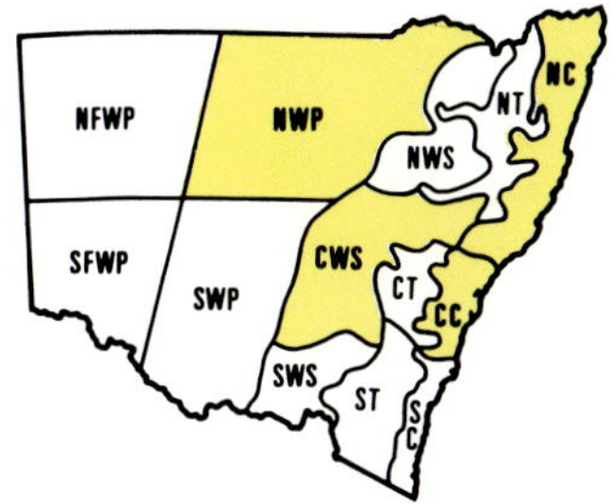

Paspalum

Rhizomatous or caespitose perennials. Ligules usually membranous. Inflorescence of (rarely 1-) 2-many 1-sided spike-like racemes along a common axis. Spikelets solitary or paired, awnless, dorsally compressed, with 1 bisexual and 1 sterile floret. Lower glume usually absent, rarely minute. Upper glume and sterile lemma subequal, as long as the spikelet. Fertile lemma tough, with inrolled margins clasping a palea as long as the lemma.

Paspalum

Paspalum dilatatum

An introduced more or less caespitose perennial with ± erect culms to 1.7 metres high. Ligule membranous, assymetrical, to 3 mm long. Blades to 45 cm long, 3-12 mm wide. Panicle ± erect or nodding, 7-25 cm long, with usually 3-7 racemes each 2.5-11 cm long. Spikelets paired, 2.8-4 mm long and about 2-2.3 mm wide. Lower glume absent. Upper glume and sterile lemma subequal, sparsely pubescent, both or either fringed with fine silky hairs. Fertile lemma 2.4-2.6 mm long. *Spikelets and florets photograph page 441.*

Growth Biology

Summer growing and flowering. Frost sensitive. Seeds float and spread rapidly with water flow.

Habitat

Almost cosmopolitan, a common roadside plant, or a planted pasture species. Can also grow and survive on banks of drains and channels in areas of low rainfall.

Economic Significance

A very useful and important pasture species, but in irrigation areas it grows on the banks of smaller supply channels, and the inflorescences drag in the water sufficiently to reduce water flow. In small (less than 25 megalitre per day flow) concrete-lined channels, inflorescences trailing in the water have been measured to reduce design capacity flow by 20 per cent. Broken inflorescences tend to block regulatory structures. Also a serious weed of sugarcane, citrus, vineyards, golf courses and lawns. Inflorescences may become infested by ergot *(Claviceps paspali)* and the sticky exudate is a nuisance where the plant grows near walkways.

Distribution

NC, CC, SC, NT, CT, ST, NWS, CWS, SWP, Queensland, Victoria, South Australia, Western Australia, South America.

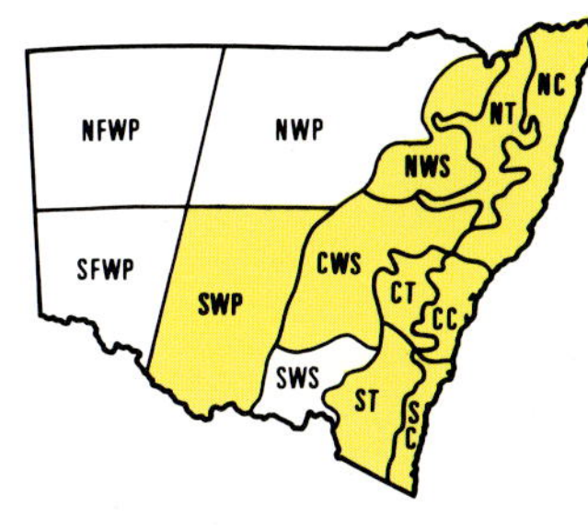

Left: *Paspalum dilatatum* overhanging a concrete channel near Yenda

Right: Inflorescences of *Paspalum dilatatum* on the left and *Paspalum paspalodes* on the right

Water Couch

Paspalum paspalodes

[Synonym ***P. distichum***]

An extensively rhizomatous and stoloniferous native perennial to 50 cm high, forming dense mats or turf. Ligule membranous, 0.5 mm long. Blades to 15 cm long, 2-7 mm wide, flat or folded. Inflorescence of 2 (rarely 3 or 4) more or less spreading racemes, each 1.5-7 cm long. Spikelets usually solitary, occasionally paired, 2.5-3.5 mm long. Lower glume usually absent, occasionally developed as a minute scale. Upper glume and sterile lemma both equal in length to the spikelet with the glume minutely pubescent. Fertile lemma 2.5-2.8 mm long.

Another native species, *P. distichum* (Saltwater Couch) [Synonym *P. vaginatum*] has both the glume and sterile lemma glabrous and grows in brackish or saline coastal habitats, but is otherwise similar to the closely related *P. paspalodes*.

Growth Biology

Flowers early to mid-summer; seeds maturing in late summer. Frost sensitive. Growth is rapid once ambient temperatures reach 20°C. Numerous seeds and floating pieces of rhizomes ensure rapid spread.

Habitat

Wet or damp situations in shallow water, or, if slowly-flowing water, floating across the water surface.

Economic Significance

Can be a useful pasture species in wet places, but its persistence makes it difficult to eradicate from supply and drainage channels. In New South Wales *P. paspalodes* is the most obstructive emergent plant infesting drainage systems. In summer it spreads quickly along the margins of drainage channels, and will encroach into water up to about 1 metre deep, severely obstructing flow and causing siltation.

Distribution

NC, CC, NT, CT, ST, NWS, CWS, SWP, NFWP, Queensland, Victoria, Northern Territory, South Australia, Western Australia, warm, temperate and tropical.

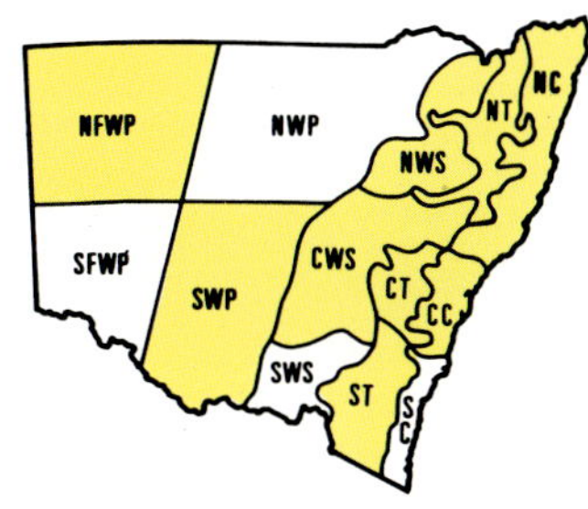

Left: *Paspalum paspalodes* growing out from the banks of a drainage channel near Leeton

Right: Inflorescences of *Paspalum paspalodes*

Reed Canary Grass

Phalaris arundinacea

An introduced robust rhizomatous perennial to 2 metres high. Ligule membranous, 2.5-16 mm long. Blades 10-35 cm long, 6-25 mm wide. Panicle erect, lanceolate or oblong, dense, but may be interrupted below, 5-40 cm long, 1-4 cm wide. Spikelets 3.5-7.5 mm long, laterally compressed, usually with obvious green nerves when young, becoming straw-coloured when older. Glumes persistent, subequal, 3-nerved, usually wingless. Fertile floret subtended by 2 equal sterile glumes, hairy above, 1.2-2.3 mm long, almost half as long as fertile lemma. Fertile lemma 5-nerved with scattered hairs above, smooth below, 2.5-4.5 mm long. Palea about as long as the lemma. Seed falling with lemma, palea and sterile florets. *Florets photograph page 441.*

Habitat

Swampy ground, channel and drain banks, creek banks.

Economic Significance

Occasionally planted as a pasture species on swampy ground. May become a problem in small supply and drainage channels. A major weed of irrigation channels in the Columbia Basin, Washington, U.S.A.

Distribution

NC, CC, SC, NT, CT, ST, Victoria, Asia, Africa, America.

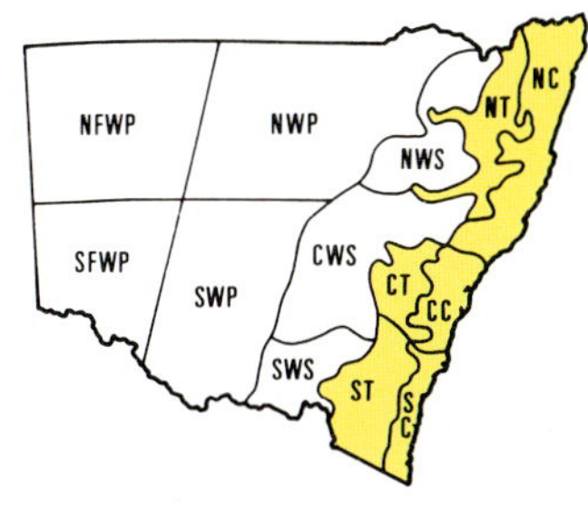

Left: *Phalaris arundinacea* near Sydney

Right: Inflorescences of *Phalaris arundinacea*

Common Reed

Phragmites australis

A native erect robust rhizomatous perennial to 3 metres high. Culms rigid, many-noded, to 1 cm diameter, hollow internodes, leaves produced all up the stem. Ligule a fringe of hairs, blades to 70 cm long, 1-3 cm wide. Panicle to 40 cm long, drooping at the tip, green with a purple tinge at first, turning white on maturity, dense, soft, many-branched. Spikelets 10-16 mm long, 2-several flowered, the lowest floret male, upper florets bisexual. Lower glume 5-6 mm long, upper glume 10 mm long. Rhachilla with long silky hairs. Fertile lemmas subulate, 1-3 nerved, up to 10 mm long, glabrous. Palea less than half the length of the lemma. *Seed photograph page 441.*

Two other closely related species are also naturalised and may be seen near wetlands. *Arundo donax* is similar to *Phragmites australis* but is a much more robust plant with broader leaves and taller culms. *A. donax* also has hairy lemmas and glabrous rhachillas. *Cortaderia selloana* ("Pampas Grass") differs from both *Phragmites* and *Arundo* in having its long (1 metre) scabrous leaves crowded at the base of the culms (caespitose) and in having unisexual spikelets (often unisexual plants). In *Cortaderia* the lemmas of the pistillate spikelets are covered with long hairs.

Growth Biology

Most stands have vertical and horizontal rhizomes. Verticals bear the aerial shoots and horizontals are the regenerating agent and main food store of the plant[1]. Horizontal rhizomes perpetuate the plant and do not directly bear aerial growth. Buds are formed on the vertical rhizome and develop throughout the year. Development is slower in cold winters. The plants may grow all year, but over most of its range *P. australis* dies back after frosting. Spread is mainly vegetative, both from growth of the rhizome and rhizome dispersal, but fertile seeds are produced and germinate to produce seedlings in either the same or the following year. It appears that germination occurs in only a narrow range of habitats.

Habitat

Grows in fresh or slightly brackish water up to 2 metres deep, mainly on a mud substrate but occasionally on sand. May also grow in seasonally inundated areas with a high watertable.

Left: *Phragmites australis* in a drainage channel

[1] HASLAM, S.M. The biology of reed *(Phragmites communis)* in relation to its control. Unpublished data.

A mature inflorescence, stem base, leaf base and cut-open internode of *Phragmites australis*

Economic Significance

In irrigation districts and flood mitigation channels it is a major pest, encroaching into water 2 metres deep, ultimately reducing the flow of the largest canals. It is also a weed of poorly drained agricultural land, particularly in high rainfall areas.

This plant has the important attributes of giving cover for animals, preventing wave or current erosion to banks and providing useful grazing for stock. Harvesting of *P. australis* in Romania has been developed to the extent that thousands of tonnes of the plant have been used for the manufacture of cardboard, printing paper, insulation and thatching materials. There are many other uses, and since the 1950's, 5000 square kilometres of wetland at the mouth of the Danube River have been managed to ensure maximum productivity.

Control

Stands are effectively reduced in size by grazing or cutting in mid-summer. Burning is less damaging than cutting. The plant requires wet conditions for maximum growth and drainage of the surrounding land is an obvious method of control. It is killed by continuous exposure to sea-strength salinity. Sea flooding of a stand for a few days will severely damage the plant.

Distribution

NC, CC, SC, CT, ST, CWS, SWS, NWP, SWP, Cosmopolitan.

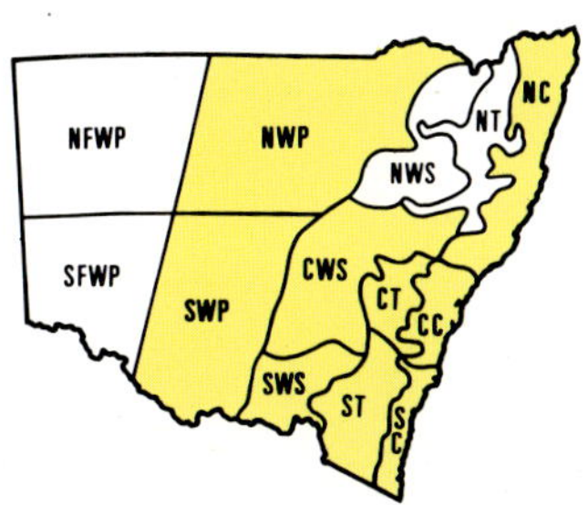

Annual Beardgrass *Polypogon monspeliensis*

An introduced decumbent or ± erect annual to 70 cm high. Blades to 20 cm long, approximately 1 cm wide. Panicle dense, spike-like, up to 10 cm long and 2 cm wide. Spikelets with 1 bisexual floret, falling entire with the glumes. Glumes 2 mm long, ciliate, notched, with a 4-7 mm long awn arising from the notch. Lemma slightly shorter than the glumes, awnless or with an awn up to 2 mm long. *Florets and seed photograph page 442.*

Growth Biology

In south-western New South Wales commences growth in winter, flowering and maturing in early summer. Numerous viable seeds are produced which germinate readily in moist conditions.

Habitat

Wet or damp ground, drains, ditches, gutters or drying mud.

Economic Significance

In early summer dense growth will block water movement in drains or channels. Occasionally a minor weed of irrigated crops.

Distribution

NC, CC, SC, CT, ST, CWS, NWP, SWP, NFWP, Victoria, Tasmania, Western Australia, Europe.

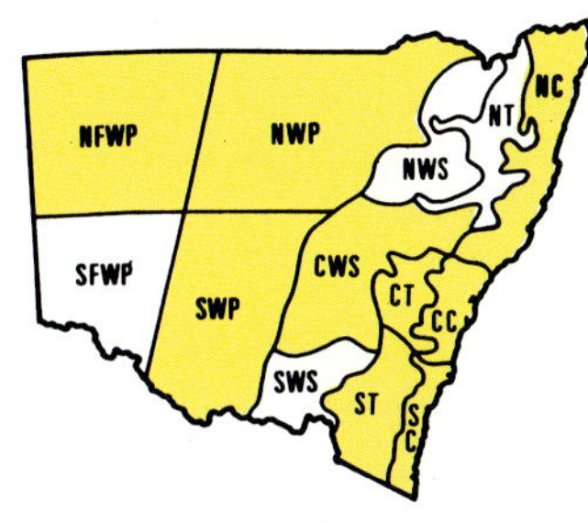

Left: *Polypogon monspeliensis* showing the fluffy green inflorescences

Right: *Polypogon monspeliensis* in a drainage channel near Hanwood

Potamophila parviflora

A native erect rhizomatous perennial to 1.5 metres high. Culms stout, to 7 mm diameter. Ligules firmly membranous, 5-15 mm long. Blades erect, to 50 cm long, 4-6 mm wide. Panicle loosely contracted, 15-45 cm long, 2-5 cm wide. Spikelets numerous, 3-5.5 mm long, 1-1.25 mm wide. Glumes reduced to an obscure or obvious persistent rim on the pedicel. Sterile lemmas 2, scale-like, one-eighth to one-quarter the length of the spikelet. Fertile lemma subtending a bisexual, or a reduced unisexual, flower subequal to the spikelet, 5-nerved. Palea subequal to the lemma.

Habitat

On the banks or beds of fast-flowing rivers. When growing in the bed it is found on stony areas that are usually exposed for about half the year.

Economic Significance

Useful for erosion control in some circumstances.

Distribution

NC.

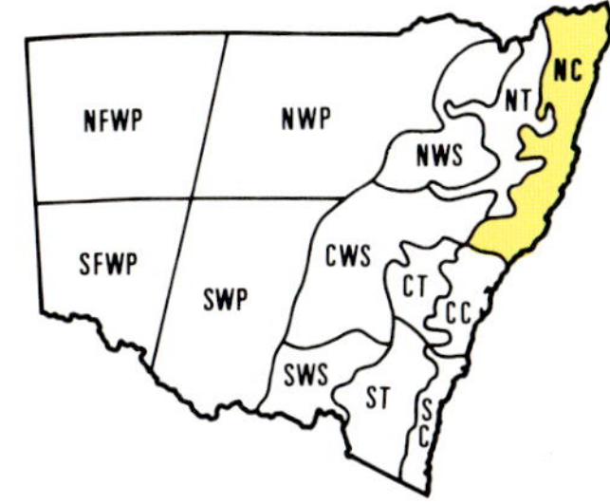

Left: Spikelets of
Potamophila parviflora

Right: *Potamophila parviflora* in the Richmond River

Inflorescences of (A)
Paspalum paspalodes
and (B) *Pseudoraphis
spinescens*

214

Spiny Mudgrass

Pseudoraphis spinescens

A native stoloniferous perennial forming a turf on mud and an intertwining mat of floating stolons on water. Ligule firmly membranous below, ciliate above, 1.5-3 mm long. Blades flat or folded with inrolled margins to 12 cm long and 2-7 mm wide. Panicle more or less pyramidal, 5-11 cm long with numerous rigid spreading branches each produced into a terminal bristle longer than the last spikelet. Spikelets distant, several along each branch, mostly 5-8 mm long (including apex of glumes). Lower glume 0.5-0.8 mm long. Upper glume tapering to a long point. Lower floret staminate, lemma shorter than upper glume. Upper floret bisexual, lemma 1.5 mm long, 0.6 mm broad. Vegetative growth may be confused with *Paspalum paspalodes*.

P. spinescens is very similar to Slender Mudgrass *P. paradoxa,* which differs by having only 1 (or sometimes 2) spikelets on each branch and in having a fertile lemma approximately 3.25 mm long.

Habitat

Both species grow in shallow water or mud beside creeks, rivers, lagoons, drains or channels and slowly-flowing water, out across the water surface.

Economic Significance

P. spinescens may reduce water flow in drains and channels.

Distribution

CC, NWS, CWS, SWS, NWP, SWP, SFWP, Queensland, Victoria, Northern Territory, South Australia, Western Australia, Asia.

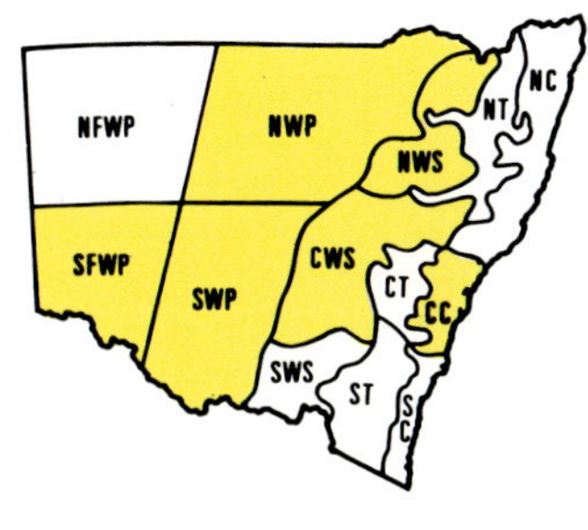

Pseudoraphis spinescens in a channel near Leeton

Water Milfoil

Myriophyllum

Mainly submerged annuals or perennials. Stems from quite short ± erect to long and flexible. Submerged leaves whorled or opposite, usually pinnately divided. Emergent leaves whorled, opposite or alternate, usually less divided than the submerged leaves. Inflorescence emergent, of several axillary clusters of 1-few flowers or spike-like. Although flowers mostly solitary, the leaves are whorled and frequently each leaf in the whorl subtends a "solitary" flower, resulting in a ring of flowers. Flowers bisexual or unisexual, frequently with male flowers at the top, female at the base, and if bisexual flowers present, these in between, each with 2 bracteoles. Male flowers with 4 minute sepals, 4 larger petals, 4 or 8 stamens. Female flowers with a calyx tube and a 4-celled ovary.

There are about 40 species around the world, about 18 in Australia, 4 common in New South Wales and sometimes pests of flowing or static water. Most have finely dissected submerged leaves which are described by the generic name *Myriophyllum* from the Greek "myrios", innumerable and "phyllon", leaf.

The genus *Myriophyllum* is currently under revision and the status of all these names is not clearly understood. A key is provided here to all the species thought to occur in New South Wales. Descriptions are provided only for the more commonly encountered and/or the more economically important species.
This key is based mainly on the emergent vegetative parts. The absence of emergent leaves can only be ascertained with surety while flowers are present.

1a. Submerged leaves once or twice divided into slender filiform branches or undivided and filiform or absent; emergent leaves entire. **2.**
1b. Submerged leaves pinnate, rarely absent and then with emergent leaves dentate or pinnate. **3.**

2a. Emergent leaves filiform; submerged leaves filiform. *M. filiforme*
2b. Emergent leaves flattened; submerged leaves filiform, once or twice divided into filiform segments, or absent. *M. pedunculatum*

3a. Emergent leaves rarely produced, inflorescence an interrupted spike with flowers subtended by bracts. Not definitely recorded from New South Wales. *M. spicatum*
3b. Emergent leaves produced (although not always present), flowers axillary. **4.**

4a. Emergent leaves entire. **5.**
4b. Emergent leaves divided or dentate. **7.**

5a. Leaves whorled. *M. elatinoides*
5b. Leaves opposite or alternate. **6.**

6a. Leaves opposite. *M. pedunculatum*
6b. Leaves alternate. *M. trachycarpum*

7a. Leaves divided (i.e. lateral lobes extending at least half way to mid vein). **8.**
7b. Leaves dentate (i.e. teeth very much shorter than the width of the entire central part), (upper younger leaves may be cylindrical ± entire). **12.**

8a. Leaves divided almost to mid-vein. **9.**
8b. Leaves divided so that entire central portion about as wide as the lobes are long. **10.**

9a. Leaves with at least 6 or more pairs of segments, pale glaucous green, feather-like. *M. aquaticum*
9b. Leaves with less than 5 pairs of segments. *M. integrifolium*

10a. Leaves glaucous, ± reddish, ovate to oblong, 2 mm or more wide. *M. verrucosum*
10b. Emergent leaves green, linear, less than 2 mm wide. **11.**

11a. Leaves whorled. *M. propinquum*
11b. Leaves alternate. *M. gracile*

12a. Leaves narrow, entire part less than 2 mm wide. *M. propinquum*
12b. Leaves broad, entire part more than 2 mm wide. **13.**

13a. Mature emergent leaves less than 1.5 cm long. *M. elatinoides*
13b. Mature emergent leaves 2-3 cm long. *M. latifolium*

Top: *Myriophyllum aquaticum* in a shallow pool in the Sydney region

Left: *Myriophyllum aquaticum* showing the whorled emergent leaves, finer submerged leaves and thick rhizome

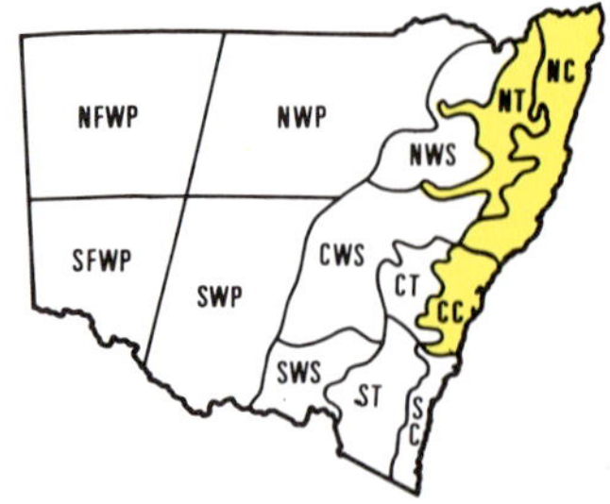

**Brazilian
Water Milfoil**

Parrots Feather

Thread of Life

Myriophyllum aquaticum
[Synonym *M. brasiliense*]

An introduced dioecious stoloniferous perennial. Stems rooting at the lower nodes, finally decumbent. Leaves whorled (4-6), a characteristic light glaucous-green, pinnate ("feathery"), 2-4 cm long. Submerged leaves rotting early, eventually leaving only the emergent leaves at the end of a long, bare, submerged stem. Female plants only recorded in Australia. Flowers axillary, solitary, with 2 bracteoles about 1 mm long. Sepals 4, about the same size, stigmas creamy, hairy. In deep water the submerged feathery leaves may be more evident, similar to *M. elatinoides*. Readily distinguished from other species of *Myriophyllum* by the pale-green emergent foliage.

Growth Biology

Flowers through most of the year in warm coastal situations. Spreads by stem fragments.

Habitat

An attractive aquatic plant that grows well in aquaria and ornamental ponds. In New South Wales observed in static or moving water to 2 metres deep, rooting in mud or gravel. Grows best in water containing high levels of nitrogen. Tolerant of a wide temperature range, thriving in water from 8°C to 30°C.

Economic Significance

A native of South America, *M. aquaticum* is a declared noxious plant for the whole of New South Wales. The classification "noxious" should be questioned. Naturalised for many years, it is seldom reported as troublesome. Dense growths can usually be attributed to high nutrient levels. Being valuable in the aquarium trade, it has spread to the United States of America, Japan, Java and New Zealand. In Java the tips are eaten as a vegetable. Care should be taken to prevent further imports as if male plants become established, seeds could be produced with as yet unknown consequences.

Distribution

NC, CC, NT, Victoria.

The broad erect emergent
leaves and finely divided
submerged leaves of
Myriophyllum elatinoides

220

Coarse Water Milfoil

Myriophyllum elatinoides

Cattail

A native procumbent monoecious perennial with stems up to 1 metre long, rooting at the lower nodes with leaves in whorls (4-6). Emergent tips procumbent on surface. Submerged leaves from 1-3 cm long, 1-2 cm wide, pinnate with 6-8 pairs of leaflets. Emergent leaves 0.5-1 cm long, less than 5 mm wide, dentate near the water surface, entire further up the stem. Flowers unisexual, axillary, solitary, sessile, with 2 white bracts about 1 mm long. Male flower 3-4 mm long: sepals 4, white, 0.3-0.4 mm long; petals white or pink to 3 mm long; stamens 8. Female flower with minute sepals and 4 petals about 0.5 mm long. The emergent leaves are paler green than *M. propinquum*.

Growth Biology

Similar to *M. propinquum*.

Habitat

Still, slowly-flowing fresh or slightly brackish water. Mainly creeks and supply channels, occasionally lagoons and anabranches. Not persistent on intermittently flooded land.

Economic Significance

Occasionally severely obstructs small to medium-sized supply channels. In the 1950's was a major weed of supply channels in Victoria. In New South Wales it has seldom persisted as an obstructive plant for more than one season.

Distribution

Scattered. Plains to above snowline. CT, ST, SWS, SWP, Victoria, Tasmania, South Australia, Western Australia.

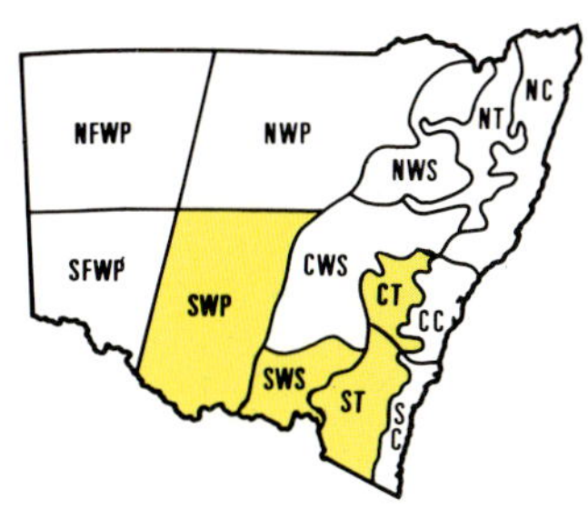

Myriophyllum elatinoides in a supply channel near Yenda

Right: Emergent shoots of *Myriophyllum propinquum* with the smaller submerged leaves visible below the water surface

Below: *Myriophyllum propinquum* in a supply channel near Yoogali

Top: The effect of the beetle *Haltica ignea* on the same area of *Myriophyllum propinquum* photographed a few weeks later

Left: The beetle *Haltica ignea* on *Myriophyllum propinquum*

Common Water Milfoil

Myriophyllum propinquum

A native procumbent perennial with stems about 2 metres long, rooting at the lower nodes and with an emergent stem up to 20 cm long. Leaves in whorls of 4-8, dark green. Emergent leaves 1-4 cm long, about 1-2 mm wide, entire or dentate. Submerged leaves 1-6 cm long, 0.5-4 cm wide, pinnate with 6-8 pairs of leaflets. Flowers unisexual, axillary, solitary, subtended by 2 bracteoles less than 1 mm long. Male flower 3-4 mm long; sepals 4, about 1 mm long; petals 4, 3-4 mm long. Female flower without calyx tube; stigmas with reddish or white hairs; seeds less than 1 mm long. *Seed photograph page 442.* Recognisable by its above-surface leaves and stems with the shape and branch arrangement of a pine tree. Differentiation of the submerged growth of *M. propinquum, M. elatinoides* and *M. aquaticum* is difficult.

Growth Biology

Main growth is in spring and summer, and flowering is sustained for long periods at this time. Flowering also recorded through most of the year in warm situations. Fruits chiefly in summer.

Habitat

Thrives in still or slowly moving fresh water up to 1 metre deep, with floating stems extending into deeper water. Also found in dampland surrounding swamps and lakes.

Economic Significance

A weed of small supply and drainage channels, frequently severely blocking flow. During the summer a small black beetle, *Haltica ignea,* is frequently found eating the emergent sections of the plant. At times this beetle or its larvae will significantly reduce plant density.

Distribution

All regions except NFWP, cosmopolitan.

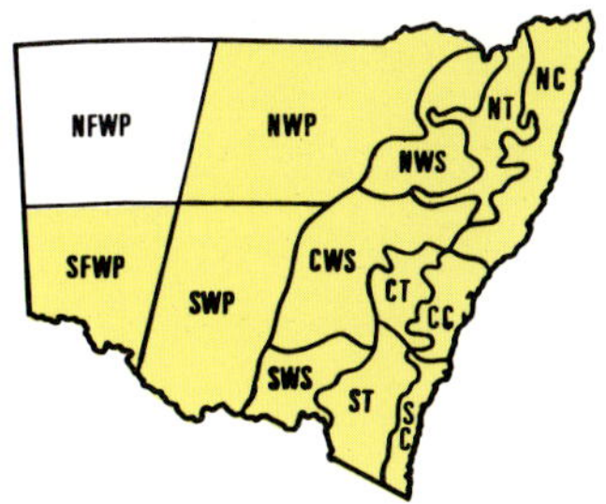

Eurasian Water Milfoil

Myriophyllum spicatum

An exotic monoecious perennial with simple or branched stems up to 2 metres long. Leaves, in whorls of 3-4, almost all submerged, 1-3 cm long, pinnate with 6-12 pairs of capillary segments. Inflorescence emergent, spike-like, the lower flowers female, the upper male; sepals 4, petals 4 about 2.5mm long. Male flower with 8 stamens.

Growth Biology

Growth is most prolific in summer. Flowering is often somewhat tardy but is most likely in late summer and early autumn.

Habitat

Enclosed water bodies to about 3 metres deep.

Economic Significance

A serious weed in many overseas lakes and water storages; not yet recorded from Australia.

Distribution

America, Asia, Indonesia.

Left: *Myriophyllum spicatum* showing the finely divided submerged leaves and emergent spicate inflorescence

A Water Milfoil

Myriophyllum trachycarpum

A native monoecious perennial with stems up to 1 metre long, rooting at the lower nodes. Emergent tips more or less erect. Leaves alternate or, when crowded, may appear to be in pseudo-whorls of 2-3. Submerged leaves pinnate or pinnatisect, 1-3cm long, segments capillary. Emergent leaves entire or slightly toothed, cylindrical or slightly flattened, 0.5-2cm long. Flowers solitary, axillary. Male flower with 8 stamens; female flower with 3-4 carpels.

Growth biology

Generally summer growing and reasonably persistent. Flowers from late summer to autumn.

Habitat

On mud and in fresh water up to 1 metre deep.

Economic Significance

This species does not spread readily and is of no economic significance

Distribution

CC, Northern Territory.

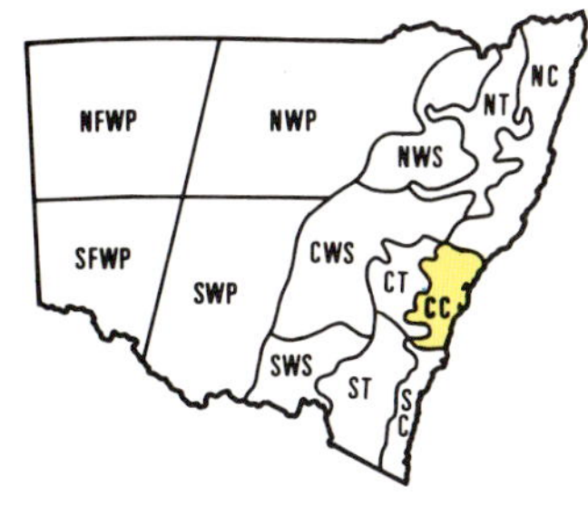

Left: An emergent shoot of *Myriophyllum trachycarpum* with axillary flowers

Right: *Myriophyllum trachycarpum* in Centennial Park, Sydney

Red Water Milfoil   *Myriophyllum verrucosum*

A procumbent native perennial with stems up to 4 metres long, rooting at the lower nodes. Leaves mostly in whorls of 3, sometimes 4. Submerged leaves 5-10 mm long, almost as wide, pinnate with 4-8 pairs of leaflets. Emergent leaves a characteristic reddish-green, curled, 4-8 mm long, 3-5 mm wide, deeply lobed. Flowers unisexual, axillary, solitary, with 2 bracteoles about 1 mm long. Male flower with 4 minute sepals, 4 petals about 2.5 mm long, 8 stamens. Female flower with 4 minute sepals and 4 minute petals. In some situations the reddish emergent leaves are not produced and the plant may be completely submerged, making differentiation from other species of *Myriophyllum* difficult. *(Fruit and stamen photograph page 442.)*

Myriophyllum verrucosum showing both emergent and submerged leaves

Growth Biology

Growth is more active in the early summer, usually reaching a peak in December in western New South Wales, after which the plant tends to break up and is not so obstructive in channel situations. It tends to persist for a few seasons and then reduce in density or disappear. Flowers are common in spring and summer and fruits plentiful at this time. A germination test at Griffith, New South Wales, had limited success, with only 1 seed out of 20 germinating. Spreads by stem fragments.

Habitat

In fresh or brackish water varying in depth from a few cm to 4 metres. Common in slowly moving or static water in inland New South Wales. Persists and flowers in dampland and may be only a few cm high.

Economic Significance

A major pest in farm dams, fouling pump inlets and generally tainting water with a fishy smell. Also obstructive in supply and drainage channels.

Distribution

NC, CC, NT, ST, CT, NWS, CWS, SWS, NWP, SWP, NFWP, SFWP, Queensland, Victoria, Northern Territory, South Australia, Western Australia.

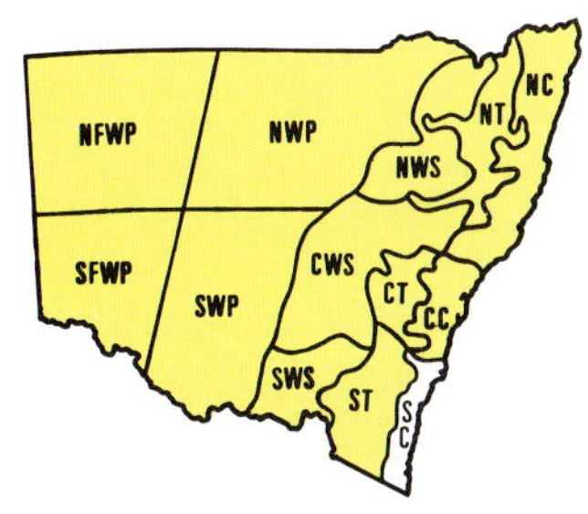

Myriophyllum verrucosum in the margins of a farm dam

Dense Waterweed

Leafy Elodea

Egeria densa

An introduced, submerged, attached, dioecious perennial with floating flowers. Stems to 1.5 metres long, rooting at the lower nodes, buoyant, so most of the growth is near the surface, upper internodes short. Leaves in whorls mostly of 4-5 (3-8) each to 4 cm long and 2-5 mm wide, with minutely serrate margins. Flowers unisexual, no female flowers recorded from New South Wales, spatheate, axillary from upper nodes, floating. Sepals 3, 3-4 mm long; petals 3, 9-12 mm long. Stamens mostly 9. Female flower similar but with an inferior ovary near the stem. *E. densa* can be confused with *Hydrilla verticillata* and *Elodea canadensis*. *E. densa* can be distinguished by its larger leaves and distinctive flowers.

Growth Biology

Flowers summer and early autumn. Spread is principally by the stems breaking into segments in the autumn. Small buds on the segments produce new plants the following spring. It is cold-tolerant and thrives in ornamental lakes in the Australian Capital Territory, where winter surface water temperatures are close to freezing.

Habitat

Native to Brazil, Uraguay and Argentina, it has spread to many temperate parts of the world and thrives in shallow lakes, ponds and slowly-flowing streams containing high nutrient levels.

Egeria densa in an ornamental pond in Canberra

Economic Significance

In New South Wales it is widespread in coastal areas and is generally not a pest. In Florida, U.S.A. it has been recorded as a menance to navigation in inland waterways containing high levels of nutrients. A common aquarium plant, care should be taken to ensure that *E. densa* is not introduced into channel systems as it is capable of rapid spread.

Distribution

CC, SWS, SWP, Queensland, Victoria.

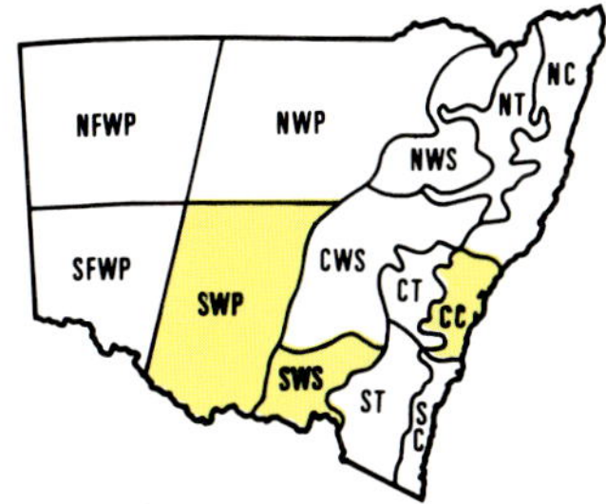

Terminal shoots of *Egeria densa* showing two axillary flowers and a node with 4 leaves

Elodea canadensis with
(A) a node subtending
3 leaves

Elodea

Canadian Pondweed

Oxygen Weed

Elodea canadensis

An introduced, submerged, attached, dioecious perennial with floating flowers. Stems up to 3 metres long with crowded internodes near the tip, producing long white threadlike roots at the lower nodes. Roots often become reddish, apparently from the deposition of iron. Leaves curled, in whorls of 3 (-4 rarely) 5-15 mm long, 2-5 mm wide. Margins minutely serrate or entire. Flowers unisexual, only male flowers recorded from New South Wales. Sepals 3, 3.5-5 mm long. Petals 3, about 5 mm long. Stamens 9.

May be confused with *Hydrilla verticillata,* common in coastal New South Wales. In *H. verticillata* the leaves grow in whorls of 3-8 and not 3, rarely 4, as with *E. canadensis.*

Growth Biology

From November-February in south-western New South Wales the inconspicuous male flowers encased in an air bubble are frequently seen rising to the surface on long white "stems". In New Zealand only female flowers have been recorded. Flowers are pollinated at the water surface. Thrives in temperate climatic zones and can survive freezing. It grows well in Lake Jindabyne in the Snowy Mountains hydro-electric scheme, where the temperature ranges from 6.5°C to 18°C. Grows vigorously once water temperatures exceed 15°C, which usually occurs in September-October in south-western New South Wales. Growth continues at a rapid rate in water temperatures around 25°C until February-March, when the growth rate declines. At this time the stems readily break into pieces, and this is the main means of spread. Fragments of stems will produce buds and these segments root at the nodes when in a suitable situation. Provided there is one node, the smallest fragment will produce a bud that will readily grow into a mature plant. Will overwinter in a leafy, semi-dormant form in water temperatures as low as 5°C. It will also overwinter under damp silt, and in such situations 4000 dormant apices per square metre have been counted. Does not thrive in water deprived of iron and appears to have a high light requirement for maximum growth.

Elodea canadensis in a supply channel near Finley

Habitat

E. canadensis has spread around the temperate world from North America, at times dominating the aquatic environment but usually existing as a balanced component of the aquatic plant population, improving water quality. In New South Wales it occurs in a wide range of habitats, from deep storages in the Snowy Mountains to irrigation supply and drainage channels.

Economic Significance

Probably an escape from aquaria and, as an attempt to prevent its wider spread, has been declared noxious in New South Wales. It infests a large percentage of irrigation canals in Victoria and New South Wales supplied from Yarrawonga Weir pool on the Murray River. In New South Wales it was first recognised in the Berriquin Irrigation District in 1958 and in 1973 spread into the Murrumbidgee Irrigation Area, at times obstructing water flow to 60 per cent of design capacity. This is one of the most obstructive and persistent of all aquatic plants infesting irrigation supply and drainage channels in temperate south-east Australia. Once established in a channel it is extremely difficult to eradicate by present methods. Mechanical removal tends to break up and spread the plant. Since 1963 annual treatments with the herbicide acrolein have given temporary relief from this weed. Each year up to 1500 kilometres of channels in New South Wales infested with *E. canadensis* are chemically treated.

Distribution

NC, CWS, SWS, SWP, Victoria, Tasmania.

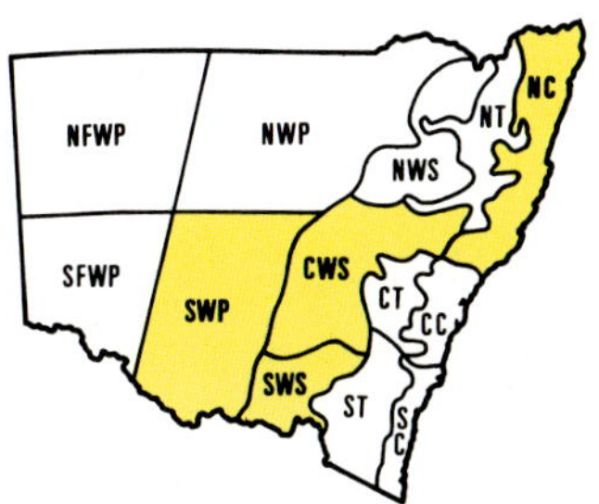

A Seagrass

Halophila

Small rhizomatous marine and/or estuarine perennials.

Seawrack

Halophila ovalis

Native dioecious perennial. Brittle rhizomes up to 3 mm diameter with 1 or 2 roots at each node. Leaves paired, glabrous with petioles usually 1-5 cm long. Blades usually ovate, 1-5 cm long. Flowers unisexual, spatheate. Male flower with 3 perianth segments 4-5 mm long, 3 stamens alternating with perianth segments. Female flower with an ovary 1-5 mm long subtended by 3 minute perianth segments on a short stalk 2-5 mm long.

Halophila decipiens

Similar, but has shorter, narrower leaves (10-25 mm long, 3-6 mm wide) often with scattered hairs. The leaves are noticeably thinner-textured than those of *H. ovalis*.

The tip of a rhizome of *Halophila ovalis* with one young pair and two expanded pairs of leaves

Growth Biology

Little is known about the germination of either *Halophila* sp. Both species spread readily by loose rhizome pieces consisting of one internode, one node and a pair of leaves, taking root and growing. In New South Wales, at least, plants flower during the summer months. Reports vary as to the frequency of flowering. In some years it can be very difficult to find any flowering material, yet in other years both species may flower quite commonly.

The rhizome pieces will grow on unstable sediments but their rate of growth is usually insufficient to stabilise these sediments, and they will frequently break up again, dispersing more rhizome fragments.

Habitat

H. ovalis grows on a wide range of sediments in the inter-tidal zone to water about 10 metres deep. It is often found growing amongst other seagrasses. *H. decipiens* usually occurs in water from 5-20 metres deep but may occasionally be found in shallower water.

Economic Significance

Neither species can be regarded as having any economic significance.

Distribution

H. ovalis occurs along the NC, CC, and SC of New South Wales and the coasts of Queensland, Victoria, Tasmania, Northern Territory, South Australia, Western Australia, the Indian Ocean and the West Pacific. *H. decipiens* occurs in the CC of New South Wales, Queensland, the tropics and in the Indian and Pacific Oceans.

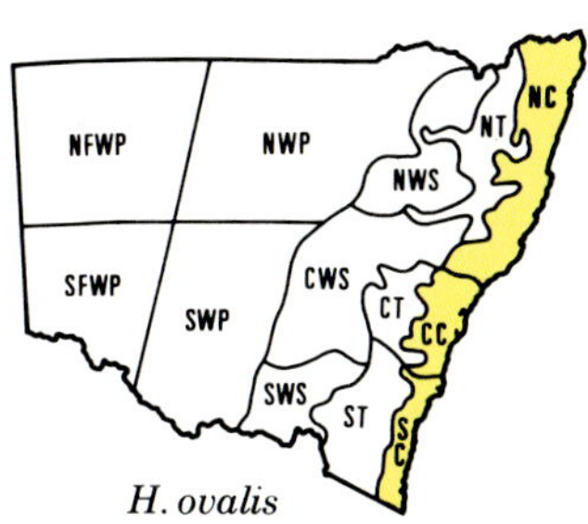

H. ovalis

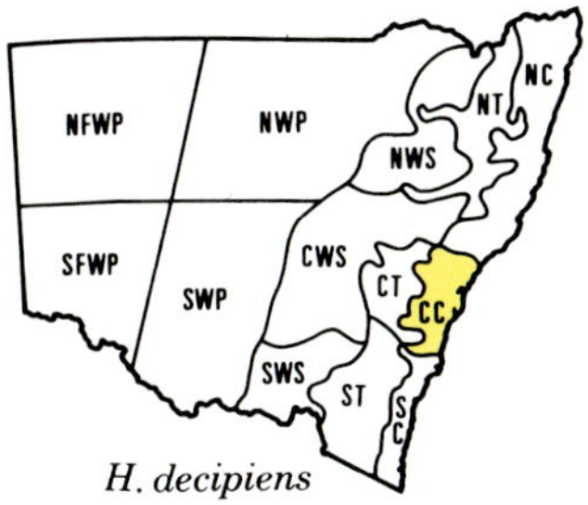

H. decipiens

Halophila ovalis in Jervis Bay

(A) Rhizome and tuber;
(B) leaf whorl;
(C) flowering piece;
(D) turion of
Hydrilla verticillata

238

Hydrilla

Water Thyme

Hydrilla verticillata

A native submerged, attached, dioecious, branched perennial with stems up to 2 metres long, usually forming a canopy immediately below the water surface. Internodes up to 5 cm apart in deeper water, much closer together near the surface. Leaves in whorls of 3-8, up to 4 cm long and 5 mm wide, with serrulate margins. Flowers unisexual, floating, solitary, axillary, with 3 sepals 1-3 mm long, 3 petals 1-2.5 mm long. Male flower initially enclosed in a spathe 1.5-3 mm long, released and floating free to the surface, with 3 stamens. Female flower remaining attached, ovary inferior, stigmas 3. Seeds 2-3 mm long. It can be confused with *Elodea canadensis* but *Hydrilla* has leaves growing in groups of more than three. *Egeria densa* is also similar, but differs in having large, 2 cm wide, 3-petalled white flowers.

Growth Biology

Roots in the mud and spreads with rhizomes and stolons. Tubers and turions are produced in profusion. Turions (see photograph) develop in the leaf axils, mainly on floating plant pieces. After a few weeks of development in the autumn the turions break off and sink to the bed where they grow. Tubers are formed on rhizomes in the autumn. They are found up to 20 cm deep in the hydrosoil. Numerous tubers are produced and up to 5000 per square metre have been recorded. Also propagates by fragmentation and a small piece of stem that includes one node will readily grow into a mature plant.

Flowering occurs in summer and autumn and both male and female flowers have been observed at Griffith, New South Wales in June with water temperatures around 10°C. The female flower reaches the surface by elongation of the thread-like "stem" (hypanthium). During its upward passage the female flower is encased in an air bubble and opens at the surface, exposing the stigmas. Male flowers detach from the parent plant as a ripe bud and rise in an air bubble, opening explosively to scatter pollen over the water surface.

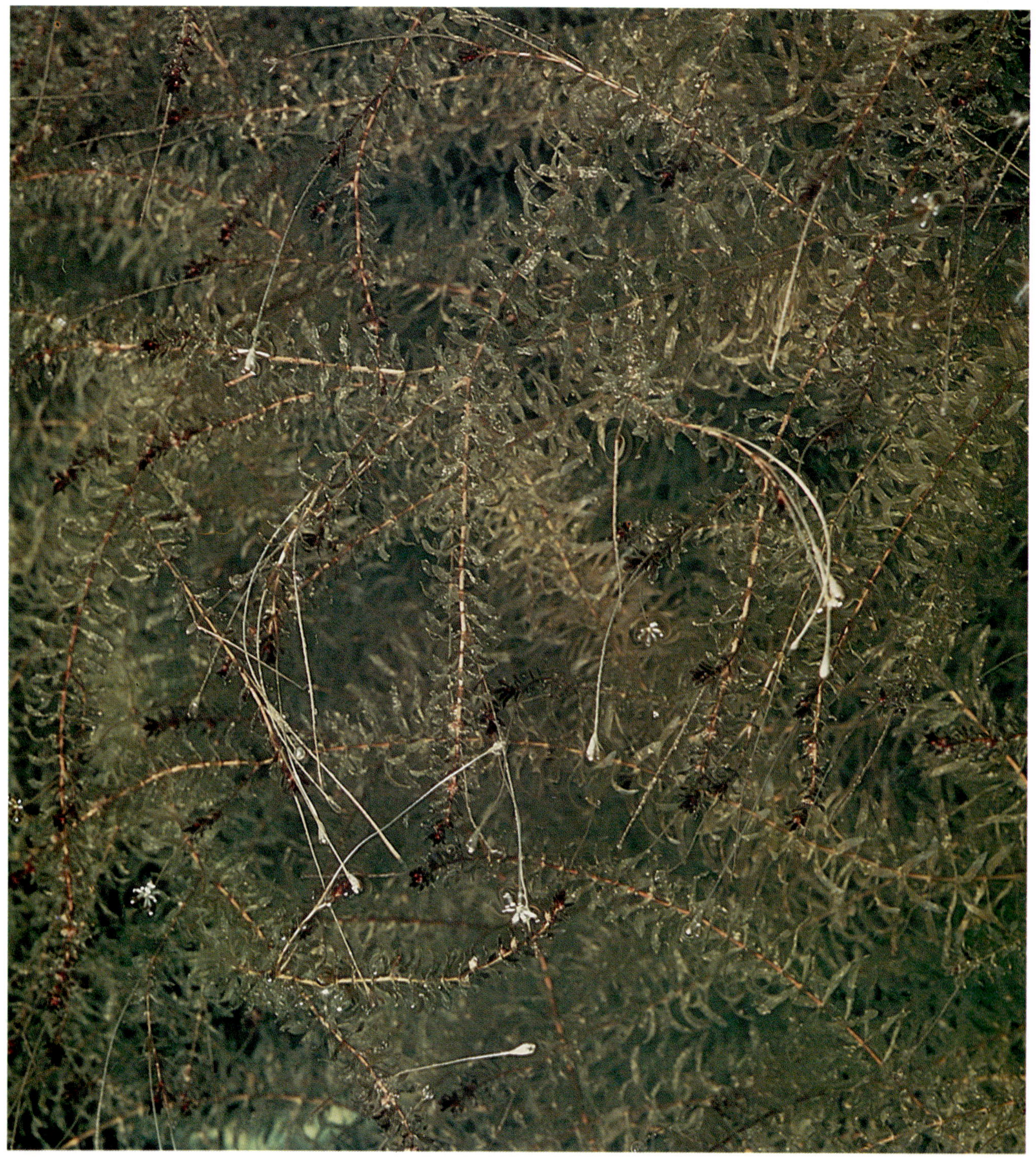

Habitat

Mainly fresh or slightly brackish creeks, small rivers, lakes, supply and drainage channels. Will grow in deep water, and rooted plants have been found in Crystal River, Florida, U.S.A. at a depth of 12 metres. Can grow at light levels of 1 per cent full sunlight.

Economic Significance

Forms a dense canopy at the water surface and crowds out other aquatic plants. Capable of rapid spread in nutrient-rich water, and in Rodman Reservoir, Florida, U.S.A., 6 hectares of *Hydrilla* in 1973 had spread to 1200 hectares by 1975. The dense mats formed by the plant reduce flow in streams and flood mitigation channels. Can restrict fishing and reduce the value of nutrient-rich recreational lakes. However, like all aquatic plants, a balanced population will improve water quality and fish production.

Distribution

Widespread in Australia and most common in Queensland and New South Wales coastal areas. *Hydrilla* is also found in Europe, Asia and Africa. NC, CC, CWS, SWS, SWP, SFWP, Queensland, Victoria, Northern Territory, South Australia, Western Australia.

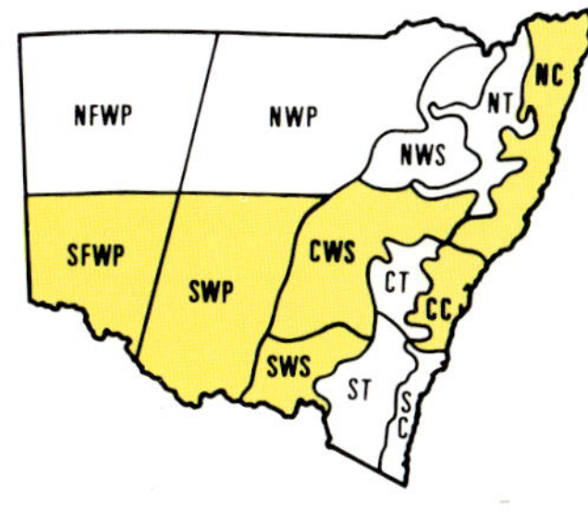

Left: *Hydrilla verticillata* showing the long slender hypanthium and terminal petals of the flowers

Frogbit

Hydrocharis dubia

A native monoecious stoloniferous perennial with emergent leaves and flowers. Stolons mostly rooted in shallow water, but will float across deeper water. Leaves erect, sometimes floating; petioles to 12 cm long, blades to 6 cm long and nearly as wide. Flowers unisexual, about 2-3 cm in diameter. Sepals 3, green, 4-5 mm long, 2-3 mm wide. Petals 3, white, 1-1.5 cm long, rounded, spatheate. Male flowers 1-4 per spathe, 12 stamens (or some staminodes). Female flower with 6 staminodes. Ovary 1-locular with 6, 2-fid styles. Fruit 5-10 mm long, about half as wide, with numerous tuberculate seeds 1-1.5 mm long.

Growth Biology

A summer growing species, flowering in late summer and autumn.

Habitat

Stationary water in lagoons or dams up to 30 cm deep, but grows mostly in very shallow water over thick mud. Will grow out into deeper water.

Economic Significance

Nil in New South Wales.

Distribution

NC, Queensland.

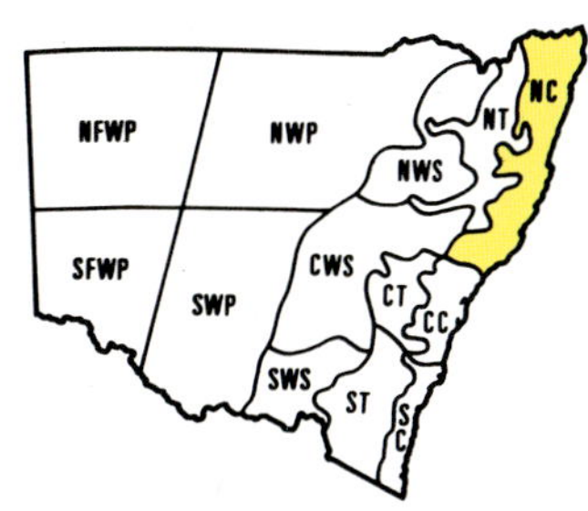

Left: Leaves and male flower of *Hydrocharis dubia*

Right: *Hydrocharis dubia* on the margins of a farm dam near Tabulam

Lagarosiphon

Lagarosiphon major

An introduced, submerged, rhizomatous, dioecious perennial with branched stems up to about 5 metres long. Leaves a long way apart below, closely spaced above, alternate, recurved, 0.5-2 cm long, 2-3 mm wide, with minutely serrate margins. Flowers unisexual, axillary, solitary. Sepals 3, 1-2 mm long; petals 3, 1-2 mm long. Male flower initially enclosed in a spathe, becoming free and floating to the surface. Stamens 3; staminodes 3, enlarged "sail-like". Female flower attached, floating, with 3 staminodes. Ovary inferior with 3, 2-branched purple styles. Similar to *Elodea canadensis* and *Hydrilla verticillata,* but the leaves are alternate, stiff and recurved. Male flowers not recorded in Australia. The species of this genus have typical spiral leaf arrangements, a useful character for identification, since plants are often sterile.

Growth Biology

Female flowers rise to the surface encased in a gas bubble on a threadlike tube (hypanthium). The ultimate length of the hypanthium depends on the species; in *L. major* it is up to 15 cm long. If the flower does not reach the surface it dies. Male flowers are released from the plant and rise to the surface to pollinate the projecting female flowers. Female flowers have been observed on aquarium material obtained from Sydney, New South Wales.

Habitat

Overseas *L. major* needs good light for maximum growth, but will grow in clear water 7 metres deep. Grows well in low nutrient levels on a silty or sandy bed.

Lagarosiphon major in an aquarium

Economic Significance

First observed in New Zealand in 1950, where it has caused problems in lakes supplying water for electricity generators by blocking intake structures. It is widely sold in Australia for use in aquaria and has the potential to cause problems in storages such as in the Snowy Mountains scheme.

Distribution

Not recorded in New South Wales except from aquaria.

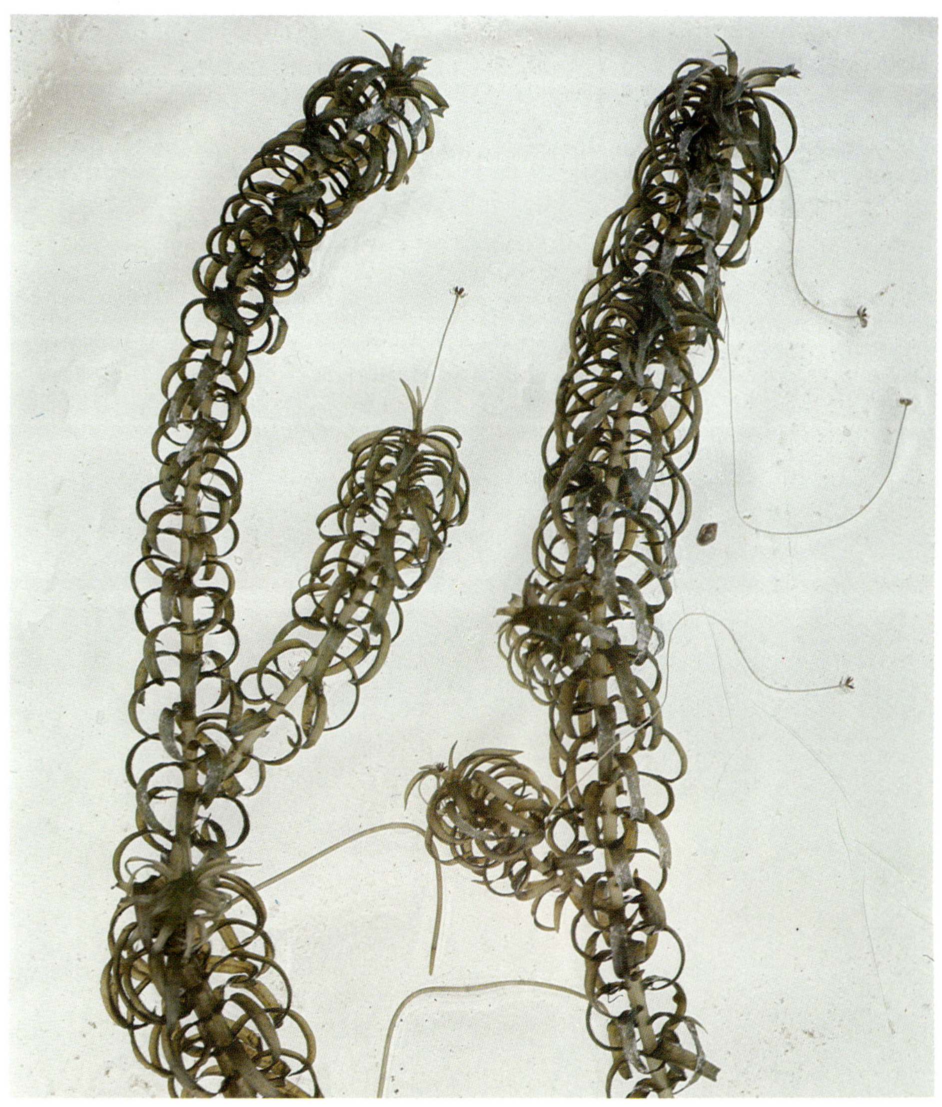

Lagarosiphon major with female flowers

Top: *Ottelia ovalifolia* showing floating
leaves, emergent flowers, submerged
buds and fruit

Left: *Ottelia ovalifolia* in a channel near
Griffith

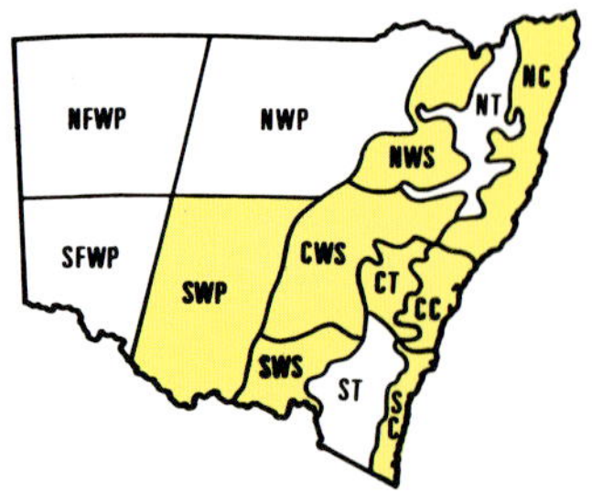

Swamp Lily   *Ottelia ovalifolia*

A native tufted perennial or annual with floating leaves and flowers. Leaves all basal on petioles up to 1 metre long, blades floating, elliptical, up to 16 cm long, 3-6 cm wide with 5-7 parellel veins and numerous transverse connecting veins. Flowers bisexual, axillary, solitary, both chasmogamous flowers (opening and being cross-pollinated) and cleistogamous flowers (not opening, remaining submerged, enclosed within the spathe, and self-pollinated) produced. Chasmogamous flowers floating, obvious, up to 6 cm across, with a thick, tuberculate, ridged, 2-lobed spathe 3-6 cm long. Sepals 3, green, 1.5-2.3 cm long. Petals 3, white with a maroon base, 4-5 cm long. Stamens in 3 groups of ± 4 stamens. Ovary inferior with 6-9 styles. Fruiting ovary withdrawn into the water when it matures, and produces numerous narrow, finely hairy seeds each 2-3 mm long. Cleistogamous flowers are not as common and have reduced sepals and petals. The specific name "ovalifolia" alludes to the oval-shaped leaves and the generic name "Ottelia" is from "ottel-ambel", the native name of an Indian species. Young submerged leaves are straplike and can be confused with submerged leaves of *Vallisneria gigantea*. Main difference at this stage is the pointed leaf tip in *O. ovalifolia* and the stolons produced by *V. gigantea*. Floating leaves can be confused with *Aponogeton distachyos*.

Growth Biology

After fertilisation the flower stem bends into the water where a bottle-shaped, prominently ridged capsule matures. Seeds germinate readily in shallow water, producing narrow, entirely submerged leaves. Plants grow in spring and summer, flowering and fruiting in summer and autumn. The seed normally germinates the following spring.

Economic Significance

Seldom troublesome, but will densely cover water surface, impeding flow or perhaps hindering stock from watering. An attractive aquarium plant.

One species, *O. alismoides* occurs in tropical Australia, is completely submerged, with only the flowers above water surface. This species is also found in shallow water in Asia, where the entire plant except roots is cooked and eaten as a vegetable.

Habitat

Thrives in still or slowly flowing fresh water up to about 1 metre deep in channels, rice crops, shallow lakes, river backwaters and stock tanks.

Distribution

NC, CC, SC, CT, NWS, SWS, SWP, found in all mainland States.

Ribbonweed

Vallisneria gigantea

Eelweed

A native tufted stoloniferous, submerged, dioecious perennial with floating flowers. Stolons white or light-brown, producing small rooted tufts at varying intervals. Leaves all basal, to 3 metres long and 5 cm wide, often much smaller, with 5-9 parallel veins. The leaves are bright-green, but in turbid inland waters are covered in slime and appear brown. Flowers unisexual, axillary, male in several-flowered inflorescences, female solitary. Male flowers numerous, less than 1 mm in diameter, produced on a central axis enclosed within a spathe 1-2 cm long on a peduncle 2-5 cm long; on maturity the spathe opens, releasing the male flowers to float freely on the surface. Perianth segments 3, 2 larger than the other, all minute; stamens 2. Female flowers 1.5-2.5 cm long on a pedicel up to 1 metre long. Flower about 5 mm in diameter when open, pink or white inside; sepals 3, 1-2 mm long; petals 3, minute. Ovary inferior cylindrical, with 3 stigmas. Seeds papillose, 1-2 mm long.

There are several forms of *Vallisneria* in New South Wales and the name of the European species, *V. spiralis,* has been used for some or all of these forms. *V. gigantea* is the more appropriate name if all the forms are treated as a single taxon as *V. spiralis* does not appear to have become naturalised. Further study is required to determine the taxonomic status of the various forms.

Plants may be confused with young leaves of *Ottelia ovalifolia* or *Triglochin procera.*

Growth Biology

Flowering takes place through the summer months with the male and female flowers produced on separate plants. The male flower, situated in an envelope near the base of the plant, floats to the surface at maturity. The female flower is on a pedicel which is initially spiral, straightens as the flower floats to the surface, then spirals again to withdraw the maturing fruit below the surface.

Spread is chiefly vegetative, the stolons extending rapidly, producing new shoots and roots at intervals. In south-western New South Wales the plant has been observed to over-winter in temperatures about 5°C. In these temperatures the leaves partially die back. If the water drains from the channel, the leaf stumps will remain green if protected from frost by silt. Maximum growth is in water temperatures around 25°C. The summer-growing *V. gigantea* often follows the spring growth of some species of *Potamogeton.*

Left: *Vallisneria gigantea*
in a channel near Griffith

Male (left) and female (right) plants of *Vallisneria gigantea*

250

Habitat

One of the most widespread and common submerged aquatic plants in New South Wales, growing prolifically in irrigation systems, rivers and lakes. Grows in a wide variety of habitats and will thrive in water containing 1500 ppm total dissolved salts. Found growing in water from a few cm to about 4 metres deep.

Economic Significance

Frequently obstructive in irrigation channels and in summer is capable of rapid growth. This plant's exceptional growth potential in irrigation channels of south-western New South Wales is shown by rapid regeneration after cutting. In an experiment at Griffith, New South Wales, leaves reaching to the surface in water 1 metre deep were cut at the base; regrowth to the original length occurred in 4 weeks during January in water temperatures varying from 22°C to 30°C, and at a time when the plant was flowering.

Vallisneria stolons appear to be a favourite food of black swans, *Cygnus atratus,* which have been observed to dig up large areas of this plant, causing large amounts of floating debris.

Distribution

NC, CC, SC, CT, ST, CWS, SWS, NWP, SWP, SFWP, all States.

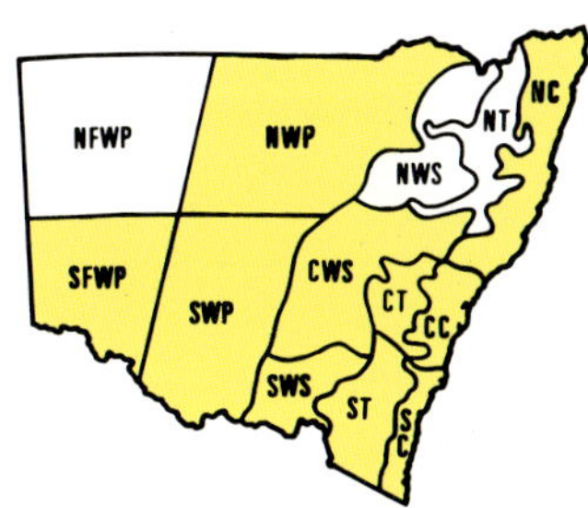

Plants of an unidentified
species of *Isoetes* showing
(1) a megasporangium, and
(2) a microsporangium

252

Quillwort

Isoetes spp.

Native, perennial, submerged, tufted grasslike ferns. Rootstock swollen. Leaves stiff, to about 20 cm long, approximately 2 mm wide, with a broad sheathing base and with 4 separate longitudinal air canals. Leaves mostly sterile, but the similar-looking fertile leaves bearing either megasporangia or microsporangia within the sheathing leaf base. The outer fertile leaves enclosing megasporangia each containing several ± globular megaspores. The inner fertile leaves with microsporangia each producing numerous microspores. Being a fern, there is a gametophyte stage as well as the sporophyte stage described above. Little is known about the gametophyte stage of *Isoetes*. It is difficult to identify *Isoetes* to species level, but this is probably of little practical significance.

Habitat

Most species are aquatic or semi-aquatic, the latter found occasionally in depressions that fill temporarily with water. The *Isoetes* photographed was found growing completely submerged in water up to 1 metre deep in an irrigation supply channel near Maffra, Victoria. The bed of the channel was covered with grasslike tufts of *Isoetes* about 10 cm high.

Distribution

About 75 species in the world. Around 8 occur scattered over most of Australia. NC, NT, ST, NWS, SWS, all States.

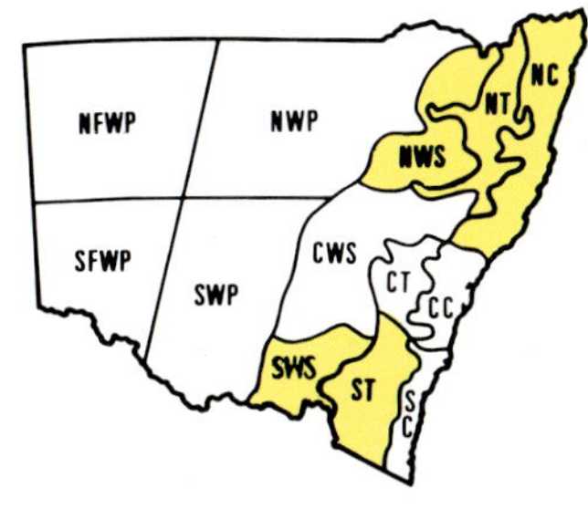

An unidentified species of Isoetes

Juncus

Caespitose or rhizomatous annuals or perennials. Inflorescence open and much branched or reduced to few-many flowered heads. Flowers mostly hermaphrodite but unisexual in some species, bracteoles 2, or 0. Perianth segments 6, in two whorls of 3, chartaceous; the outer whorl usually longer and more robust than the inner whorl. Stamens 3-6, consistent or variable. Ovary usually partially triseptate, seeds numerous.

Most species of *Juncus* occur in periodically damp areas but vary greatly in their tolerance to the intervening dry periods. It is very difficult to effectively classify *Juncus* spp. as "aquatic" or "non-aquatic" and, as well, they are a very difficult group to identify. A key is provided here to the major sections of the genus likely to be encountered in or around aquatic areas. Descriptions are provided for some of the more frequently encountered species but identification is best left to experts.

1. Leaves present and clearly distinct from stems. **2.**
1a. True leaves absent (reduced to chaffy scales) or similar in appearance to flowering stems. **3.**

2. Plants with flattened grass-like leaves; usually caespitose with terminal inflorescences. Auricles absent; flowers without bracteoles.
sect. **Graminifollii**
includes *J. planifolius*
2a. Plants with cylindrical or somewhat flattened, septate leaves, either cauline or basal. Auricles present; flowers without bracteoles.
sect. **Septati**
includes *J. prismatocarpus*
J. holoschoenus
J. articulatus

3. True leaves absent, reduced to chaffy scales at base of stems; rhizomatous perennials.
sect. **Genuini**
includes *J. usitatus*
J. ingens
J. aridicola
J. australis
3a. True leaves present (but basal chaffy bracts also present), arising singly from underground rhizomes (rhizomes often very short) and similar in appearance to the flowering stems. **4.**

Left: A clump of *Juncus acutus* near Rochester, Victoria

4. Robust perennials with stiff to pungent-pointed leaves and bracts. Perianth (and hence inflorescence) red-brown from an early stage.

sect. **Juncus**
includes *J. acutus*
J. kraussii

4a. Slender perennials with soft leaves. Inflorescence greenish until mature.

sect. **Poiphylli**
includes *J. cognatus*

The brown stem bases, large fruit and needle-sharp bracts typical of *Juncus acutus*

Spiny Rush

Juncus acutus

An introduced shortly rhizomatous perennial to 1.5 metres high. Stems and stem-like leaves rigid and pungent, to 5 mm diameter, non-septate and with continuous pith. Inflorescence apparently lateral, to 15 cm diameter, panicle-like, with several dense clusters on spreading branches, subtended by a much longer, apparently terminal, bract which appears to be a continuation of the stem. The individual cluster may be up to 3 cm diameter and is composed of several reddish-brown flowers. Perianth segments 2.5-4 mm long, acute, much shorter than the capsule. Stamens 6, seeds usually tailed at both ends. *Fruit photograph page 442.*

Growth Biology

Flowers mainly in spring or summer, producing numerous small seeds which presumably germinate in either autumn or spring. More common in saline areas or on weakly structured clays (i.e., those with a higher proportion of monovalent cations than divalent cations).

Habitat

In drains and ephemeral creeks or swamps and on channel banks and wasteland where there is adequate subsurface moisture.

Economic Significance

A very spiny plant which is very difficult to eradicate once established and can spread to impede water flow in drains and channels. Has the potential to be a troublesome weed in some areas and should be controlled as early as possible.

Distribution

NC, CC, SC, NT, ST, SWS, SWP, SFWP, Victoria, South Australia, Western Australia, Europe, Asia, Africa.

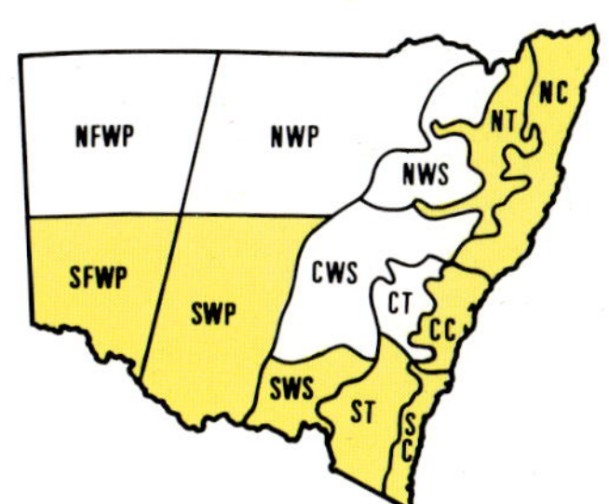

Juncus aridicola

A native shortly rhizomatous perennial to 1.5 metres high. Stems to about 6 mm in diameter, non-septate with interrupted pith. Inflorescence apparently lateral, open and spreading, with numerous more or less regularly spaced flowers, subtended by an apparently terminal bract which appears to be a continuation of the stem. True leaves absent but some "sterile stems" may be present. Perianth pale-coloured, 2-3 mm long. Stamens 3-6. Capsule subequal to or longer than the perianth. *Seed photograph page 442.*

Growth Biology

Growth and flowering dependent on standing water, but usually in late autumn to spring.

Habitat

Permanently or periodically inundated areas in semi-arid or arid areas.

Economic Significance

Useful as a soil stabilising plant and provides cover for wildlife.

Distribution

CC, CT, NWS, CWS, SWS, NWP, SWP, NFWP, SFWP, Queensland, Victoria, Northern Territory, South Australia, Western Australia.

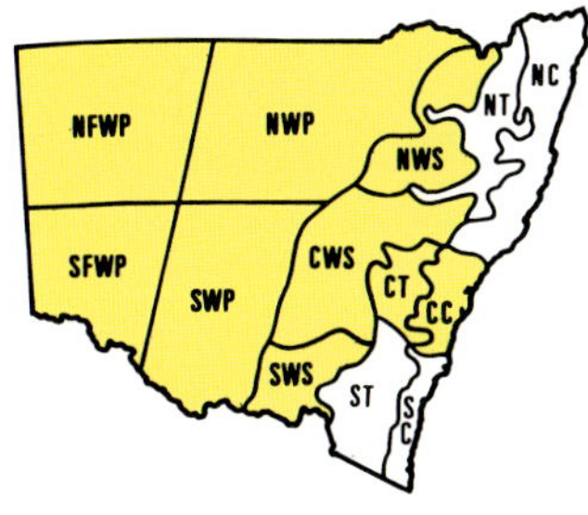

Left: An inflorescence of *Juncus aridicola*

Right: *Juncus aridicola* on the margins of a billabong on the Darling River

Jointed Rush

Juncus articulatus

An introduced rhizomatous perennial to 50 cm high. Leaves mostly on the stem, sheaths auriculate, blades about 2 mm wide and to 10 cm long, terete or ± laterally compressed, hollow with regular complete septa. Inflorescence terminal, a loose open arrangement of flower clusters subtended by leafy bracts very much shorter than the inflorescence. Each cluster usually less than 1 cm diameter with up to 15 flowers. Perianth segments 2-3 mm long. Stamens 6, capsule longer than the perianth. *Seed photograph page 442.*

Growth Biology

Growth and flowering mainly in spring and summer.

Habitat

In wet places under a wide range of conditions.

Economic Significance

Will grow along channel margins and in drains, causing minor obstruction to water flow.

Distribution

NC, CC, SC, NT, CT, ST, NWS, CWS, SWS, SWP, Queensland, Victoria, Tasmania, South Australia, Western Australia, Europe, Asia.

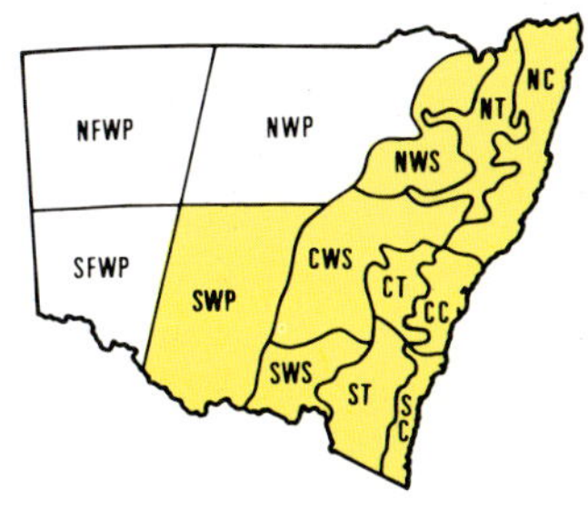

Left: *Juncus articulatus*

Right: *Juncus articulatus* along a drainage channel near Hanwood

Giant Rush

Juncus ingens

A native dioecious rhizomatous perennial to 4 metres tall. Stems to 1 cm diameter, non-septate with very interrupted pith. Inflorescence apparently lateral, open and spreading, to 20 cm long with numerous, more or less regularly spaced flowers, subtended by an apparently terminal bract up to 1 metre long which appears to be a continuation of the stem. True leaves absent, but some "sterile stems" may be present. Flowers unisexual. Perianth segments 1-2 mm long. Stamens 6, reduced to staminodes in female flowers. Ovary absent in male flowers. Fruiting capsule subequal or shorter than perianth. *Seed photograph page 442.*

Growth Biology

Very little is known about this species; apparently seeds are not commonly formed.

Habitat

Lagoons and swamps along the Murray River.

Economic Significance

Useful for bank stabilisation and as habitat for waterbirds.

Distribution

SWS, SWP, Victoria.

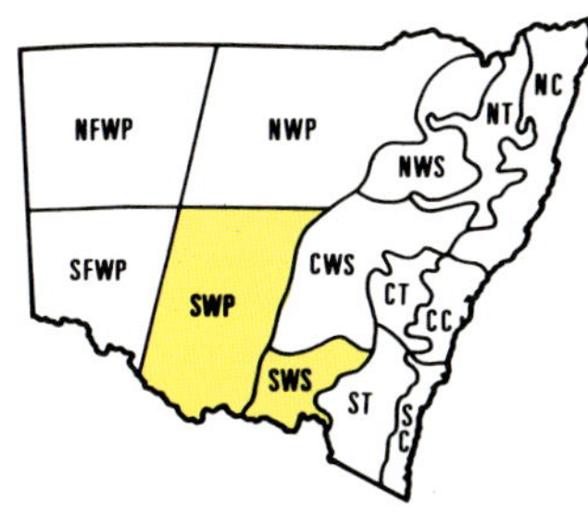

Left: *Juncus ingens* on Gulpa Creek near Mathoura

Right: Female (left) and male (right) inflorescence of *Juncus ingens*

Inflorescences and base of
Juncus usitatus

Common Rush

Tussock Rush

Juncus usitatus

A native, shortly rhizomatous, almost tufted perennial to 1 metre high. Stems usually 1-2 mm diameter, non-septate but with interrupted pith. Inflorescence apparently lateral, open and spreading with numerous flowers, subtended by an apparently terminal bract which appears to be a continuation of the stem. True leaves absent but some "sterile" stems may be present. Flowers evenly spaced, not clustered. Perianth segments usually 1.5-2 mm long. Stamens 3. Capsule subequal to or longer than the perianth. *Seed photograph page 442.*

Growth Biology

Flowers mainly in spring and summer, though old flowers usually present at any time of the year.

Habitat

Damp places under a wide range of conditions.

Economic Significance

This species serves a useful purpose in irrigation supply channels by providing competition on the channel margin for less desirable, more obstructive plants. Rarely obstructive to water flow.

Distribution

NC, CC, SC, NT, CT, ST, NWS, CWS, SWS, NWP, SWP, Queensland, Victoria, South Australia, Western Australia.

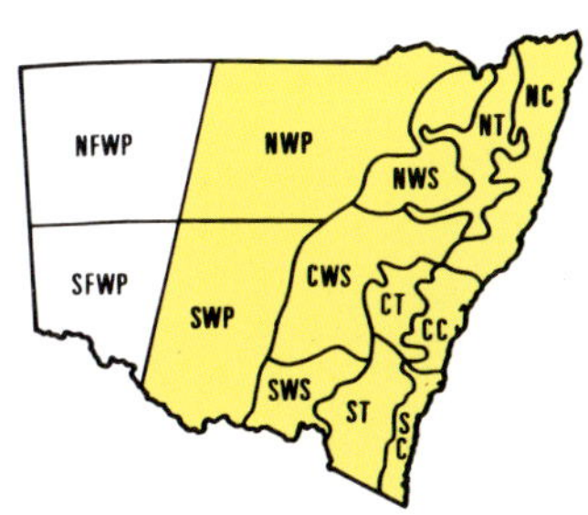

Juncus usitatus on the margin of a farm dam near Rankins Springs

Immature fruiting inflorescence of *Maundia triglochinoides*

Maundia triglochinoides

Native rhizomatous perennial. Rhizomes approximately 0.5 cm thick. Leaves spongy, inflated, triangular in cross-section, to 80 cm long and 5-10 mm wide. Inflorescence a spike-like raceme to 10 cm long and 2.5 cm wide.

The floral structure has been interpreted rather differently by various people and the actual structure needs further clarification. There are 2 bracts (perianth segments?), the carpels are surrounded at the base by 12 bilocular stamens, each opening outwards by a slit. Perianth appears to be absent. Carpels usually 4, fused almost to the top. Each carpel is surmounted by a broad spreading stigma. Fruit to 1 cm long, 8 mm wide, each carpel with a short beak. The flower structure has also been interpreted as consisting of equal numbers of perianth segments and bilocular stamens (4-6) or as 4-6 perianth segments subtending twice the number of unilocular stamens. Neither of these latter interpretations seems to be satisfactory, as the 12 structures all appear identical and the pollen is released to the outside.

Habitat

Grows in freshwater swamps and shallow streams.

Economic Significance

Nil.

Distribution

NC, CC, Queensland.

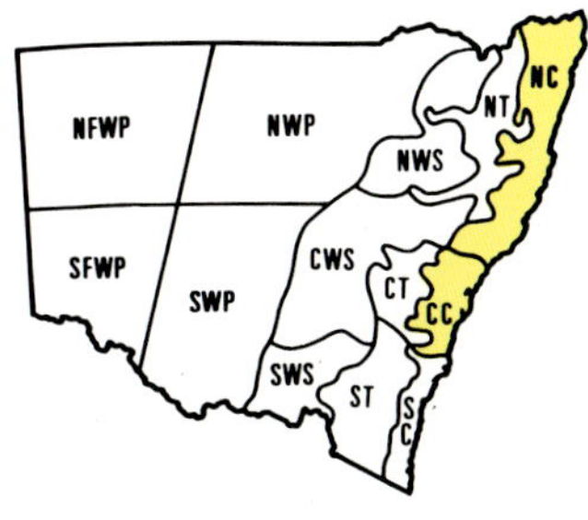

Maundia triglochinoides in a swamp near Wyong

Triglochin

Native perennial or annual rhizomatous herbs. Tubers often present. Stems reduced, leaves mostly basal. Inflorescence a spike or spike-like raceme. Perianth segments 6, falling off early. Stamens 6. Carpels 6, although not necessarily all fertile. Five species of *Triglochin* have been recorded in New South Wales; most, unlike *T. procera,* are small and inconspicuous.

Water Ribbons

Triglochin procera

Robust rhizomatous perennial with roots frequently ending in tubers. Leaves basal, more or less flat, 0.5-2 metres long, 1-3 cm broad, erect, semi-erect or floating. Inflorescence a fairly dense spike-like raceme to 30 cm long and 3 cm broad. Flowers shortly pedicellate, 6 green perianth segments in 2 whorls; each segment approximately 2 mm long, deciduous. Stamens 6, each attached to the base of a perianth segment. Ovary superior, consisting of 6 fertile carpels fused below the middle and with no central axis. Stigmas 6, recurved, prominent, with long white hairs. Fruit consists of the 6 elongated carpels (5-10 mm long), each with a single seed. *Fruit photograph page 443.* Eventually the carpels separate and fall. The plant has a spongy root system which at times carries tubers about 12 mm in diameter and 4 cm long. There is considerable variation in leaf morphology, and this appears to be correlated with variations in carpel shape. The taxonomic status of these variants requires clarification. Plants without flowers may be confused with *Vallisneria gigantea.* The leaves of *T. procera* are usually more spongy than *V. gigantea* and the latter does not produce root tubers.

Triglochin procera in a drain near Koo-wee-rup, Victoria

Growth Biology

Seeds germinate readily in shallow water and the small plants survive winter. Growth and flowering take place in spring and summer, with fruiting in late summer or autumn.

Habitat

Stationary or slowly-flowing water in streams or swamps. Will grow on wet mud or in water to 2 metres deep.

Economic Significance

Occasionally grows in sufficient density to be obstructive, but is of minor importance in channel systems. Component of habitat for water-birds and fish.

Distribution

NC, CC, SC, NT, CT, ST, CWS, SWS, SWP, all States.

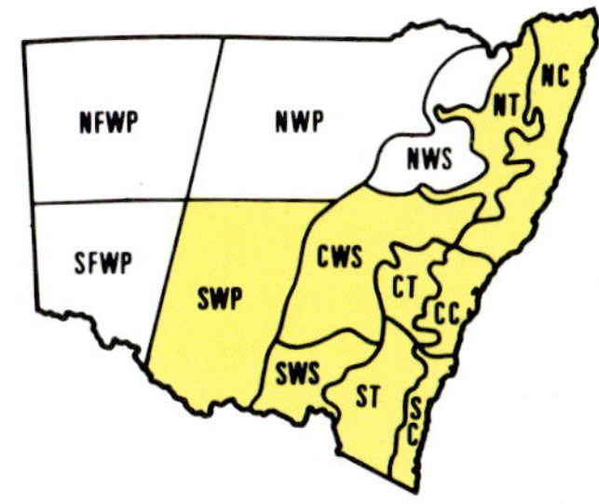

Triglochin procera with drooping leaves and erect inflorescences

Streaked Arrow-grass

Triglochin striata

Native slender, erect, low-growing rhizomatous perennial. Leaves to 30 cm long and 1-3 mm wide arising in distinct tufts separated by several nodes along an extensive rhizome. Sheath developed with a distinct obtuse ligule. Inflorescence a spike-like raceme usually 2-15 cm long and less than 1 cm broad. Flowers approximately 2 mm long on short pedicels. Perianth segments 6, in 2 whorls, each segment approximately 1 mm long. Stamens 6, but mostly not all developing. Ovary of 3 fertile and 3 sterile carpels. *Fruit photograph page 443.*

Habitat

Tidal and freshwater swamps or on muddy creek banks.

Economic Significance

Useful stabilising species.

Distribution

NC, CC, SC, all States, South Africa, America.

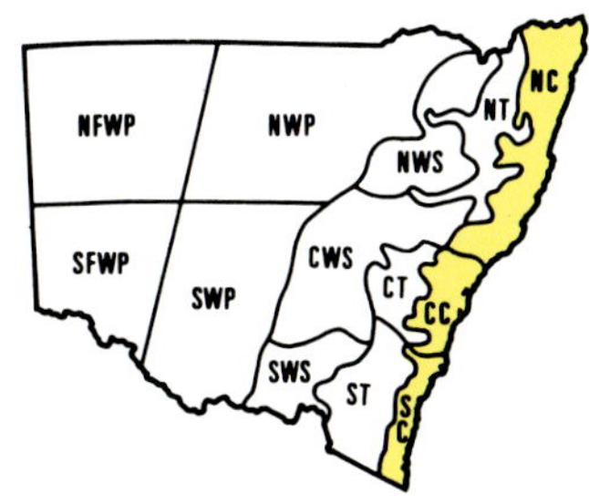

Left: *Triglochin striata* in a salt marsh on the lower Hunter River

Right: *Triglochin striata* showing the open stand of narrow erect leaves and inflorescences

A mixture of *Lemna trisulca* (left below), a species of *Wolffia* (centre below) and a species of *Spirodela* (right below)

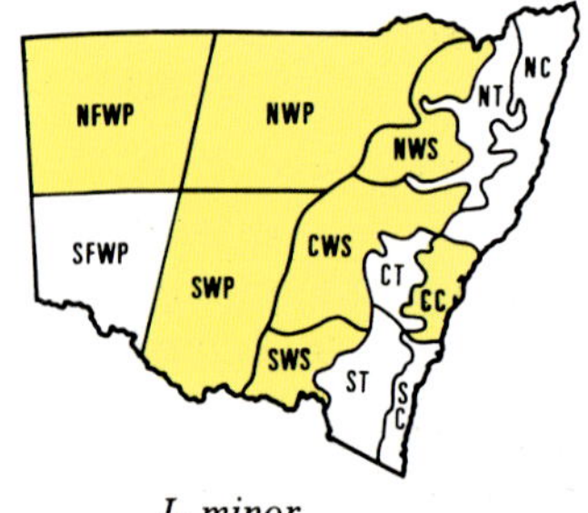

L. minor

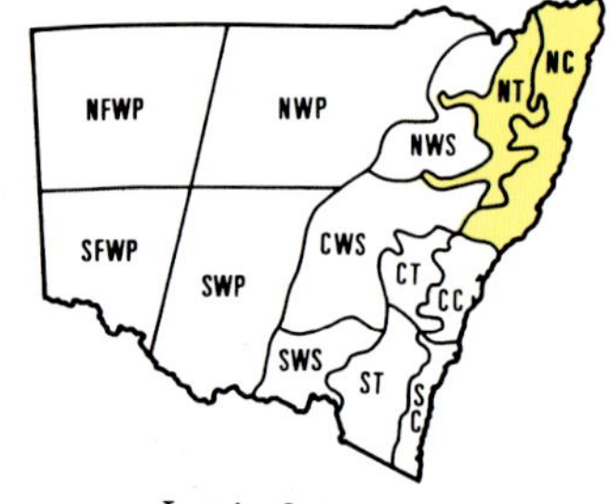

L. trisulca

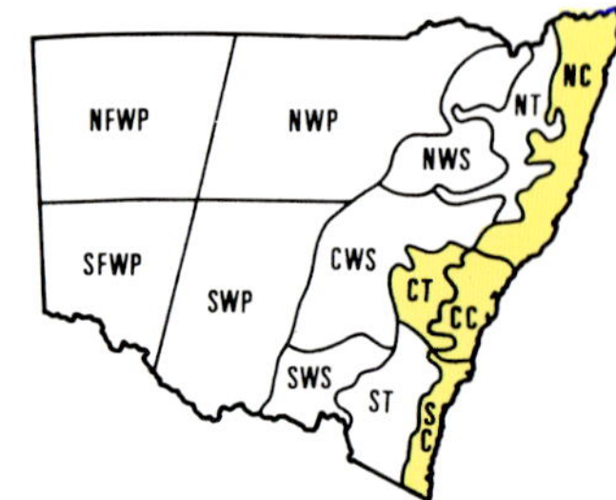

S. oligorrhiza var. *oligorrhiza*

Duckweeds

Lemna spp.

Wolffia spp.

Spirodela spp.

Very small floating native monoecious perennials. The plants consist of flattened or inflated thalli looking like small leaves, with 0-several hair-like roots from the undersurface. Thallus with 1 or 2 lateral pockets. New thalli arising from the lateral pockets, the inflorescence developing either from the lateral pocket or through the surface of the thallus. Inflorescences each with 1 female and 1 or 2 male flowers together. Flowering rarely observed in some species. Inflorescence spatheate in *Lemna* and *Spirodela,* espatheate in *Wolffia*. Male flowers of a single stamen with a 2-celled or 1-celled anther. Female flower consists of a solitary ovary with a terminal style.

This family consists of 4 genera of which 3, *Lemna, Spirodela* and *Wolffia,* occur in Australia.

They are not well understood botanically and identifications to species are still not consistent. For the purposes of this book, and for most practical applications, it is usually adequate to identify the plants to genus and a key is provided to aid identification to this level.

1. Thallus minute, usually less than 1 mm long, inflated; globose with 1 lateral pocket; rootless; inflorescence breaking through the top surface of the thallus, espatheate and consisting of 1 female flower and 1 male flower with one unilocular anther.
Wolffia
1a. Thallus usually more than 1 mm long, flattened, with 2 lateral pockets, with or without roots; inflorescence emerging from a lateral pocket; spatheate, consisting of 1 female and 2 male flowers, each with 1 bilocular anther.

2. Thallus with deciduous dorsal and ventral scales (present when thallus is young); roots 1-18; nerves 3-15.
Spirodela (mostly S. oligorrhiza)
2a. Thallus never with dorsal and ventral scales; roots absent or 1 per thallus; thallus nerves obscure, 1-3.
Lemna

3. Thallus floating on the surface, opaque ± elliptic; no obvious stalk; 1 root.
L. minor
3a. Thallus suspended below water surface, translucent, conspicuously stalked; usually rootless.
L. trisulca

The species of this family are amongst the simplest and smallest of flowering plants and are usually relegated to the category of botanical

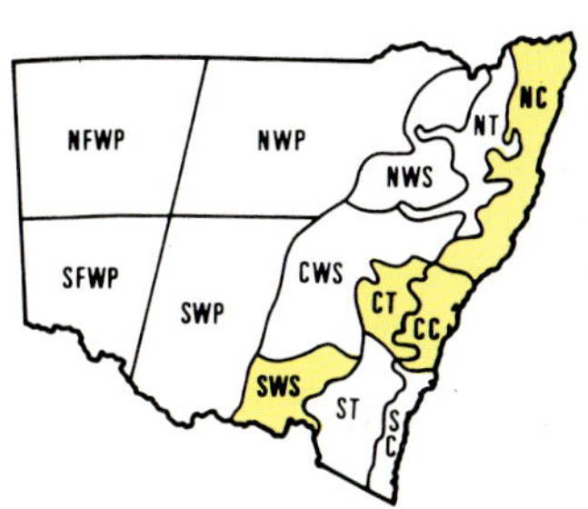
S. oligorrhiza var. *pleiorrhiza*

273

A species of *Spirodela* in eutrophic water in the Georges River

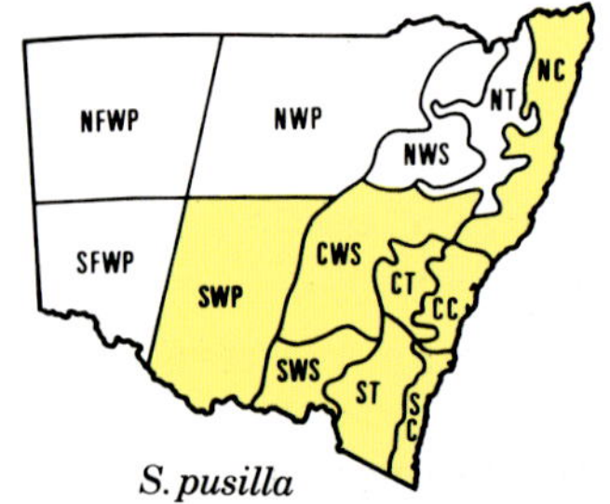

S. pusilla

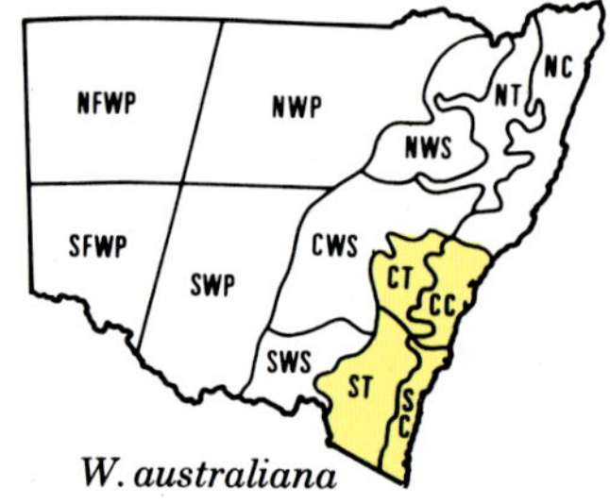

W. australiana

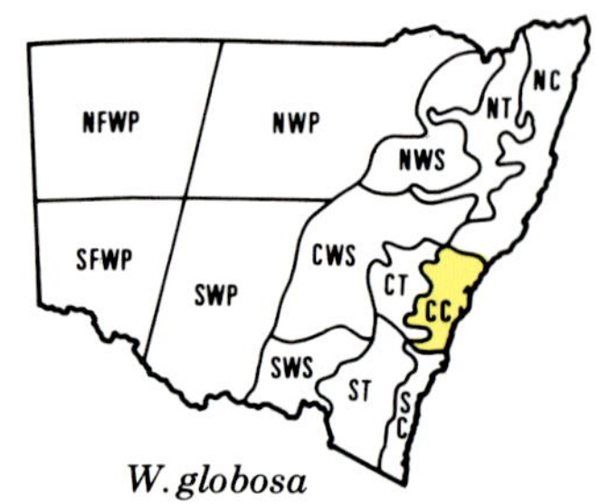

W. globosa

curiosities. The plants in the genus *Wolffia* are so small as to appear merely as a greenish tinge on the water surface, and species of this genus are the smallest flowering plants.

Growth Biology

Plants in this family propagate mainly by vegetative means. Daughter plants are budded from adult plants and by this method some species can double plant numbers in 3 days.

Flowering generally takes place throughout the summer or in late spring. Extent of fruiting is not known, but reportedly prolific in *Lemna minor*. The plant over-winters by seed and vegetatively by small plants that sink to the mud and rise to the surface the following season.

Habitat

On or slightly beneath the surface of static water or harboured in backwaters by other floating plants such as *Potamogeton, Ludwigia* and *Azolla*. When scattered, plants are hard to see. When growing densely a green mat covers the water. Dense mats are a sure indication of eutrophic conditions. Wave action will break up and disperse plant masses.

Economic Significance

An important source of food for humans, and *Wolffia arrhiza* has been used as food in Thailand and Burma for centuries. In those countries plants are harvested at about 4-day intervals from ponds enriched by animal manure. Dried plants contain 20 per cent protein and 40 per cent carbohydrate and are a nutritious food. This family also shows promise for recovering nutrients from waste water. In addition they have potential as experimental organisms for research due to their small size, rapid growth and structural simplicity. Although Duckweeds do not obstruct water flow in channels to any extent, they can block pump inlet structures and cause a nuisance in farm dams used for domestic purposes.

Distribution

L. minor: CC, NWS, CWS, SWS, NWP, SWP, NFWP, Victoria, Tasmania.

L. trisulca: NC, NT, ST, Queensland, Victoria, Tasmania, Western Australia.

S. oligorrhiza:
Var. *oligorrhiza:* NC, CC, SC, CT.
Var. *pleiorrhiza:* NC, CC, CT, SWS.

S. pusilla: NC, CC, SC, CT, ST, CWS, SWS, SWP, Victoria.

W. australiana: CC, SC, CT, ST, Victoria, South Australia.

W. globosa: CC, Queensland, Victoria, South Australia, Western Australia.

W. sp. CC, SWP, Victoria, South Australia, Western Australia.

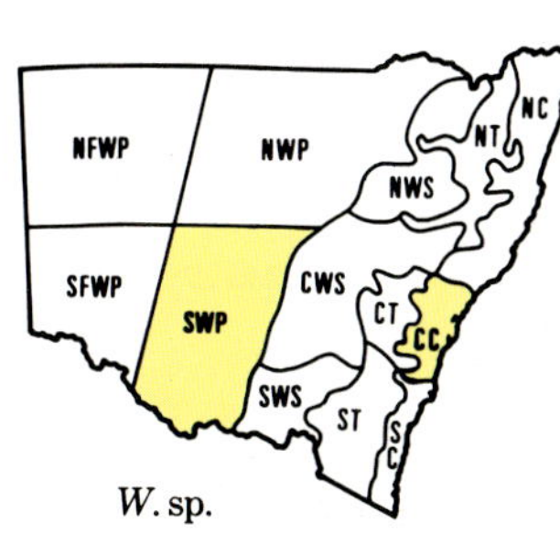

Plants of *Utricularia australis* in a swamp near Putty

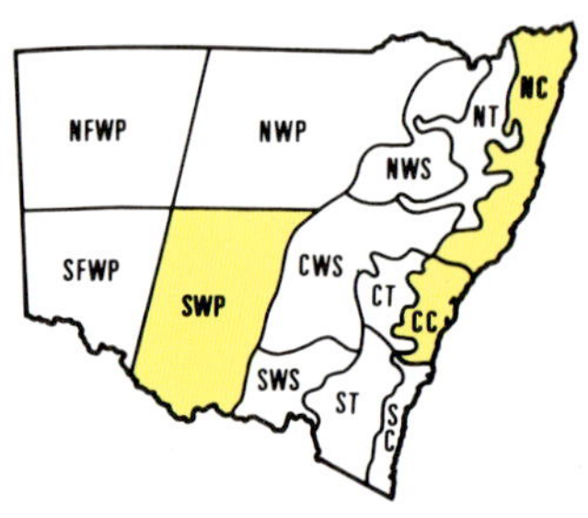

Bladderworts *Utricularia*

Annual or perennial herbs, usually aquatic or swamp plants. Leaves alternate or basal, usually dimorphic with the submerged leaves much branched and forming small traps for aquatic insects or other life forms. The traps are subglobular bladders with a hinged flap and projecting bristles. The trap has a lower internal pressure and when the bristles are triggered the valve opens and the victim is carried in with the water.

In aquatic species the plants are usually submerged, free floating with no floating or emergent leaves. Inflorescences emergent, 1-few flowered, bisexual, irregular. Calyx 2-lobed; corolla 5-lobed; 2-lipped, tubular, spurred at the base. Stamens 2 ($\pm$ staminodes). Ovary superior, fruit a capsule.

There are two truly aquatic species in New South Wales:

1. Leaf segments few, corolla 4-6 mm across, the traps usually replacing one of the leaf segments at a fork. **_U. gibba_ ssp. _exoleta_**
1a. Leaf segments numerous, corolla 9-10 mm across, the traps borne near the base of a leaf segment. **_U. australis_**

Yellow Bladderwort *Utricularia australis*

A native submerged, unattached perennial. Stems up to 70 cm long, readily fragmenting, often wrapped around submerged obstructions. Leaves alternate, divided in two at the base, each branch divided into 10-14 alternate pinnae which are in turn divided into 4-8 fine segments. Traps 1.5-4 mm long. Inflorescence emergent, 3-6 flowered. Pedicels reflexed in fruit. Corolla yellow with brownish markings, 9-10 mm across.

Growth Biology

Basically summer growing. Seeds apparently rarely develop and propagation is mainly by fragmentation of individuals and turions or winter buds formed at branch ends. Although flowering occurs in summer, it apparently does not occur every year in any one locality. The plants photosynthesise but also derive a substantial nutrient supply from their traps.

Habitat

Stationary or slow-flowing water up to several metres deep and usually floating within the top metre in acid lagoons, small lakes or swamps.

Economic Significance

Nil.

Distribution

NC, CC, SWP, Tasmania.

Bladderwort

Utricularia gibba ssp. ***exoleta***

A native, submerged unattached perennial. Stems up to 50 cm long readily fragmenting, often wrapped around submerged obstructions. Leaves alternate, divided in 2 at the base, each branch with only a few, rarely divided, segments. Traps replacing a leaf segment on the branch. Corolla yellow with brown markings, 4-6 mm across.

Growth Biology

Virtually the same as *U. australis* as far as is known, and the two are frequently found together in New South Wales.

Habitat

As for *U. australis*.

Economic Significance

Nil

Distribution

NC, CC, Queensland, Northern Territory.

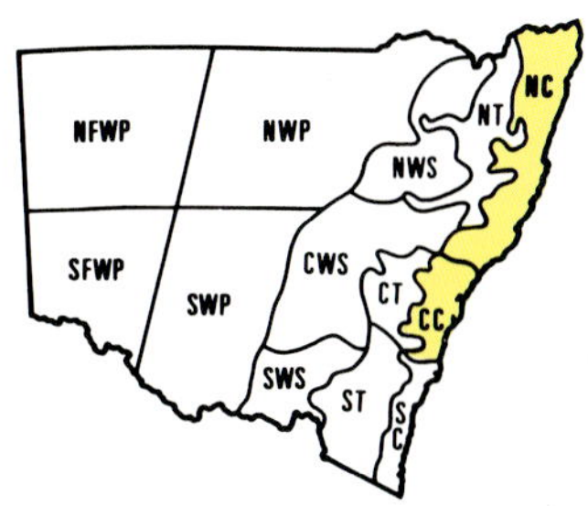

Left: Submerged growth of *Utricularia australis* (left) and *U. gibba* ssp. *exoleta* (right)

Right: Flowers of *Utricularia gibba* ssp. *exoleta*

Lythrum hyssopifolia
with the comparatively
inconspicuous flower
near the base of the
central piece

Loosestrife *Lythrum*

Perennial or annual herbs. Leaves opposite or alternate. Flowers bisexual, mostly regular, solitary or in many-flowered inflorescences. Calyx with 4-6 lobes, shorter than the corolla, usually with appendages present; corolla tube with 4-6 lobes; stamens twice as many or as many as petals; ovary superior, 2-locular.

1. Leaves opposite; flowers large, approx. 1 cm diameter.
L. salicaria

1a. Leaves alternate; flowers small, approx. 0.5 cm diameter.
L. hyssopifolia

Hyssop Loosestrife *Lythrum hyssopifolia*

A native decumbent or weakly ascending annual. Leaves mostly alternate, 5-10 mm long, 1-5 mm wide. Flowers axillary, solitary, ± sessile. Corolla 2-4 mm long, fused for about one-third of the length, blue. *Seed photograph page 443.*

Habitat

Drains or damp places, banks of channels or margins of water a few cm deep, commonly growing in damp areas around rice crops.

Economic Significance

Nil.

Distribution

NC, CC, SC, NT, CT, ST, CWS, SWS, NWP, SWP, NFWP, SFWP, Queensland, Victoria, Tasmania, South Australia, Western Australia.

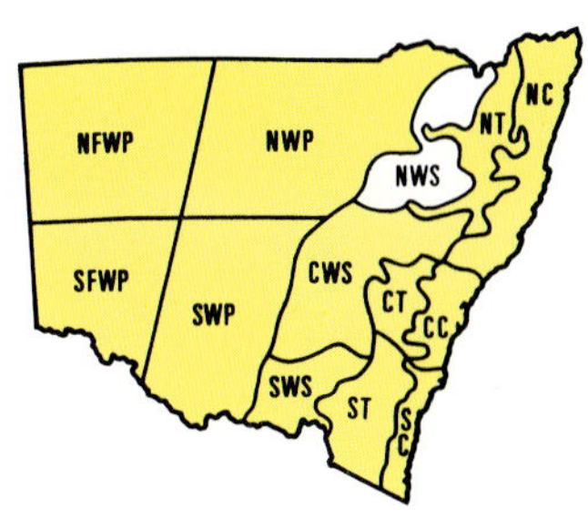

Lythrum hyssopifolia on the margins of a rice crop

The showy pink flowers
on an inflorescence of
Lythrum salicaria

Purple Loosestrife

Lythrum salicaria

An erect, frequently pubescent native perennial up to 1.5 metres high. Leaves opposite 2-5 cm long, about 1 cm wide. Inflorescence a terminal leafy spike. Corolla pink or purple, about 8 mm long, fused for about 5 mm. Stamens usually 12, 6 longer or exserted.

Habitat

Drains or damp ground as well as some dryland habitats, such as small islands in streams.

Economic Significance

An attractive species for cultivation.

Distribution

NC, CC, SC, NT, CT, ST, SWS, NWP, SWP, NFWP, Queensland, Victoria, Tasmania.

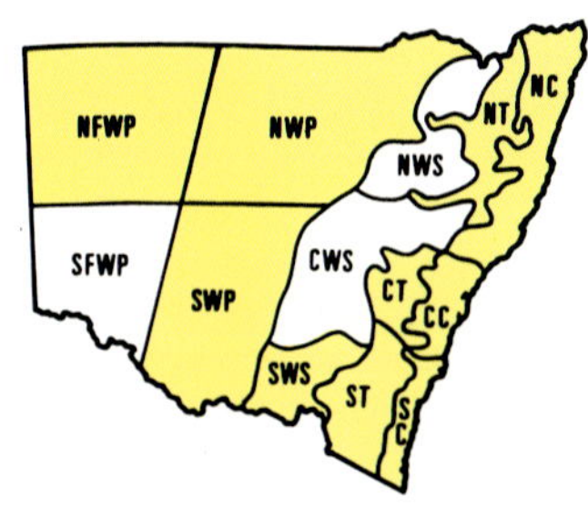

Lythrum salicaria in a swamp on the Coxs River

"Mangroves" is a name given to a group of ecologically related plants from many different plant families. They are usually excluded from treatments of aquatic plants and treated as a group by themselves. "Mangrove", like "aquatic" can be difficult to define rigidly, although most people understand the term. "Mangroves" are usually defined as the trees and larger shrubs that normally only grow in plant communities that are periodically flooded by tidal waters. As there are only five species generally regarded as true mangroves in New South Wales it is comparatively simple to include at least a brief treatment and key.

Only two species, viz., *Avicennia marina* and *Aegiceras corniculatum* are relatively common in New South Wales; the other three species are variously restricted to parts of the North Coast, but even there occur in only a small proportion of the mangrove vegetation.

1. Milky-white latex present.   *Excoecaria agallocha*
1a. Milky-white latex absent.   **2.**

2. Stilt roots or buttresses absent.   **3.**
2a. Stilt roots of buttresses present.   **4.**

3. Leaves opposite.   *Avicennia marina* var *australasica*
3a. Leaves alternate.   *Aegiceras corniculatum*

4. Stilt roots present, produced from branches.
     Rhizophora stylosa
4a. Buttresses present at base of trunk.   *Bruguiera gymnorrhiza*

Left: *Avicennia marina* forest
on the shores of Botany Bay

Aegicerus corniculatus
on the shores of Botany
Bay

River Mangrove

Aegiceras corniculatum

A native shrub up to 4 metres tall with irregularly arranged, obtuse, ± fleshy leaves up to 7 cm long and 5 cm wide. Inflorescence a lateral or terminal umbel. Calyx with 5 stiff lobes each 4-6 mm long. Corolla tube shorter than the calyx lobes, corolla lobes about 5 mm long, white. Stamens 5. Ovary superior, 1-locular. Fruit elongated, germinating on the plant.

Growth Biology

Plants flower in spring, early summer, producing fruit in summer and autumn. The fruit germinate on the plant, drop off, then float in the tide until lodging in a suitable habitat.

Habitat

Periodically inundated margins of estuaries and saline or brackish rivers. Usually grows in shallower water, and further upstream than *Avicennia* although mixed stands are frequent.

Economic Significance

Important for stabilising estuary and river banks and providing a suitable habitat for young fish. A honey source. A protected plant.

Distribution

NC, CC, SC, Queensland, Northern Territory, Western Australia.

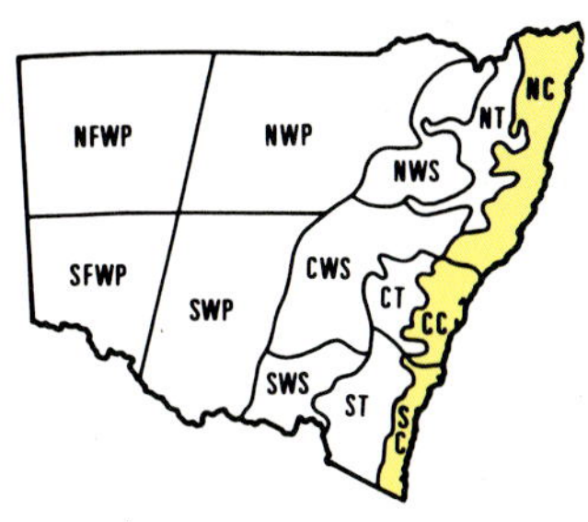

Inflorescence of *Aegicerus corniculatus*

Grey Mangrove

Avicennia marina var. *australasica*

A native tree up to 8 metres high. Vertical pneumatophores, cylindrical, unbranched, up to about 1 cm diameter. Bark grey with very fine cracks. Leaves opposite, glossy above, slightly hairy below, up to 8 cm long, 5 cm wide. Inflorescence axillary or terminal, small, dense. Calyx with 5 lobes, divided almost to the base, corolla 4-lobed, orange, tube shorter than the calyx. Stamens 4. Fruit laterally compressed, about 3 cm diameter, with a hairy seed coat enclosing a single large seed consisting of 2 closely folded cotyledons. The seed coat usually splits (i.e. germinates) on the plant.

Growth Biology

Flowering is in spring with fruit formed in late spring, early summer. The seed coats are usually split on the plant, dropping the young seedling. The seedlings float for up to a week, being dispersed by currents and tide. The radicle is usually slowly elongating during this time. Eventually the seed sinks, or becomes stranded by the higher tides. If in a suitable habitat the plant will continue to grow.

Habitat

Mud or sandy-mud alluvium along periodically inundated margins of estuaries and saline or brackish rivers.

Economic Significance

Stabilises banks and channels and provides suitable habitats for young fish and water birds.

Distribution

NC, CC, SC, Queensland, Victoria, Tasmania, Northern Territory, South Australia, Western Australia.

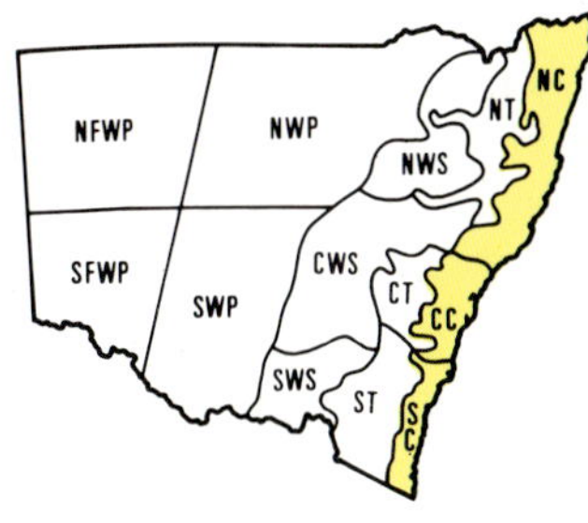

Left: *Avicennia marina* on the banks of the Lower Hunter River

Left: Fruit of *Avicennia marina*

Milky Mangrove

Excoecaria agallocha

A native tree to 14 metres high with an irritant milky-white sap. Bark with rows of corky lenticels. Leaves usually alternate up to 7 cm long, 4 cm wide. Inflorescence an axillary spike.

Distribution:

NC, Queensland, Northern Territory.

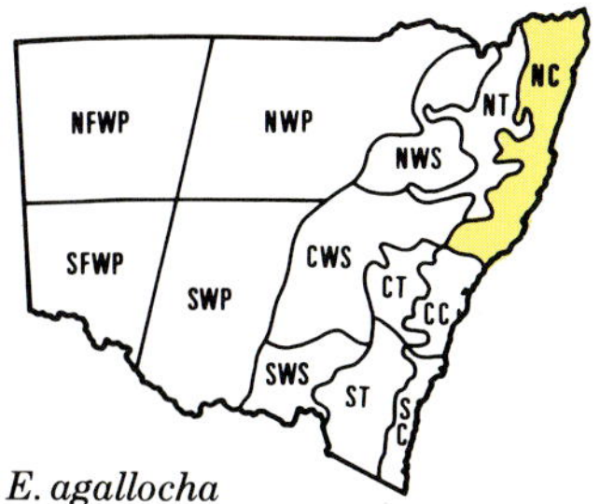

E. agallocha

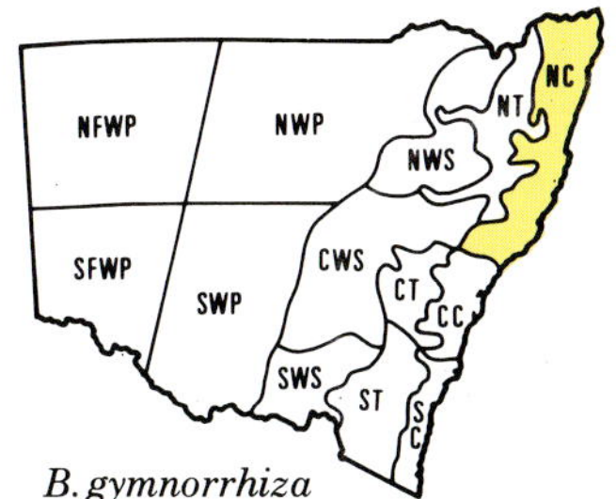

B. gymnorrhiza

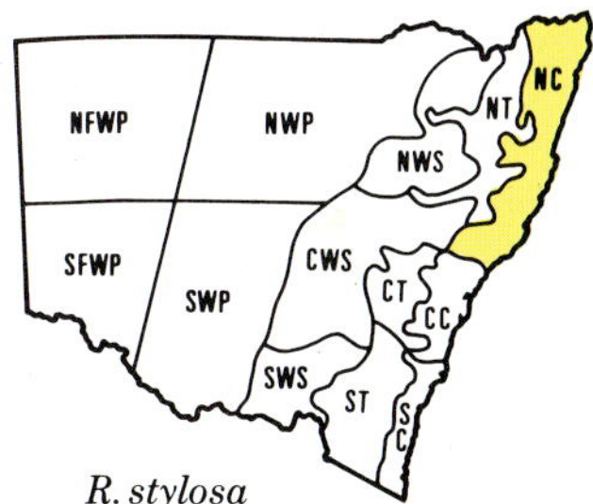

R. stylosa

RHIZOPHORACEAE

Bruguiera gymnorrhiza

A native dioecious tree up to 12 metres tall. Stem buttressed at base, bark rough. Pneumatophores knobbly, bent. Leaves opposite, glabrous, up to 20 cm long, 7 cm wide. Flowers solitary, axillary.

Distribution

NC, Queensland, Northern Territory.

Red Mangrove

Rhizophora stylosa

A native tree to 4 metres high with long aerial, basally divided prop roots produced from the lower branches. Pneumatophores bowed, looped roots growing away from the base. Leaves opposite, up to 12 cm long and 6 cm wide. Inflorescence open, spreading.

Distribution

NC, Queensland, Northern Territory.

Left: A community of mangroves, including *Excoecaria agallocha*, in North Queensland

Thalia dealbata

An introduced rhizomatous perennial to 2 metres tall. Sheaths closed, rather loose, with a distinct petiole and a lanceolate blade up to 60 cm long and 25 cm wide with a whitish bloom, especially on the lower surface. Inflorescence up to about 30 cm long and 10 cm wide, an open arrangement of bracteate spikes. Flowers paired, sessile, purple, initially enclosed in a deciduous bract. Perianth 2 whorls of 3; staminodes 2, 1 petaloid, stamens 1, ovary 1-locular, inferior. *Fruit photograph page 443.*

Growth Biology

Grows in spring/summer, flowering and fruiting in summer.

Habitat

Usually only cultivated, in standing water less than 1 metre deep.

Economic Significance

An ornamental. Could prove troublesome on North Coast.

Distribution

Not recorded as naturalised in New South Wales. South and Central America.

Left: *Thalia dealbata* in an ornamental pond in Canberra

Right: Inflorescence of *Thalia dealbata*

MARSILEACEAE

Marsilea spp.

Perennial rhizomatous ferns bearing characteristic erect, floating, submerged or emergent fronds with 4 terminal leaflets (looking like a 4-leaved clover). Some species basically terrestrial. Sporocarp separate from the leaflets, attached near or on the base of the frond close to the rhizome. Both megasporangia and microsporangia borne together in the hard, ± hairy, flattened sporocarp; megasporangia producing a solitary megaspore, microsporangia producing numerous microspores.

Like all ferns, *Marsilea* has a gametophyte stage as well as the sporophyte stage described above. The megaspores germinate to produce prothalli, which in turn produce archegonia and microspores. The microspores germinate into prothalli bearing antheridia.

Another uncommonly recorded species, *Pilularia novae-hollandiae,* is apparently related to *Marsilea* but differs in the absence of the terminal leaflets *Pilularia* also grows in or around ephemeral water bodies.

1. Terminal leaflets absent, plants grasslike, with basal sporocarps present.   ***Pilularia novae-hollandiae***
1a. Terminal leaflets present.   ***Marsilea* 2.**

2. Fronds small, glabrous, usually about 2 cm in diameter with the leaflets often not perpendicular to the stalk, particularly in submerged or young fronds; leaflets about 1 cm long, half as wide.   ***M. angustifolia***
2a. Fronds 1-5 cm diameter, usually fronds of 3 cm diameter present; hairy or glabrous. All leaflets perpendicular to frond stalk.   **3.**

3. Leaflets usually hairy, not bicoloured; sporocarps borne singly.   ***M. drummondii***
3a. Leaflets usually glabrous, with a yellow-brown zone on lowest third; more than one sporocarp borne on branched pedicels.   ***M. mutica***

Growth Biology

If the plants are subject to flooding after a prolonged dry period, as frequently occurs in inland watercourses, the sporocarps will open within hours and spores begin development into prothalli. The old plants are capable of reabsorbing water and apparently dead plants may resume growing. This plant makes vigorous growth in the hot weather and does not die back in the winter.

Habitat

Common and widespread in damp situations in New South Wales.

Economic Significance

At times *M. drummondii* is obstructive in small supply and drainage channels and is a minor problem in the Coleambally Irrigation Area, New

South Wales. On several occasions *M. drummondii* has been suspected of poisoning sheep and cattle. The aborigines are known to have collected the sporocarps for food, preparing them by grinding and mixing with water before cooking.

M. angustifolia infrequently occurs as a submerged plant in rice.

M. mutica, a large-leafed Nardoo, is occasionally cultivated as a decorative plant. The leaves have a distinctive brown line across the leaf. This species grows in an irrigation supply channel near Maffra in Victoria in water to 1.5 metres deep.

Thin-leaf Nardoo   *Marsilea angustifolia*

A delicate native perennial with submerged or floating leaves. Frond stalks occasionally emergent to 10 cm long. Leaflets usually glabrous, when young usually deeply lobed or divided rather than clover-like, each to about 1 cm long by 0.5 mm broad, producing a mature clover-like leaf 1-3 cm diameter. Sporocarps usually on unbranched pedicels. Widespread, but not as common as *M. drummondii.*

Distribution

M. angustifolia: CC, CT, NWS, SWS, NWP, SWP, NFWP, all mainland States.

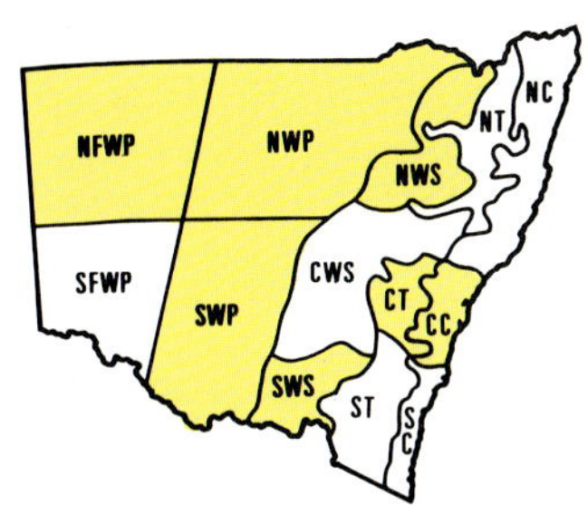

Top: *Marsilea angustifolia* on the margins of a rice crop

Left: *Marsilea angustifolia* showing the range of leaf shapes found in this species

Marsilea drummondii
with sporocarps (A)

296

Common Nardoo *Marsilea drummondii*

A native perennial rhizomatous fern with either erect, emergent or floating cloverlike fronds to 30 cm high. Fronds to 6 cm in diameter, usually hairy, but not obviously so on some floating fronds. Sporocarps *(photograph page 445)* solitary, less than 1 cm in diameter, on stalks usually less than 5 cm long.

Distribution

M. drummondii: NWS, CWS, SWS, NWP, SWP, NFWP, SFWP, Queensland, Victoria, Northern Territory, South Australia.

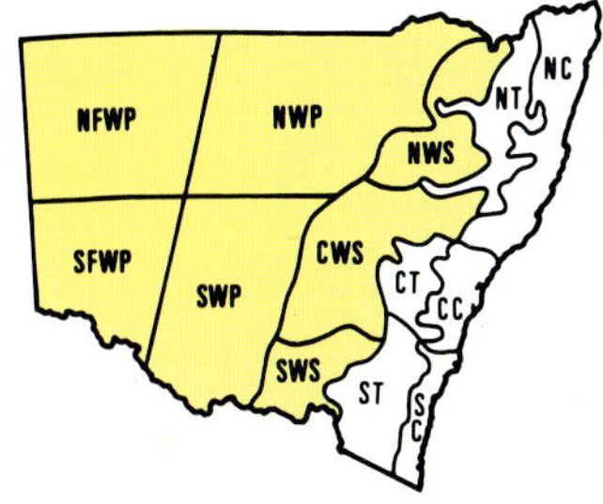

Marsilea drummondii in a channel near Coleambally

Marsilea mutica (left,1) and *Marsilea drummondii* (right, 2) showing (3) sporocarps and (4) undersurface of leaf

298

Nardoo

Marsilea mutica

A native rhizomatous perennial with the clover-like fronds produced on ± flexuous stalks to about 1 metre long. Fronds to about 10 cm in diameter, usually glabrous. Individual leaflets with a yellowish-brown band on lowest third. Sporocarps usually 2-4 on a branched pedicel usually less than 2 cm long.

Distribution

M. mutica: NC, CC, SC, CT, ST, NWS, NWP, Queensland, Victoria, Tasmania, Northern Territory.

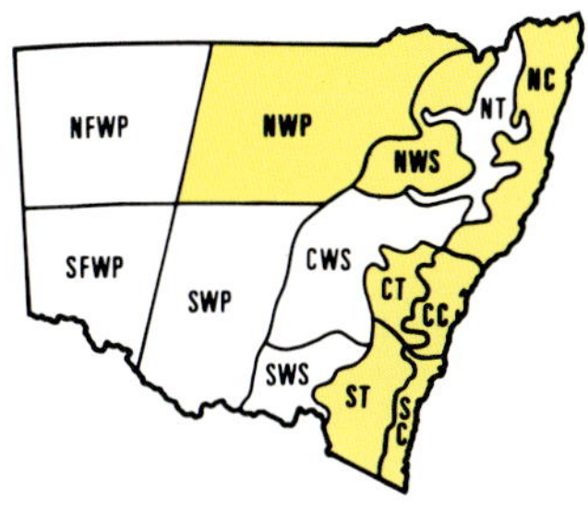

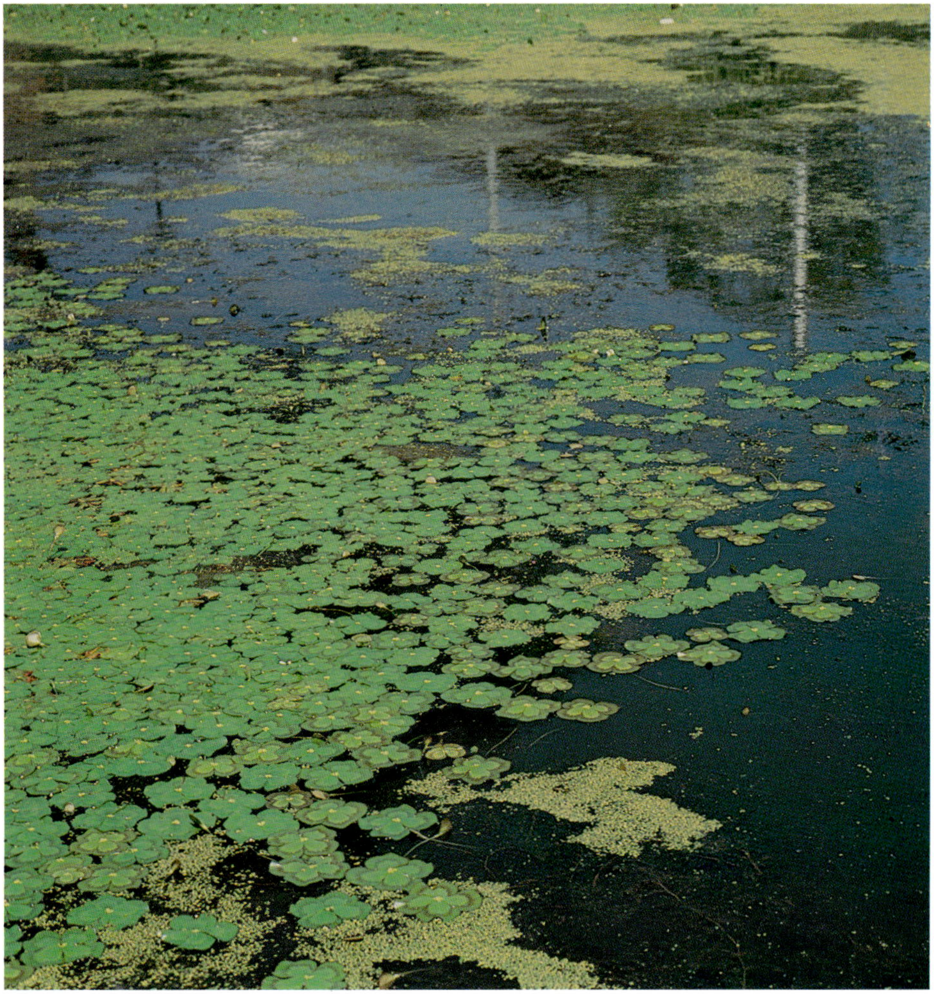

Marsilea mutica in an ornamental pond in Canberra

Nymphoides

Stoloniferous attached perennials, or occasionally annuals. Plants rooting in substrate, rarely floating until taking root, stolons floating. Leaves floating, ovate to rounded, strongly cordate, long or short petiolate. Inflorescences either produced virtually behind the leaf, or on separate peduncles, emergent. Flowers submerged in bud, emergent when open, submerged when fruiting. Flowers only open for one day, opening in the morning, usually dishevelled by afternoon. Flowers often heterostylous. Calyx fused, 5-lobed. Corolla fused at the base, mostly 5-lobed, yellow or white, often winged or hairy on inner surface. Stamens 5. Stigmas equal in number to ovary cells, often of different lengths (2 or 3) in one species. Seeds variously smooth or sculptured.

1. Flowers white, leaves shortly petiolate, flowers borne immediately behind leaf. *N. indica*
1a. Flowers yellow, leaves long petiolate, inflorescences separate from leaf bases. **2.**

2. Leaves crenate, corolla lobe with a median fringed wing inside. *N. crenata*

2a. Leaves entire (rarely slightly crenate), corolla lobe without an inner wing. *N. geminata*

Left: *Nymphoides crenata* in a supply channel near Yanco

The delicate flowers of *Nymphoides crenata* with the median petal wing. The three upper flowers are short-styled and the three lower long-styled

Wavy Marshwort

Nymphoides crenata

Robust native perennial with long-petioled floating leaves and $\pm$ floating stolons up to 2 metres long. Basal leaves ovate, strongly cordate, crenate, 5-12 cm long. Leaves on stolons smaller with shorter petioles subtending few to many flowers with pedicels 3-8 cm long. Flowers heterostylous, calyx 7-10 mm long, corolla mostly 5-lobed, yellow with a fringed wing on inner midline, margins fringed. Seeds numerous, 0.6-1.2 mm long, straw coloured, smooth and shiny. The waxy leaves have crenate margins, from which the specific name "crenata" has been derived.

Habitat

Slowly-flowing water in creeks and rivers, lagoons and channels in water to 1.5 metres deep, usually on a mud substrate. Can persist for some time on drying mud.

Economic Significance

Rarely blocks drains or channels. Alone it is not a pest, but growing with other aquatic plants it will impede water flow. Used as a horticultural subject.

Distribution

NWS, CWS, NWP, SWP, NFWP, Queensland, Victoria, Northern Territory, Western Australia.

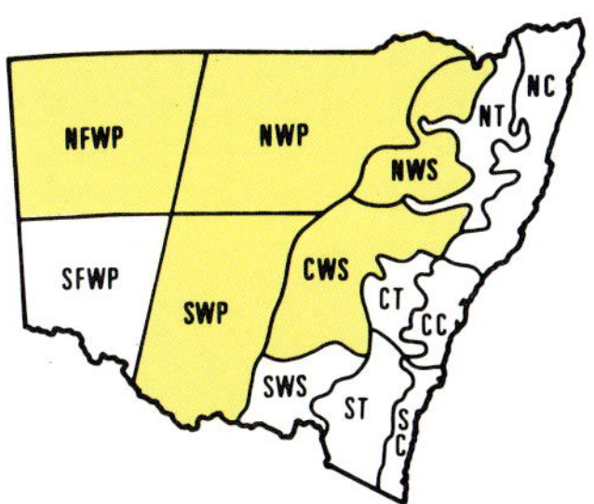

Nymphoides geminata

Similar to *N. crenata* but has mostly entire leaves 4-10 cm long and produces a leafless axillary inflorescence 5-20 cm long from the ± floating stolon. Corolla yellow with only lateral fringed wings. Seeds 1.2-1.7 mm long, straw coloured, smooth.

Habitat

In still water of swamps to fast-flowing fresh water streams common to high country at altitudes of up to about 1000 metres.

Economic Significance

Occasionally used in garden pools as an ornamental.

Distribution

NC, CC, NT, ST, Queensland.

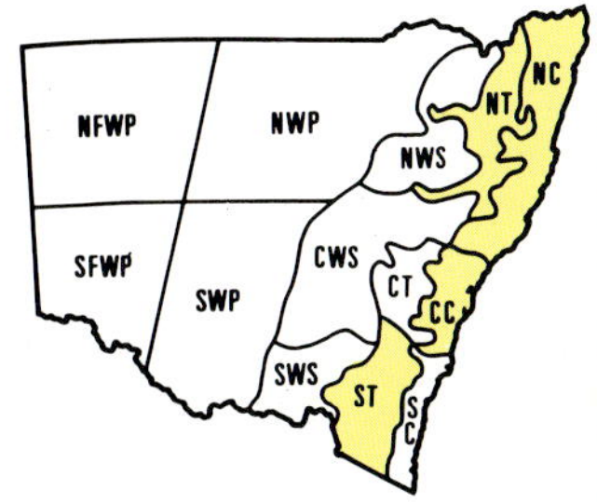

Left: Flower of *Nymphoides geminata* with no median petal wing

Right: A cultivated specimen of *Nymphoides geminata*

Water Snowflake

Nymphoides indica

A native robust attached perennial with floating stolons and leaves. Leaves and inflorescence terminal, but the stolons are occasionally branched, producing a second leaf and inflorescence. Flowers numerous in each cluster, each with a pedicel 2-8 cm long. Corolla lobes white, yellow or orange at the base, bearded inside. Heterostylous. Seeds 1-2 mm long, smooth, straw coloured.

Habitat

Rivers and creeks, farm dams, channels and drains in still and flowing water to 2 metres deep, on a variety of substrates from mud to sand. Can survive for some time on drying mud.

Economic Significance

Can grow in sufficient density to block drains and channels. Commonly cultivated as an ornamental.

Distribution

NC, Queensland, Northern Territory, Western Australia.

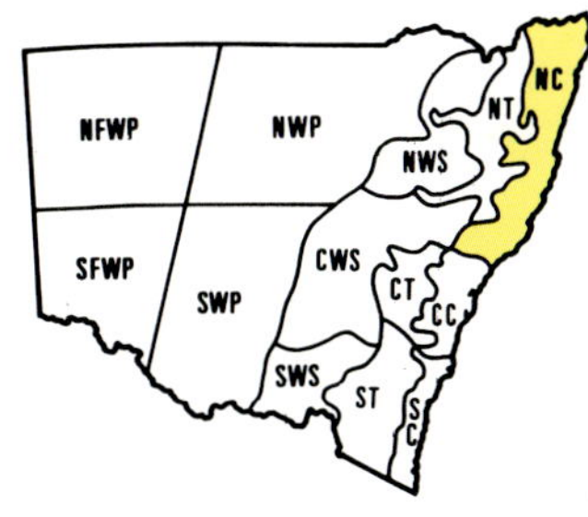

Left: *Nymphoides indica* in the Coldstream River

Right: Short-styled flowers of *Nymphoides indica* with white hairs on the inner petal surface

Top: *Villarsia exaltata* in a spring near Sydney

Left: A short-styled flower and leaf of *Villarsia exaltata*

Villarsia

Tufted or stoloniferous perennials with emergent inflorescences and emergent and/or floating leaves. Inflorescence an open panicle. Flowers heterostylous or homostylous. Calyx fused, usually 5-lobed. Corolla united at base, mostly 5-lobed, yellow and in New South Wales species non-bearded and with undulate margins.

Two species in New South Wales:

1. Leaves emergent, erect, longer than broad, plants tufted; flowers heterostylous. Corolla 9.5-18 mm long; 16-30 mm diameter.
V. exaltata

1a. Leaves emergent and/or floating, as long as broad, plants usually stoloniferous in water, flowers homostylous. Corolla 12-23 mm long, 19-40 mm diameter.
V. reniformis

Villarsia exaltata

Erect tufted native perennial. Petioles 13-40 cm long. Blades 6-15 cm long, mostly longer than broad, rounded or shallowly cordate at base ± similar on both surfaces. Stems 1-few, to 1.5 metres high, leafy below, terminating in an open panicle to 40 cm long. Flowers heterostylous. Calyx lobes 4.5-8 mm long. Corolla yellow, 9.5-18 mm long, 16-30 mm across. Seed 1.7-3 mm long, light or dark coloured with whitish caruncle, surface tuberculate.

Habitat

Stationary or slowly-flowing water to 50 cm deep, swamps or ephemeral pools. Survives on drying mud for some time.

Economic Significance

An attractive native species.

Distribution

NC, CC, SC, CT, ST, Queensland, Victoria, Tasmania.

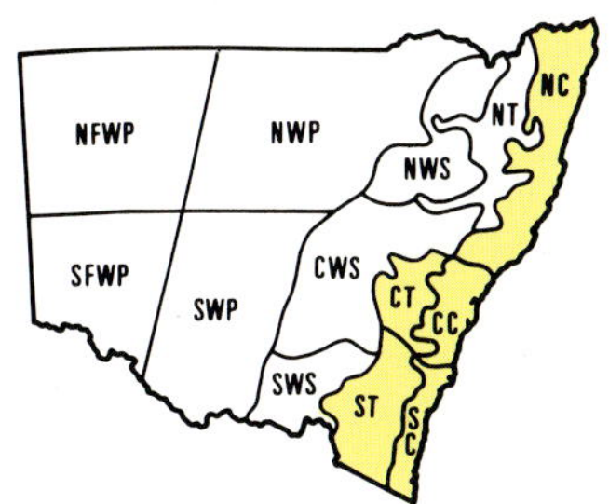

Villarsia reniformis

Stoloniferous native perennial with floating and/or emergent cordate dorsiventrally differentiated leaves. Leaves 4-8 cm long, roughly as long as broad. Culms 1-few, up to 1 metre high, ± erect, bearing an open panicle to 30 cm long. Flowers homostylous. Calyx lobes 6.5-10 mm long. Corolla yellow, 12-23 mm long, 19-40 mm across. Seed 1-1.8 mm long, pale, non-carunculate, smooth.

Habitat

Stationary or slowly-flowing water up to 80 cm deep in swamps or bogs.

Economic Significance

An attractive native plant.

Distribution

CC, SC, CT, ST, Victoria, Tasmania, South Australia.

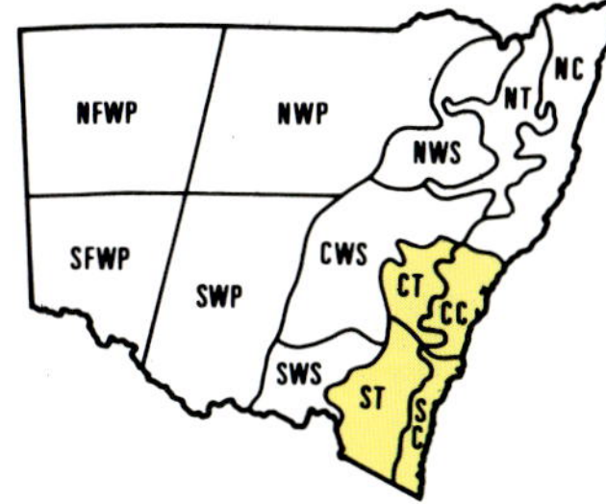

Left: *Villarsia reniformis* (×0.3)

A species of moss

Mosses require free water for fertilisation and for completion of their reproductive cycle. Consequently mosses generally grow in wet or damp places, or in places that are seasonally wet. No Australian mosses are regarded as aquatic, although there are some truly aquatic species in other countries (e.g. *Fontinalis* spp.). Mosses may, however, survive inundation, specially in areas where the water is rich in dissolved oxygen, such as attached to rocks in shallow turbulent streams. As it is not practical to select some species for treatment here, it is suggested that the excellent book, "The Mosses of Southern Australia" by G. A. Scott and I. G. Stone (Academic Press, 1976) be used as an identification guide for New South Wales mosses.

A species of moss. Intruding into the illustration on the right are pieces of a species of *Riccia*.

Paperbark

Melaleuca quinquenervia

A native tree to 15 metres high with an open to dense crown. The bark consists of many layers of thin papery sheets ("paperbark") around a trunk usually less than 1 metre diameter and unbranched for up to 5 metres above ground. Leaves 4-10 cm long, 2-15 mm wide, lanceolate or elliptic, not necessarily 5-nerved (usually 3-5). Inflorescence a white spike 2-5 cm long, to 2 cm wide. Flowers with 5 small white petals and numerous stamens with long white filaments connate into 5 bundles. Fruit a small woody capsule up to 5 mm diameter. *Seed photograph page 443.*

Growth Biology

This species seeds freely. After the seeds have been released, and particularly after a fire, many seedlings can establish and grow close together forming a dense stand. Flowering is usually in summer and seeds may be released the following spring or summer or following a fire. Seedlings need shallow, almost stationary water, or damp soil, to establish.

Habitat

Stationary or slowly-flowing fresh or brackish waters in swamps, creeks, drains or channels. Saplings and small trees can survive quite fast-flowing water. Trees can grow permanently in water up to 1 metre deep.

Economic Significance

This tree is extensively planted for windbreaks and erosion control and in gardens all over the world. It can block irrigation and flood mitigation channels which are used intermittently. In Florida, U.S.A., it has spread into the Everglades region, establishing dense thickets, and is a major problem.

Distribution

NC, CC, Queensland.

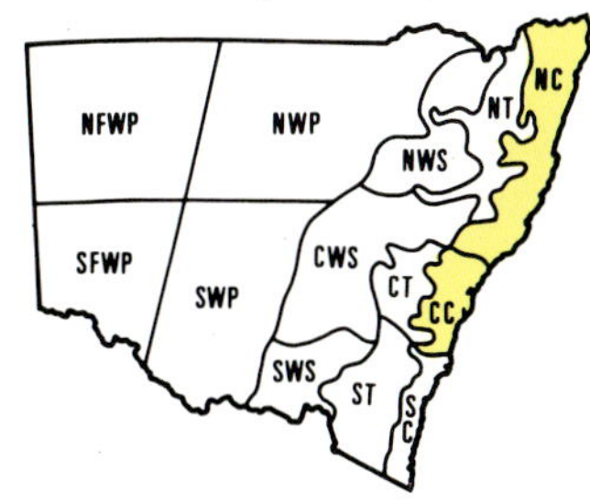

Left: Flowers and fruits of *Melaleuca quinquenervia*

Right: A dense stand of *Melaleuca quinquenervia* in a swamp in Centennial Park

Najas

Submerged monoecious or dioecious, much-branched, diffuse annuals; leaves opposite or whorled with a sheathing base and dentate margins. Flowers unisexual, axillary, 1-few together. Male flowers spatheate, with 1 stamen. Female flower espatheate; ovary solitary with 2-4 stigmas.

Worldwide there are about 50 species found in a wide range of aquatic situations to depths of up to about 5 metres. Many species are found in rice fields. Some can withstand high water temperatures. In Australia, there are doubtfully 5 species. Only *N. marina* is distinctive, the other species being very similar in appearance.

Flowering and fruiting *Najas marina;* (A) an axillary fruit

Prickly Waternymph

Prickly Naiad

Najas marina ssp. *armata*

A native submerged dioecious annual up to 3 metres high. Leaves up to 5cm long and 0.5 cm wide with numerous large spine-tipped teeth along the margin, with sheathing base. Leaf surface and/or internodes may have similar teeth. Flowers unisexual, axillary, solitary. Male flowers spatheate 3-4 mm long, with a single stamen. Female flowers espatheate, 3-4 mm long, mostly with 3 stigmas.

Growth Biology

Small flowers are borne at the leaf-stem junction for a short period in mid-summer. Flowering and fruit development takes place beneath the water surface. When the fruit is mature the plants become detached, floating around in masses, breaking up and dispersing the seed. Grows vigorously in the mid-summer when water temperatures are consistently above 20°C. Since 1965 this species has spread throughout the lakes at Lake Wyangan, Griffith, New South Wales, thriving in slightly saline water typically containing 700 parts per million total dissolved salts. Observations of submerged plant growth in these lakes (area 300 hectares) indicates that *N. marina* will compete with and significantly smother *Potamogeton pectinatus* in water up to 3 metres deep. *N. marina* behaves as an annual, disappearing in the winter months.

Habitat

In lakes containing fresh or brackish water.

Economic Significance

It may form large masses just below the surface of the water, interfering with boating and fishing.

Distribution

NC, CC, SWP, Victoria.

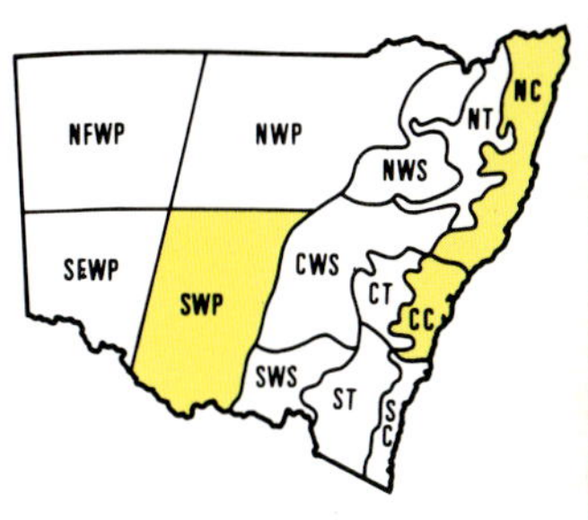

Najas marina in a lake near Griffith

Najas tenuifolia with
axillary flowers and thin,
rapidly wilting leaves

Waternymph

Thin-leafed Naiad

Najas tenuifolia

A native diffuse monoecious annual or short-lived perennial with stems up to 1 metre long. Leaves opposite or ± whorled, up to 5 cm long, 1-3 mm wide with a sheathing base, margins with several fine teeth. Flowers unisexual, axillary, 1-few together, the male flower produced towards the tip of the branches, the female lower down. Male flower spatheate, 1-3 mm long, consisting of a single stamen. Female flower espatheate, 1-4 mm long, a solitary ovary with 2 stigmas.

N. tenuifolia may be confused with other species in the genus, except for the more robust *N. marina*.

Growth Biology

Main growth takes place when water temperatures exceed 20°C. Flowering and fertilisation occur underwater.

Habitat

This is an uncommon but widespread species found in static, fresh or slightly saline water, particularly in the northern coastal part of New South Wales. It is occasionally seen as a dark mass under the water in earth tanks, lakes and shallow drainage channels.

Economic Significance

Generally, species of *Najas* are not considered to be of great economic importance, although they may be a nuisance in irrigation ditches.

Distribution

NC, CC, CT, NWS, CWS, SWS, SWP, Queensland, Victoria, Northern Territory, South Australia, Western Australia.

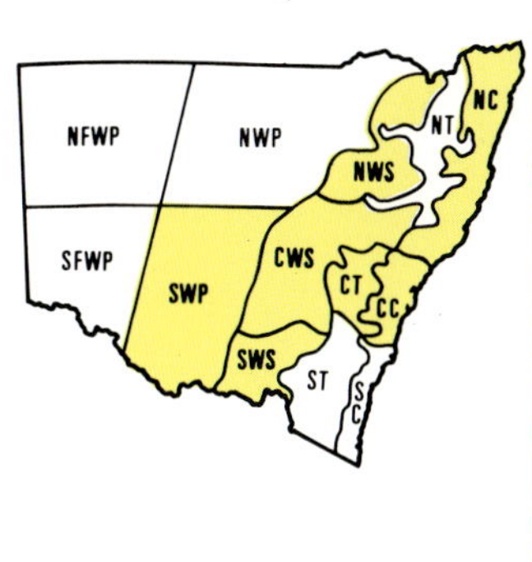

Najas tenuifolia in the
Coldstream River

CM

American Lotus

Yellow Nelumbo

Nelumbo lutea

A rhizomatous exotic perennial with floating and/or emergent leaves and emergent flowers. Rhizomes either thin and spreading vegetatively or thick and buried, so-called "storage-rhizomes" with large longitudinal air canals. The rhizomes can be many metres long. Leaves peltate, usually 40-80 cm diameter, on spiny petioles up to 2 metres long. Flowers bisexual, solitary, axillary up to 15 cm diameter. Perianth segments numerous, yellow. Stamens numerous with a terminal appendage. Carpels 10-20, solitary in a characteristic pithy conical receptacle.

Nelumbo nucifera is a pink-flowered native Lotus found in northern Australia in fresh water lagoons. This is the same species as the Sacred Lotus of India.

Growth Biology

Summer growing and flowering. Young leaves float on water surface and as the petioles strengthen, are held erect.

Habitat

Usually found on the margin of ponds and lakes in shallow water up to about 1 metre deep.

Economic Significance

Grown as an ornamental in Australia, but has the potential to be a troublesome weed of waterways of the North Coast. It provides a favourable habitat for mosquitoes. Not recorded as naturalised.

Distribution

Cultivated, not naturalised.

Left: A cultivated specimen of *Nelumbo nucifera*

Right: *Nelumbo lutea* in a Tennessee Valley impoundment, U.S.A.

Nuphar lutea

An introduced perennial with a large tuberous rhizome, floating and/or emergent leaves and emergent flowers. Petiole attached near centre of blade, which has a radial gap. Blade to 15 cm diameter. Flowers emergent, solitary, axillary. Sepals 6, the outer green, the inner yellow. Petals numerous, reduced and scalelike or stamenlike. Stamens numerous. Ovary superior, 5-20 united carpels. Seeds not carunculate.

Growth Biology

Fruit ripens above water, but seeds do not seem to germinate or seedlings survive in Australia. The massive tuber overwinters, sending up new leaves in spring.

Habitat

Permanent water to about 1 metre deep.

Economic Significance

An uncommon ornamental species. A species of *Nuphar* is recorded as a weed in Florida, U.S.A.

Distribution

Not recorded as naturalised in New South Wales.

Left: *Nuphar lutea* in an
ornamental pond in Canberra

Waterlilies

Nymphaea

Emergent perennials with floating leaves, tubers or massive rhizomes and various coloured large floating or emergent flowers. Roots thick, white, fleshy up to 5 mm diameter. Leaves sub-peltate or cordate, rising from a short, erect stem. Mature leaves mostly floating with various distinctive margins. Petioles long, dependent on water depth, up to 1 cm diameter. Seedling leaves usually oval, mature leaves circular with radial slit, up to 75 cm diameter.

Flowers axillary, solitary, either floating with flexuous pedicels or stiffly emergent on more rigid pedicels. Sepals commonly 4, usually greenish. Petals numerous and characteristically coloured. Stamens many, either grading into petals or clearly disjunct from them. Anther sacs either dorsally or laterally joined. Anther appendages absent, present or deciduous. Ovary globular, many-celled, multi-ovulate, topped with lobes which differ in length and with varying amounts of stigmatic surface. Fruit submerged, enclosed by the discoloured perianth on the end of the now spirally twisted pedicel. Seed variously coloured and ornamented.

A world-wide genus of approximately 50 species with approximately 7 native Australian species and 2 naturalised species and hybrids cultivated. One native and 2 naturalised species in New South Wales.

1a. Plants with horizontal rhizomes or stolons; leaf margins usually entire, may be repand; flower colours yellow, white, pink, red, orange or similar. **2.**

1b. Plants with short vertical rhizomes only; leaf margins dentate or irregularly sinuate; flower colours blue, white, rarely blue-pink. **3.**

2a. Plants with horizontal rhizomes producing roots all along; flower colours cream, white, pink, red, orange or similar, rarely yellow.

ornamental ***Nymphaea*** hybrids, usually with **N. *alba*** or **N. *tuberosa*** as one of the parents.

2b. Plants with vertical rhizomes and horizontal stolons only producing roots underneath clumps of leaves; flowers yellow; leaves often with brown markings. **N. *mexicana*** or ornamental hybrids with **N. *mexicana*** as one parent.

3a. Mature leaves dentate with acute, regularly spaced teeth; perianth segments obtuse, blue to white (fading with age); anthers without terminal appendages. **N. *gigantea***

3b. Mature leaves irregularly sinuate; perianth segments acute, blue, white or blue-pink (not changing with age); anthers with an obvious terminal appendage. **N. *capensis***

Left: Flower of *Nymphaea gigantea* showing the obtuse perianth segments

Cape Waterlily

Nymphaea capensis

An introduced perennial with small tubers which may develop into short vertical rhizomes. Mature leaves up to 30-40 cm diameter with sinuate margins. Flowers emergent, up to 15 cm diameter and standing about 30 cm above the water surface. Sepals 4, green with margins white,or white tinged with blue. The green areas of the sepals may be flecked with very dark blue spots, especially in cultivated forms. Petals blue, rarely white. Outermost series of petals obtuse, inner whorls acute. Outer stamens petaloid, only slightly differentiated from the innermost petals and with no obvious gap between them. Anther appendages the same colour as the petals, large on the outer stamens to small on the inner stamens. Ovary lobes short but distinct with stigmatic surface on the lower ½-⅔. Seed small, egg-shaped, up to 2 mm long, olive-green and yellow under the microscope, appearing brown *en masse*. The specific epithet "capensis" refers to the South African distribution of the species.

Growth Biology

Seeds germinate from spring to autumn on mud in water up to about 1 metre deep. Seedling leaves submerged, lanceolate, then elliptic, then ovate with a radial slit, then floating almost circular leaves with a radial slit.

Flowers during late spring to early winter. Flowers remain open for 4 to 5 days, closing at night. As the flower matures the pedicel becomes spiral and the spiral gradually tightens drawing the young fruit below water level. After a maturation period under water the seeds are released from the fruit. The seeds may germinate immediately but can survive drying out, germinating after re-wetting.

The rhizomes overwinter fairly readily and, after several years, may be up to 10 cm long. Several leaves usually survive through winter and growth is usually retarded rather than interrupted. This species spreads spatially only by seeds, which are produced in very large numbers.

Habitat

This species grows in shallow water usually up to about 1 metre deep, but can tolerate a subsequent level change up to about 2 metres. It can grow in soft mud or in heavier clays or in poorer soils with a higher nutrient input, e.g., stock dams.

Economic Significance

Grows in stock dams where it is probably aesthetically advantageous and only likely to cause minor problems in very shallow dams. It may reduce flow rates in smaller drainage channels and may warrant control in some cases. Care should be taken to differentiate between this species and the native *N. gigantea* which is far less likely to cause a problem.

Distribution

NC, CC, South-eastern Queensland.

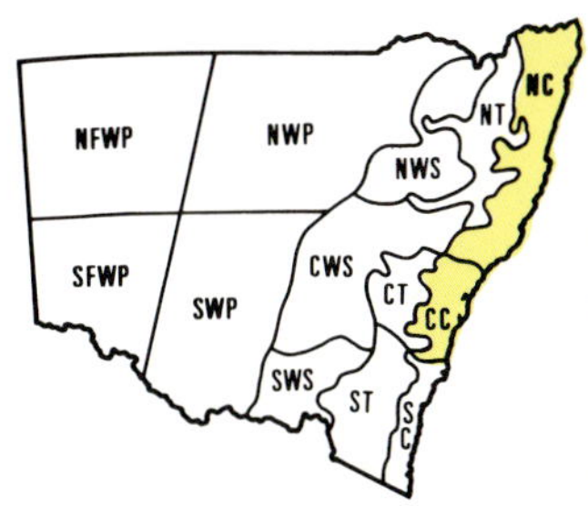

Left: The crenate-margined leaf and blue flowers with acute perianth segments typical of *Nymphaea capensis*

Left: Submerged and floating juvenile leaves of *Nymphaea capensis* seedlings

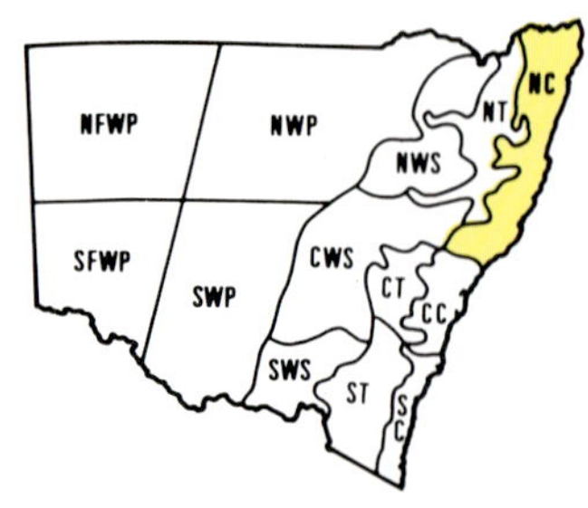

Nymphaea gigantea in a deep pool of a North Coast river

Giant Waterlily

Native Waterlily

Nymphaea gigantea

A large native perennial with small tubers. Mature leaves up to 75 cm diameter with margins of regularly spaced teeth each approximately 5 mm long. Flowers emergent, up to 25 cm diameter and standing up to 50 cm clear of the water surface. Sepals 4, green with bluish tinge on margins. Petals obtuse, usually lilac blue when young, gradually fading to light blue, almost white, joined to the receptacle below the ovary. Stamens arising from near the top of the receptacle, no obvious gap between petals and stamens. Filaments non-petaloid, anther appendages absent or minute and deciduous. Ovary crowned with short, yellow lobes, each totally covered with stigmatic surface. Seed red, carunculate and with rows of white papillae. The specific epithet "gigantea" refers to the large flowers and leaves produced by this species.

Growth Biology

Seeds germinate in early summer on mud in shallow water. Seedling leaves initially almost lanceolate, submerged, followed by floating ovate leaves with a radial slit, in turn followed by mature circular, cordate leaves. Flowers produced during summer and autumn. The flowers open during the day and close at night for about a week. Insects are often trapped in the closed flowers at night. As the flower ages the pedicel becomes spiral and gradually tightens the spiral, drawing the young fruit below water level. After a maturation period underwater the seeds are released from the fruit, float to the surface and usually stay afloat for 2-3 days before sinking. Seeds may germinate immediately but if the water temperature is low they can survive till the following summer.

The mature plant produces small tuberous rhizomes which overwinter. The rhizome initiates the next season's growth and during the season is replaced by a new rhizome immediately above it. The species spreads spatially by seed but survives dry periods with both seed and rhizome.

Habitat

Grows in lagoons or slowly-flowing creeks in water 1-3 metres deep in soft mud which may be a further 1 metre deep.

Economic Significance

May slightly reduce flow in ultimate drainage channels in far North Coast, but probably of little economic significance.

Distribution

NC, Eastern and Central Queensland. Some populations in the Northern Territory may be referable to this species.

Yellow Waterlily

Nymphaea mexicana

An introduced perennial with vertical rhizomes up to 30 cm long and 4 cm thick with long anodal stolons produced at the apex of the rhizome. Mature leaves broadly elliptical up to 25 cm long with a radial slit. Upper surface may have brown blotches up to 1 cm in diameter. Margins slightly sinuate. When uncrowded the leaves are floating, but usually the leaves are crowded and semi-erect.

The yellow-flowered *Nymphaea mexicana* near Sydney

Flowers up to 12 cm diameter, mostly produced among the semi-erect leaves. Sepals 4, green. Petals yellow, acute, grading slowly into petaloid stamens. No obvious gap between petals and stamens. Ovary lobes large with stigmatic surface on lower ⅔. Seeds large, apparently dark-brown, but rarely formed in New South Wales. The specific epithet "mexicana" refers to the Mexican specimens which were first described.

Growth Biology

New growth and spread is mainly initiated vegetatively. Stolons are produced during spring to autumn, specially following flooding. Each stolon terminates in a new plant which in turn may produce further stolons the same season. Juvenile leaves of these small plants tend to be more elliptical than the mature leaves. Flowers are produced during late spring to early autumn. The flowers open during the day and close at night for about 4 or 5 days. Fertile seeds seem to be very uncommon, although the young fruit is still pulled under water by the tightening spiral in the pedicel as with the other two species of *Nymphaea*. The plant may overwinter with several leaves. Spatial spread is from the stolons or from the small plants produced at the end of the stolon breaking off and washing downstream.

Habitat

Deep lagoons or dams or slow-flowing creeks on coastal areas. As this species spreads by stolons from the parent plant, it can gradually spread into much deeper water than the other two *Nymphaea* species. It has been found in water 2 metres deep and presumably can grow in even deeper water. As little viable seed is produced, this species is usually confined to a particular watercourse after it has been introduced there, either intentionally by man or via seeds.

Economic Significance

Probably slight at present because of the limitations of spread between water systems. Its ability to grow in deep water, to form very dense growth, and its hardiness, does suggest that some care may be necessary to prevent its establishment in coastal dams or lagoons used as water sources or as recreational areas, specially if these water bodies have a high nutrient input.

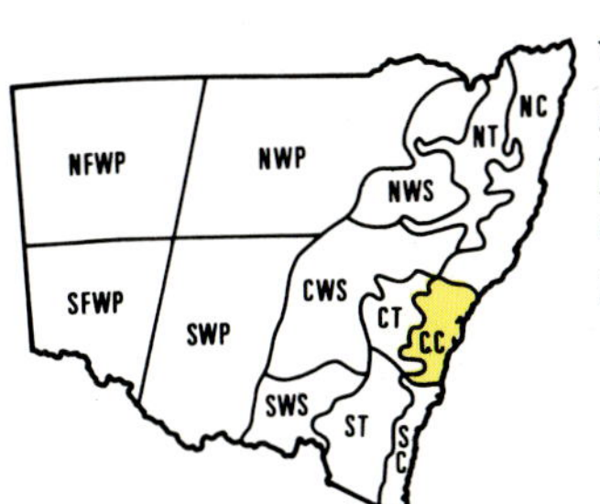

Distribution

CC, Queensland.

Among the most beautiful cultivated waterplants in New South Wales are *Nymphaea alba* hybrids. These are three of the colour variants

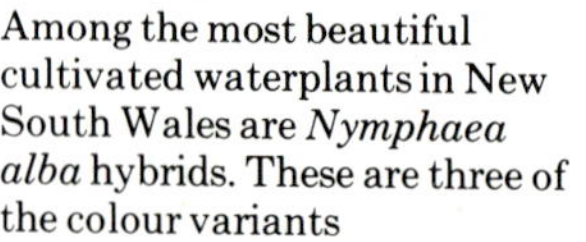

Nymphaea spp. Cultivated

Many hybrids and cultivars of _Nymphaea_ spp. are grown, and some may be planted in stock dams, irrigation channels or even natural lagoons. The most common hybrids are with _N. tuberosa_ and/or _N. alba_ as one of the parents. Most of these plants have white, red or pink flowers and usually have strong, vigorous rhizomes. The hybrids rarely produce seed and the few seeds that may be produced are rarely viable. Mostly these hybrids are of little concern, but under some conditions they may be able to produce sufficient vegetative growth to reduce water flow.

Right: A hybrid of _Nymphaea tuberosa_ planted in a lake near Griffith

Ludwigia

Erect or stoloniferous perennials. Leaves opposite or alternate, mostly entire. Flowers bisexual, regular, solitary, or arranged in an inflorescence. Sepals 4 or 5, fused below, petals yellow, same number as sepals. Stamens as many or twice as many as petals. Ovary inferior with the same number of locules as sepals.

1a. Plants herbaceous with decumbent or floating stems; branches glabrous.
L. peploides
1b. Plants shrubby, up to 3 metres tall, hemispherical; young branches hairy.
L. peruviana

Water Primrose

Clovestrip

Ludwigia peploides ssp. *montevidensis*

A prostrate native perennial with creeping or floating stems up to about 4 metres long, roots at the nodes either attached to soil or hanging in water. Leaves alternate up to 6 cm long and 3 cm wide, petiolate with small stipules at the base of the petioles. Flowers bisexual, axillary, solitary on pedicels shorter than the leaves. Calyx lobes 5, 7-9 mm long; petals 5, yellow, about 12 mm long; stamens 10; ovary 5-celled, producing numerous seeds about 1 mm diameter.

Growth Biology

Prolific flowering through summer.

Habitat

Edges of lakes and streams, supply and drainage channels. Common in permanent water.

Economic Significance

At times obstructive in supply and drainage channels. The long runners extend from the shore into deep water and often grow with other floating plants. It is suspected of causing gastro-enteritis in cattle.

Distribution

NC, CC, SC, NWS, NWP, SWP, NFWP, Queensland, Victoria, South Australia.

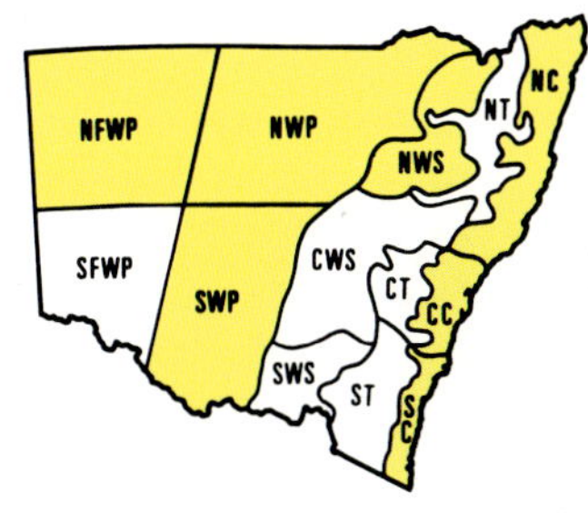

Left: Prostrate floating stems and leaves typical of *Ludwigia peploides* ssp. *montevidensis*

Right: *Ludwigia peploides* ssp. *montevidensis* in a channel near Griffith

Ludwigia peruviana

An introduced perennial, open, hemispherical shrub to 3 metres tall. Branches, particularly the young ones, hairy. Leaves alternate (rarely opposite), 5-10 cm long, 1-3 cm wide, hairy. Flowers bisexual, axillary, solitary, on pedicels about half the length of the leaves, 2-4 cm diameter. Sepals 4, 8-12 mm long, pale green. Petals 4 (sometimes 5), yellow, 1-3 cm long and wide. Stamens 4. Ovary inferior, 4-celled with numerous light-brown seeds 0.6-0.8 mm long.

Growth Biology

Leaves die back in winter (in Sydney), leaving bare branches. New leaves produced in spring and flowers from late summer to autumn. Seeds are apparently spread along the watercourse by water, and between watercourses by birds.

Habitat

Stationary or moving water to 1 metre deep in lagoons or swamps or in the margins of flowing creeks.

Economic Significance

Can fill slowly-flowing watercourses, substantially reducing water flow. At the moment it is restricted to the Sydney region, but has the potential to spread and cause problems in several areas of the State. Seeds eaten by waterbirds.

Distribution

CC, America.

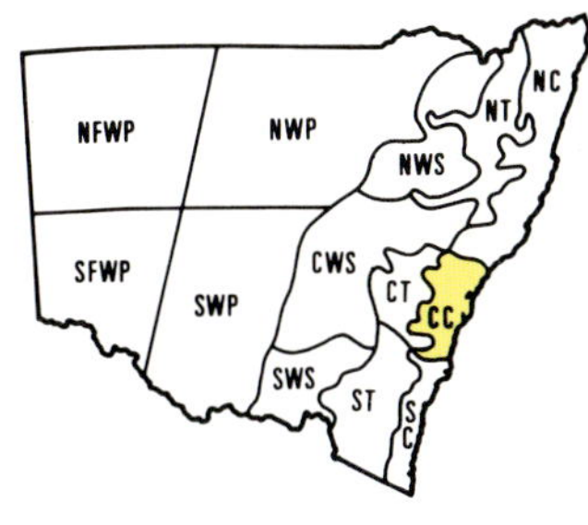

Left: Flowering and fruiting branches of *Ludwigia peruviana*

Right: *Ludwigia peruviana* in a creek near Botany

CM 1 2 3 4 5 6 7 8 9 10 11 12

Swamp Fern

Ceratopteris thalictroides

A native free-floating or attached and submerged perennial fern. Fronds erect, spongy, simple to tripinnatifid. Sterile fronds stalked, 4-30 cm long, 2.5-8 cm wide, submerged, floating or emergent. Fertile fronds only emergent, 20-60 cm long, ultimate segments narrower than on sterile fronds. Sporangia sessile, enclosed by the revolute margins of ultimate segments. Plantlets may be produced in the axils of fertile pinnae.

Growth Biology

Plants spread vegetatively by the plantlets; prothalli grow either submerged or on mud. Basically summer growing, it does not survive low winter temperatures.

Habitat

Stationary or slowly-flowing fresh water in the tropics.

Economic Significance

C. thalictroides is sold widely for use in aquaria. This species is common in South-East Asia, growing wild in swamps and rice fields. The uncurled new fronds are eaten raw or cooked; the whole plant except the roots may be cooked as greens.

Distribution

NC, Queensland, Northern Territory.

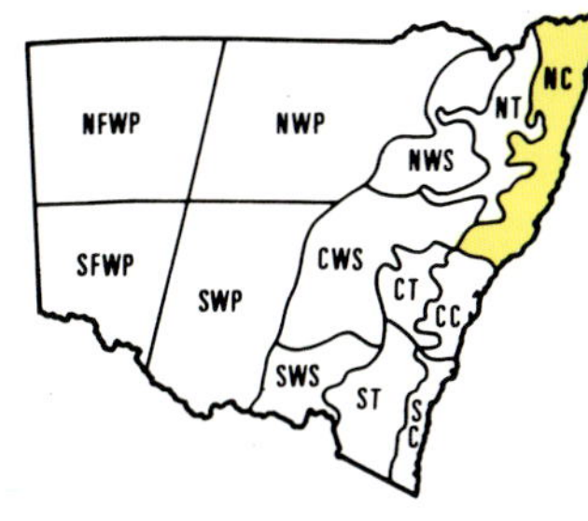

Left: Sterile fronds and young plants of *Ceratopteris thalictroides*

Right: *Ceratopteris thalictroides* in an aquarium

Philydrum lanuginosum
in a creek near Kempsey

Frogsmouth

Philydrum lanuginosum

Erect native perennial to 2 metres or more high. Leaves hairy, to 60 cm long and 2 cm wide, passing gradually into the smaller inflorescence bracts. Inflorescence a simple or branched spike to 1 metre long. Flowers solitary and sessile, subtended by a bract almost as long as the flower. Perianth segments yellow, 4 in 2 whorls of 2, the outer whorl about 12-15 mm long and 10 mm wide, the inner whorl approximately 8 mm long and 2 mm wide. Stamen 1, with a flattened filament. Ovary superior. *Seed photograph page 443.*

Habitat

Margins of ponds, swamps, farm dams, in stationary or slowly-flowing water to 20 cm deep. Normally grows in water, but often may flower and fruit on drying mud.

Distribution

NC, CC, SC, NT, CT, CWS, Queensland, Victoria, Northern Territory, Malesia.

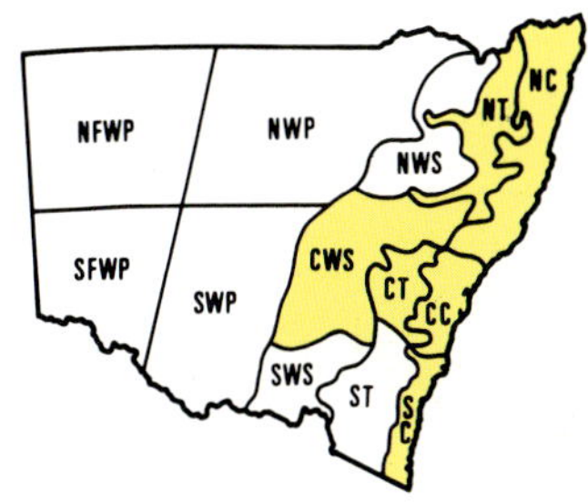

Flowers of *Philydrum lanuginosum*

Lignum

Muehlenbeckia cunninghamii

A native, intricately branched dioecious shrub to 3 metres high and 2 metres diameter. Stems usually leafless, striate, branches often ending in a spine. Leaves permanently drought deciduous, to 4 cm long and less than 1 cm wide when present. Inflorescence an interrupted raceme. Perianth segments 5, white or tinged with green. Male flowers with 8 stamens. Female flowers with a 1-locular ovary and 3-branched style. Fruiting perianth enlarging and enclosing a 3-angled nut. *Seed photograph page 443.*

Habitat

Ephemeral swamps and slowly-flowing ephemeral creeks and rivers in drier areas.

Economic Significance

May be a problem in some low-lying flood-irrigated pastures, but generally does not survive cultivation. An ideal habitat for both native and introduced birds and animals.

Distribution

NT, NWS, CWS, SWS, NWP, SWP, NFWP, SFWP, Queensland, Victoria, Northern Territory, South Australia, Western Australia.

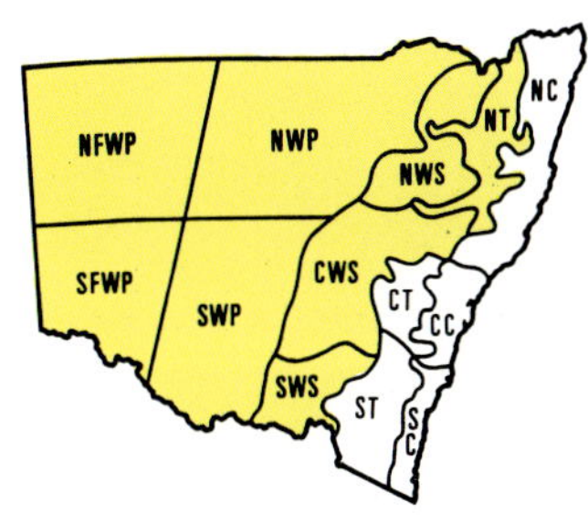

Left: *Muehlenbeckia cunninghamii* in a depression near Whitton

Right: The intricate branches and small white flowers of *Muehlenbeckia cunninghamii*

Polygonum

Erect or floating annual or perennial herbs. Nodes may be swollen or cylindrical. Leaves alternate, petiolate, ochreate. Inflorescence of axillary clusters of 1-several flowers or a variously reduced panicle. Flowers regular, bisexual, perianth segments usually 5, subequal, completely or partially enclosing the fruit. Stamens 5-8, styles (or style branches) 2-3.

1. Ochreae glabrous or mostly glabrous with a ring of stiff retrorse bristles at base. **2.**
1a. Ochreae with scattered hairs or bristles (may be deciduous from upper ochreae in *P. hydropiper*). **7.**

2. Ochreae with a ring of stiff retrorse bristles at base. **3.**
2a. Ochreae glabrous. **4.**

3. Internodes with stiff hairs along angles. *P. strigosum*
3a. Internodes mostly glabrous or with few scattered stiff bristles. *P. dichotomum*

4. Peduncles with glandular hairs. *P. praetermissum*
4a. Peduncles with no glandular hairs. **5.**

5. Upper leaf surface glabrous or, if hairs present, these restricted to along the larger veins. *P. lapathifolium*
5a. Upper leaf surface hairy. **6.**

6. Leaves sessile or almost so. *P. subsessile*
6a. Leaves with a petiole more than 1 cm long. *P. attenuatum*

7. Plant prostrate, mat-forming perennial, leaves narrow, 1 cm wide, less than 4 cm long. *P. prostratum*
7a. Plants erect or decumbent, annual or perennial, with at least some leaves 4 cm long or more. **8.**

8. Leaves hairy. *P. orientale*
8a. Leaves glabrous. **9.**

9. Perianth dotted with oil glands; green, white or pink. *P. hydropiper*
9a. Perianth without oil glands; pink. *P. decipiens*

Left: *Polygonum decipiens* in a channel near Tharbogang

Polygonum decipiens
with the dark leaf patch
common in this species

346

Slender Knotweed

A native procumbent or erect perennial (sometimes behaving as an annual), usually less than 1 metre tall. Leaves subtended by ochreae which are loosely hairy on the back with hairs about 3 mm long. Leaves petiolate with a glabrous blade to 15 cm long and 2 cm wide, often with a dark blotch in the centre. Inflorescence of one or more lightly crowded, flexuous spikes to 10 cm long and 0.5 cm wide. Perianth pink, without oil glands. Nut trigonous. *Seed photograph page 443.*

Growth Biology

Summer growing and flowering, dying back in winter.

Habitat

Creek and river banks, margins of lagoons, swamps and channels.

Economic Significance

In summer vigorous growth forms dense mats along the beds or margins of drainage channels, blocking flow.

Distribution

NC, CC, SC, NT, CT, ST, NWS, SWP, all States.

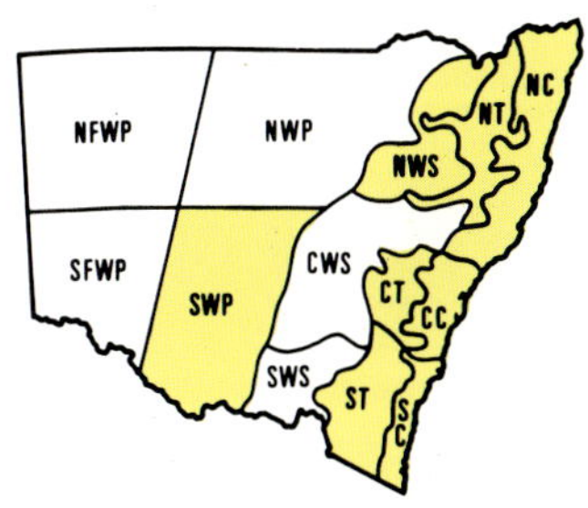

A plant of *Polygonum
lapathifolium* with
delicate drooping
inflorescences

Pale Knotweed

Willow Smartweed   *Polygonum lapathifolium*

A (?) native erect annual herb to about 1 metre high. A variable species which can, at times, be difficult to distinguish from some other species. Stem robust, brown, to 2.4 cm diameter. Ochrea truncate, not hairy, the top often becoming torn and laciniate with age, frequently more than 1 cm long, occasionally less. Petiole 0.5-2 cm long, usually hairy. Leaf blade to 25 cm long, 5 cm wide, but frequently much smaller, upper surface ± hairy. Hairs, if present, only along larger veins. Lower surface usually covered with glandular hairs. Inflorescence a terminal arrangement of one or more flexuous or stiff, lightly to densely crowded spikes to 10 cm long and 0.5 cm wide, often dotted with glands. Perianth pink. *Seed photograph page 444.*

Growth Biology

Similar to *P. decipiens*.

Habitat

Swamps or lagoon and billabong margins, creek and river banks. Frequently found with *P. decipiens*.

Economic Significance

Usually not a pest.

Distribution

All regions of New South Wales, Queensland, Victoria, Tasmania, Northern Territory, South Australia.

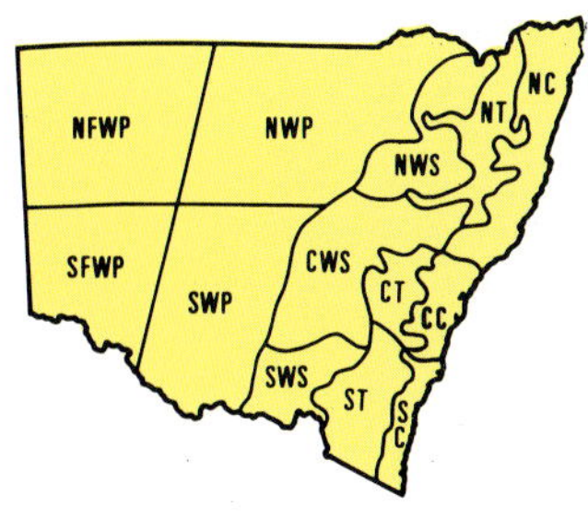

Polygonum lapathifolium in a flood channel near Kempsey

Top: Inflorescences of *Polygonum strigosum* (left) and *Polygonum decipiens* (right)

Left: *Polygonum strigosum* near Botany

Prince's Feather

Polygonum orientale

An introduced annual up to 2 metres high. Stems robust, to 2 cm diameter. Ochrea hairy, with terminal hairs, frequently dilating with age. Petioles usually less than 4 cm long, hairy. Blades to 20 cm long and 8 cm wide, both surfaces usually hairy. Inflorescence a terminal arrangement of one or more usually dense spikes to 10 cm long and 2 cm wide. Perianth pink to crimson.

Habitat

In and near swamps, creeks, etc., frequently near disturbed areas.

Economic Significance

Nil.

Distribution

NC, CC, SC, NWS, CWS, SWS, NWP, SWP, Queensland, Northern Territory.

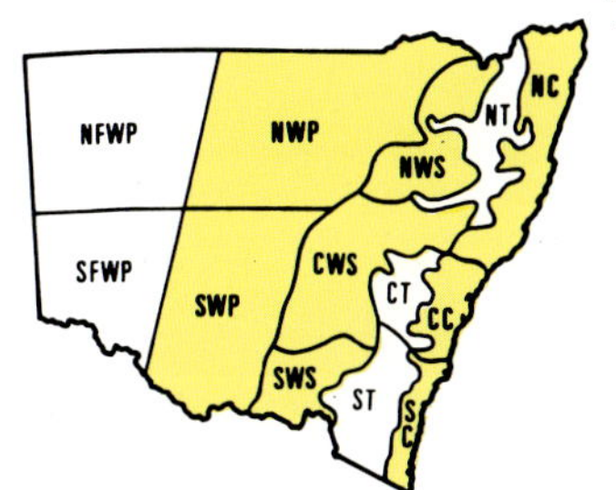

Polygonum strigosum

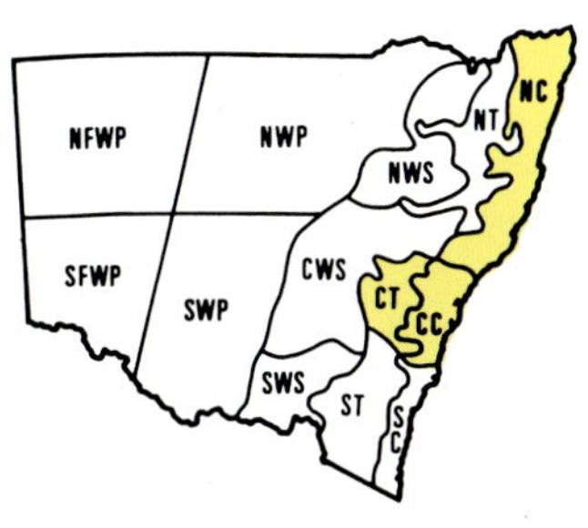

A native decumbent or straggling perennial with stem to 2 metres long. Stem angles lined with stiff retrorse bristles. Ochrea 1-2 cm long, mainly glabrous but often with a line of stiff retrorse bristles near base. Petioles also with retrorse bristles. Leaf blade to 8 cm long and 2.5 cm wide, midveins with retrorse bristles, lamina ± hairy. Inflorescence a loose arrangement of one or more short, not very dense spikes of 1-10 flowers, to about 3 cm long. Inflorescence branches covered with glandular hairs and with an understorey of shorter, stiffer hairs. Perianth white, occasionally with a pinkish tinge or pink.

Habitat

Swamps, margins of slowly-flowing water in creeks or rivers.

Economic Significance

Nil.

Distribution

NC, CC, CT, Queensland.

Rumex

Perennial or annual tufted or decumbent herbs. Leaves alternate with ochreous stipules. Flowers green when young, often turning red, in whorls or clusters, in panicles or leafy racemes, bisexual or unisexual. Perianth segments 6 in 2 whorls of 3. The outer whorl smaller than the inner whorl, which enlarges and closes over the fruit. Stamens 6. Styles 3. Nut triquetrous.

Several species can grow in damp places, but only *R. bidens* is truly aquatic. *R. crispus* can grow in waterlogged soil and survive inundation.

1. Stems with inflated internodes, prostrate or floating, rooting at the nodes. Inflorescence a reduced panicle, leafy at the base.

R. bidens

1a. Plants erect with a large panicle. Stems not inflated nor rooting at the nodes.

R. crispus

Rumex bidens

Mud Dock
Swamp Dock

A native prostrate or floating perennial herb with prostrate, hollow stems rooting at the nodes, terminally erect when flowering. Leaves alternate, to 30 cm long and 4 cm wide with undulate margins. Petiole to 7 cm long and with a basal ochrea. Inflorescence elongated, open, leafy at the base. Flowers unisexual with upper clusters mostly male, lower clusters mostly female. Both flowers 2.5-4 mm long. Male flower with 6 stamens. Nut 3-3.5 mm long, light brown, falling enclosed in perianth.

Habitat

Freshwater swamps and creek banks in water up to about 20 cm deep or on mud, floating out over deeper water.

Economic Significance

May have some potential as a weed of drains in cooler areas of the State.

Distribution

SWP, Victoria, Tasmania, South Australia.

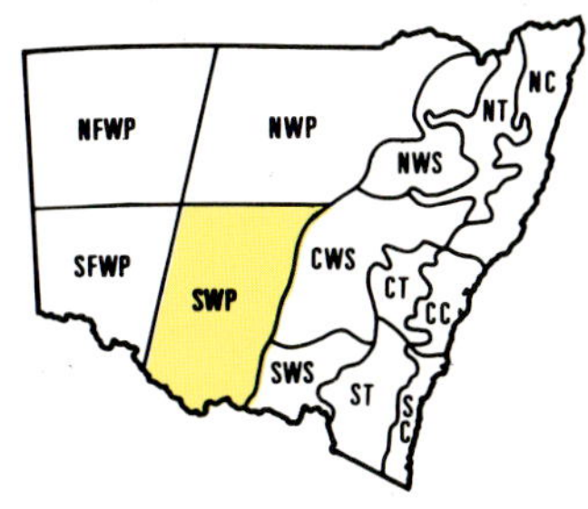

Left: Prostrate stem, leaves and inflorescence of *Rumex bidens*

Right: *Rumex bidens* on the margin of Lake Leake, South Australia

Top: *Rumex crispus* on the margin of a lake near Griffith

Left: *Rumex crispus* showing the large brown inflorescence

Curled Dock

Rumex crispus

An erect, introduced perennial herb to 1.5 metres tall. Tap root thick, fleshy, to 40 cm long. Leaves mostly basal, to 30 cm long and 4-10 cm wide, with undulate margins. Petiole to 5 cm long with an ochrea. Inflorescence a dense panicle, 20-50 cm long, with a few leaves, mainly towards the base. Flowers bisexual, outer perianth segments small, inner segments 3.6 mm long, cordate, one or more tuberculate. Nut trigonous, about 2 mm long, smooth, brown. *Seed photograph page 444.*

Growth Biology

The seed produced is polymorphic, i.e., it varies in size and weight, and on one plant seeds are produced covering a range of optimal germination conditions for light, temperature and daylength. One plant may produce up to 60,000 seeds in a year, and under experimental conditions the seed can survive up to 80 years in soil. The plant starts growing in winter and spring and flowers in response to increasing daylength. Under ideal conditions the plant may flower 9 weeks after germination. The top dies off in summer.

Habitat

Perennial plants grow in drains or intermittently flooded channels or swampy ground. May behave as an annual weed of cultivation.

Economic Significance

Native to Europe, it has spread through much of the world except the tropics, and is a serious weed of crops. In south-west New South Wales it thrives on the margins of drainage channels that remain dry during winter months. *R. crispus* is the most significant weed of the genus found in Australia. The plant is not palatable to stock and is toxic to poultry.

Distribution

NC, CC, SC, NT, CT, ST, CWS, SWS, NWP, SWP, NFWP, SFWP, Victoria, Tasmania, Western Australia, Europe, Asia.

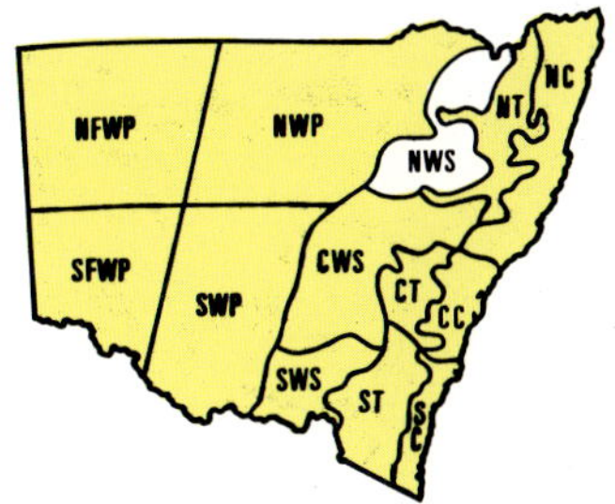

CM

Water Hyacinth

Eichhornia crassipes

An introduced erect free-floating stoloniferous perennial, up to 1 metre tall. Roots black to purple, feathery and up to 1 metre long. Leaves all basal, with spongy inflated petioles up to 75 cm long, blades up to 25 cm long, apiculate, to 20 cm wide. Inflorescence axillary, slightly exceeding the leaves, to 20 cm long. Flowers bisexual, irregular, 3-6 cm in diameter. Perianth segments 6, blue, 3-4 cm long, fused below, upper lobe with a darker blue patch with a yellow centre. Stamens 6, 3 long, almost exserted, 3 shorter. Ovary superior, 3-celled with a curved style. Fruiting inflorescence recurving, becoming submerged. *Seed photograph page 444.*

Growth Biology

This species spreads vegetatively by stolons rooting at nodes to produce new plants, eventually developing into mats covering large areas. In New South Wales flowering occurs in the summer months, each flower opening for 1 to 2 days and then withering. After all the flowers in an inflorescence have finished flowering, the flower stalk then bends into the water, with seeds being shed in about 18 days. Many seeds are produced and they mature in a capsule at the base of the flower. In New South Wales seeds are not always common in flowering stands. The seeds can germinate in 3 days or remain dormant for at least 15 years. Winter frosts will cause leaves to die off, leaving crowns of the plants to overwinter and commence new growth the following spring. Vegetative propagation is rapid. In one growing season a few plants can multiply through daughter plants to cover a small dam 30 metres in diameter. One hectare of *E. crassipes* may yield as many as 2,000,000 plants and weigh as much as 400 tonnes. Rapid growth only occurs in nutrient-rich water. It will barely survive when free-floating in fresh, clean water. Explosive growths are a symptom of eutrophication. Plants lying in mud grow at a reduced rate. Long roots may indicate low nutrient levels, short roots may indicate nutrient-rich water.

Habitat

Grows world-wide in fresh static or slowly-flowing water in tropical and temperate climatic zones to latitudes of 40°. Man has been responsible for its spread, as the flowers are prized for cultivation in garden pools.

Economic Significance

Floating plants obstruct water movement, restrict boating and interfere with fish and waterfowl. Dense mats will reduce oxygen levels in water and harbour mosquitoes that are vectors of serious diseases. Masses may build up at bridges, creating a dam and endangering the structure. Increased loss of water through transpiration has been reported.

Left: *Eichhornia crassipes*

Immature fruit (A), seeds
(B) and seedlings of
Eichhornia crassipes

E. crassipes has been used experimentally in waste water treatments. The plant is harvested from the waste water to remove some of the nutrients. It has also been experimented with by the National Aeronautical and Space Administration, U.S.A., to manufacture methane gas. However, none of the processes are economically viable, the major problems being the cost of harvesting, transport and drying the material which is usually more than 95 per cent water.

It is also useful in making compost and is used on a small scale for pig food in Indonesia. Despite its usefulness in some situations, this is a declared noxious plant in New South Wales, and its spread should be prevented.

Common in coastal areas of New South Wales, its inland spread has been limited, but suitability to this environment is exemplified by an infestation formerly covering thousands of hectares in the Gwydir Valley west of Moree, New South Wales. Once established in timbered, inaccessible swamp areas, the only economical means of control at present is by draining the swamp, although biological control agents such as the recently introduced weevil, *Neochetina eichhorniae* can limit growth *(see section on Biological Control, page 479)*.

The plant is rapidly killed by sea-strength salinity. Salinity levels of around 4000 ppm total dissolved salts will limit growth after a few months' exposure and eventually the plant will die. A number of herbicides will kill *E. crassipes,* the most effective being 2, 4-D and diquat, which must be used strictly according to label instructions. Regeneration from seed will negate the effect of a herbicide program.

Distribution

NC, CC, NWP, Queensland, Victoria, South Australia, Western Australia.

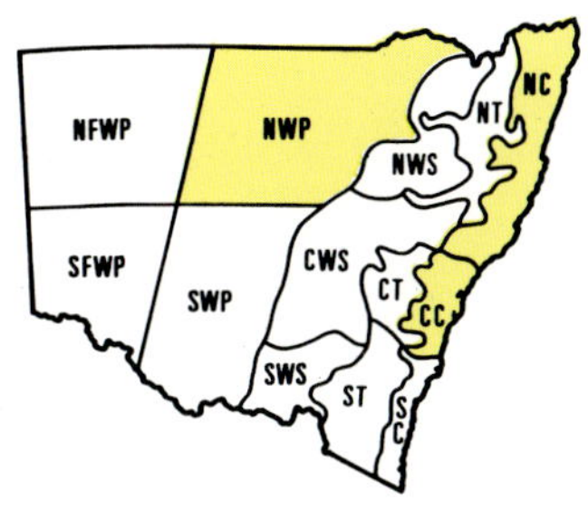

Eichhornia crassipes in a creek near Gosford

Monochoria cyanea

A native decumbent annual or perennial with floating and/or emergent leaves and emergent inflorescences up to 80 cm tall. Stems decumbent, rooting at the lower nodes. Blades 4-15 cm long and 1-6 cm wide, ovate. Inflorescence 5-10 cm long, 3-several flowered on axillary peduncles surrounded at the base by the sheath of a subtending leaf. Flowers bisexual with 6 ± equal, free, blue perianth segments each 1-2 cm long. Stamens 6, more or less equal. Seeds numerous.

Habitat

Stationary or slowly-flowing shallow water to about 40 cm deep in swamps, lagoons or dams. Plants root in mud but can grow out from the margins floating over the water. They can survive for a while on drying mud. When in permanent water this species usually behaves as a perennial, but in ephemeral waterholes or in New South Wales, where it occasionally occurs, it behaves as an annual.

Economic Significance

Nil in New South Wales. Another species, *Monochoria vaginalis,* is a weed of rice in parts of Asia.

Distribution

NWP, Queensland, Northern Territory, Western Australia.

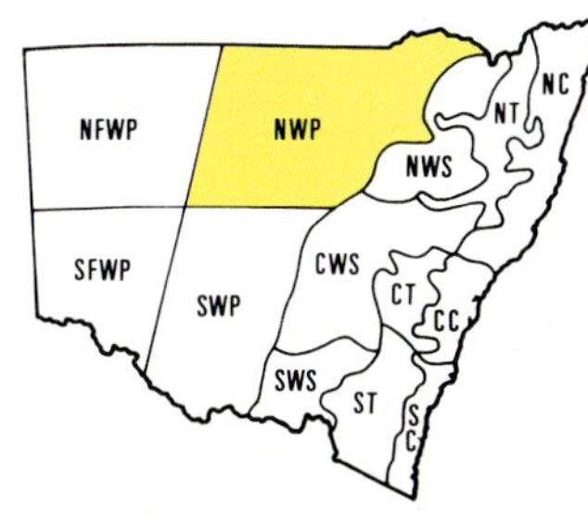

Left: Flower of *Monochoria cyanea*

Right: *Monochoria cyanea* (blue flowers) in a lagoon near Rockhampton, Queensland

Pickerel Weed   *Pontederia cordata*

Introduced erect, emergent rhizomatous perennial to 1 metre tall. Leaves basal with long petioles and mostly emergent blades, although first-produced leaves may be submerged or floating. Mature blades to 20 cm long and 10 cm broad, cordate or lanceolate. Inflorescence subtended by a shortly petiolate stem leaf which clasps the stem when cordate. The spike-like panicle is dense, 5-15 cm long, blue. Flowers bisexual. Perianth blue, glandular, hairy with yellow markings in the throat, 3 segments fused for half their length and 3 segments fused near their base, 1-2 cm long. Stamens 6, 3 with long, 3 with short filaments. This species contains two varieties, *P. cordata* var. *lanceolata,* with lanceolate leaves and ± naturalised in southern New South Wales, and *P. cordata* var. *cordata,* which is cultivated as an ornamental and has cordate and/or wider leaves. In New South Wales material, the perianth is commonly glandular-hairy in both varieties.

Habitat

Stationary or slowly-flowing water less than 1 metre deep.

Economic Significance

Of little significance at present. The area of the State in which it is currently naturalised is not suited to its rapid spread. If introduced to the North Coast it could grow and retard water flow in drains and channels.

Distribution

SWS, Queensland, America.

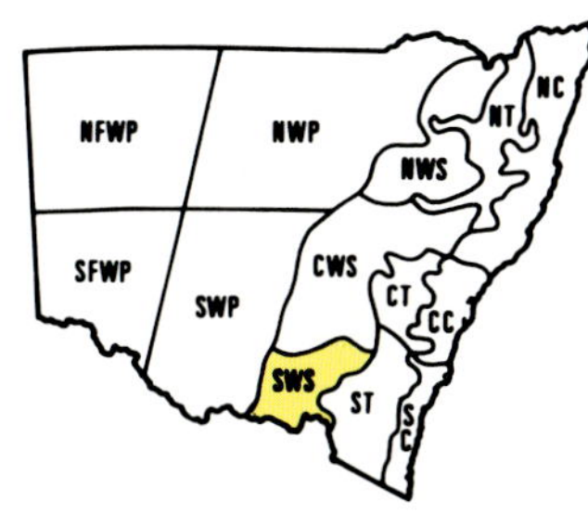

Left: Inflorescences of
Pontederia cordata

Right: *Pontederia cordata* in
an ornamental pond in Canberra

The white flowers and flattened fleshy leaves characteristic of *Montia australasica*

White Purslane

Montia australasica
[Synonyms *Claytonia australasica*
Neopaxia australasica]

A native decumbent perennial rooting at the nodes. The stolons up to 1 metre long but usually much less. Leaves alternate, fleshy, with a winged sheath-like base; blades linear to spathulate, flattened, 1-10 cm long, usually less than 1 cm wide. Inflorescence 1-few flowered, open with the flowers on pedicels 1-4 cm long. Flowers 1-2 cm diameter; calyx 2-lobed, green, 2-3 mm long; corolla white with 5 petals joined at the base, each 6-8 mm long. Stamens 5. Ovary superior, style 1, stigmas 3.

Growth Biology

Prolific spring flowering. Does not die-back in winter.

Habitat

Emergent aquatic in stationary or slowly-flowing water or on adjacent wetlands. More common at higher altitudes.

Economic Significance

May have some value as an ornamental.

Distribution

NT, CT, ST, CWS, SWS, Victoria, Tasmania.

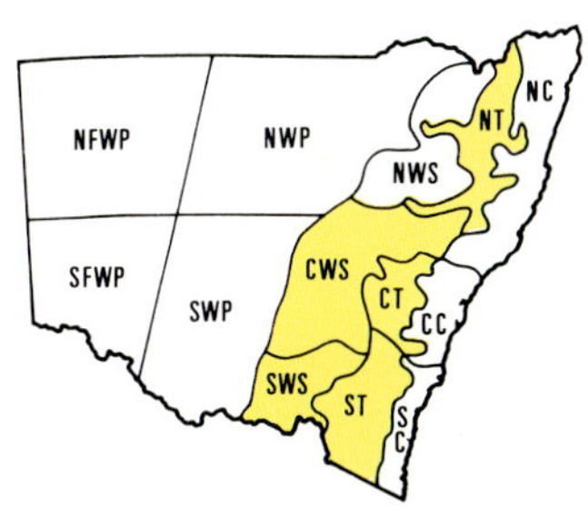

Montia australasica in an ornamental pond in Canberra

A Seagrass

Posidonia australis

Fully submerged marine or estuarine rhizomatous native species with grasslike leaves and submerged flowers. Rhizomes short, thick, to 10 cm long and 1-2 cm thick, enclosed in fibres from decaying leaf sheaths. Leaves linear, flat, normally about 50 cm long (occasionally longer) with a basal sheath and small ligule at the junction of sheath and blade. Leaf apex rounded-obtuse but normally eaten or worn truncate with marginal fibres slightly protruding. Inflorescence a terminal cluster of spikes on a leafless stem, usually the same length as the leaves. Lower spikes subtended by leaflike bract and 2 or more bladeless bracts. Each subsequent spike subtended by outer bracts with progressively shorter blades as well as the 2 or more bladeless bracts. Each spike approximately 5 cm long with, usually, 3, 4 or 5 bisexual flowers. Each flower consists of 3 stamens and 1 ovary only.

Growth Biology

The seeds usually germinate in late summer or autumn on either a sandy or silty bottom. The plants are very slow-growing and it is not known how old plants are before they flower. The rhizomes can grow vertically, and the plants are capable of gradually accumulating deposits, slowly raising the bottom. Mature plants flower about October-November, but in any one population (at least in New South Wales) the few flowers are produced over only a short period of time.

Flowers are entirely submerged. Fruits are formed in November-December and the seeds released. The seeds float for a considerable time until the outer layers disintegrate.

Posidonia australis in Jervis Bay

Habitat

P. australis grows in estuaries or sheltered bays on sandy bottoms in water (1-) 2-10 metres deep. The less sheltered the habitat, the deeper the minimum depth.

Economic Significance

Of great value as fish breeding and nursery areas. Stabilises bottom sediments. No significant disadvantage known.

Distribution

NC, CC, SC, Victoria, Tasmania, South Australia, Western Australia.

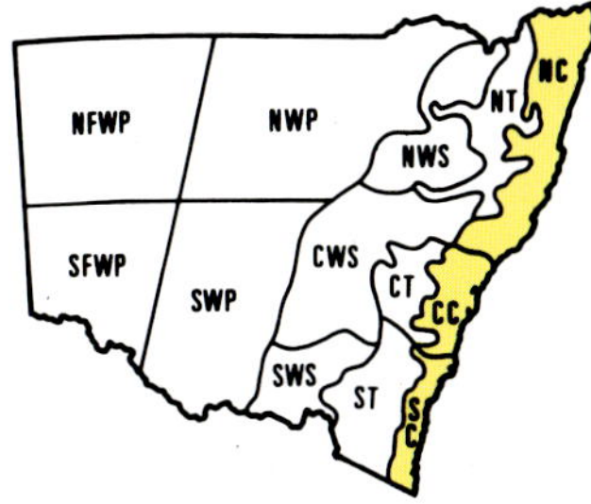

Posidonia australis showing the thick rhizomes and strap-like leaves

Pondweeds

Potamogeton

Submerged or attached floating annuals or perennials with long straggling stems and emergent inflorescences. Often producing either an extensive rhizome system and/or specialised overwintering buds (turions). Leaves alternate, submerged and/or floating with stipular ± sheathing bases. Inflorescences mostly emergent (± emergent in *P. pectinatus*), spicate. Flowers bisexual, 3-4 mm diameter, regular. Perianth segments 4; stamens 4; ovary superior with 4 free 1-locular carpels.

Species of this genus occur in a wide range of habitats and some are tolerant of strongly brackish water. *Potamogeton* spp. are an important food for waterfowl. About 100 species worldwide and many cosmopolitan. The generic name *Potamogeton* is derived from "potamos" a river; "geiton", neighbour.

1. Plants with upper leaves floating and dissimilar to the submerged leaves.   **2**

1a. Plants with all leaves submerged and similar.   **3**

2. Floating leaves usually greater than 1 cm wide (2-5 cm), several; common.   ***P. tricarinatus***
2a. Floating leaves usually less than 1 cm wide, few; not common.
                                                                             P. javanicus
3. Leaves less than 3 mm wide with an obvious ligule and sheath loosely or not enclosing stem.   ***P. pectinatus***

3a. Leaves more than 3 mm wide, no obvious sheath but with a stipule $\pm$ resembling a ligule.   **4.**

4. Leaves linear, 3-6 mm wide, margins entire, flat, leaf tip obtuse, stipule entire, membranous, becoming laciniate.   ***P. ochreatus***

4a. Leaves not linear, usually more than 6 mm wide and/or with undulate margins.   **5.**

5. Leaf bases stem clasping, leaf ovate, cordate, margins entire.
                                                                             P. perfoliatus

5a. Leaves not obviously stem clasping, linear oblong not cordate, margins entire.   ***P. crispus***

Curly Pondweed   ***Potamogeton crispus***

A native submerged rhizomatous annual or perennial with trailing flattened stems up to 3 metres long. Leaves alternate, opposite when subtending an inflorescence, up to 10 cm long and 0.5-1.5 cm wide, translucent with undulate and minutely serrulate margins. Stipular sheath enclosing stem above the node. Modified leaves or turions, when present, formed terminally on the branches. Inflorescence emergent with few (up to 10) flowers. Flowers bisexual; perianth segments 4; stamens 4; carpels 4. Fruiting carpel 5-7 mm long with a conical $\pm$ curved beak and $\pm$ tuberculate body.

Growth Biology

Flowers spring to early summer. Turions or winter buds (see photograph) are formed during the summer and autumn in the leaf axils and branch tips. When the parent plant is mature, usually in autumn, these hard, burr-like buds fall to the mud where they overwinter to probably commence growth in the new season. Great numbers of winter buds are produced and these can be readily felt in the silt surrounding the plant. Growth period varies, but usually rapid growth commences in the spring months when water temperature reaches about 15°C.

Left: *Potamogeton crispus* in a drainage channel in Tharbogang

Habitat

Common in the slowly-flowing water in drainage channels and thrives in those channels in the Murrumbidgee Irrigation Areas, New South Wales, fed with saline tile drainage discharge and where constant flow levels are maintained. It is also common in the irrigation channels that carry permanent flows in the Lachlan and Macquarie valleys.

Economic Significance

Problems with restricted flows are frequently reported in October-November in western New South Wales coinciding with water temperatures in the vicinity of 20°C. Growth rate is rapid, severe water flow obstruction occurring in less than 2 weeks. Spring blooms of filamentous algae, principally *Cladophora* spp., entwine *P. crispus,* adding to the obstruction. In these situations it has become sufficiently obstructive to require treatment in channels of up to 400 megalitres per day capacity. By December, with flowering completed, the leaves have died back or been eaten by aquatic insects to the extent that this plant is usually not a significant problem.

Distribution

Cosmopolitan and probably the most frequently mentioned *Potamogeton.* NC, CC, CT, ST, NWS, CWS, NWP, Queensland, Victoria, Northern Territory, South Australia, Western Australia.

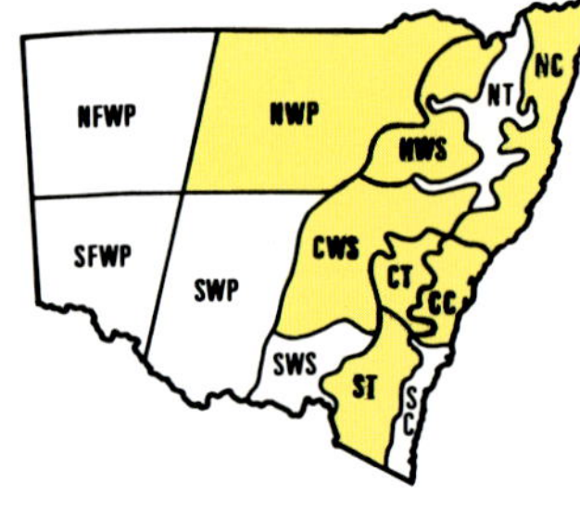

Potamogeton crispus showing a turion on the left and fruiting inflorescences in the centre

POTAMOGETONACEAE

Potamogeton javanicus

A native perennial with numerous submerged and a few acute floating leaves 2-4 cm long and less than 1 cm wide with 5-7 veins. Submerged leaves translucent, 5-10 cm long and 2-3 mm wide. Inflorescence emergent, dense, up to 2 cm long. Flowers bisexual; perianth segments 4; stamens 4; carpels 4. Fruiting carpels with a prominent beak, ± keeled. The long slender submerged leaves enable a simple and positive differentiation from small-leafed forms of *P. tricarinatus*.

Growth Biology

Main growth and flowering during summer.

Habitat

Shallow water of lakes and creeks.

Economic Significance

Reported obstructive to water flow in the Ord River irrigation scheme of Western Australia.

Distribution

NC, CC, Queensland, Western Australia.

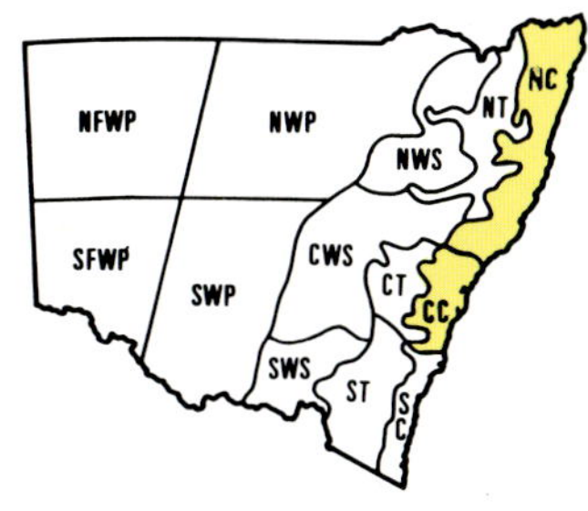

Potamogeton javanicus
in the Coldstream River

Blunt Pondweed

Potamogeton ochreatus

A native submerged rhizomatous annual or perennial with terete stems to about 4 metres long. Leaves to about 10 cm long, 6 mm wide, with entire, flat margins. Stipular sheath initially up to 2 cm long, membranous and ± sheathing the stem above the node, eventually becoming torn and laciniate and free from the stem. Inflorescence emergent, dense, to 3 cm long. Flowers bisexual; perianth 4; stamens 4; carpels 4. Fruiting carpel 3-4 mm long, compressed, rounded, beaked. *Seed photograph page 444.*

May be confused with *P. crispus.* However, the leaf edge of *P. ochreatus* is flat, not wavy.

Potamogeton ochreatus showing the typical obtuse leaves, stipular sheath and dense inflorescence

Growth Biology

Commences to grow vigorously in September in inland New South Wales, and reaches maturity in early summer, when it rapidly disintegrates or is destroyed by aquatic insects. A second burst of growth and flowering can occur in the early autumn. The flowering spike is projected above water for pollination. After pollination the spike curves into the water where the fruits mature. *P. ochreatus* occurs erratically and is not a long-lasting perennial plant.

Habitat

Usually found growing in the deeper water of channels and lakes. It will grow in deep silt or gravel, preferring still or slowly-flowing water up to 5 metres deep. Tolerant of slightly saline water containing around 2000 ppm total dissolved salts.

Economic Significance

In the 1960's the plant was common in the Murrumbidgee and Coleambally Irrigation Areas, New South Wales, severely obstructing channels during periods of vigorous growth. In these areas in the 1970's no major channel blockages have been attributed to this species.

Distribution

NC, CC, SC, NT, CT, ST, CWS, SWS, SWP, all States.

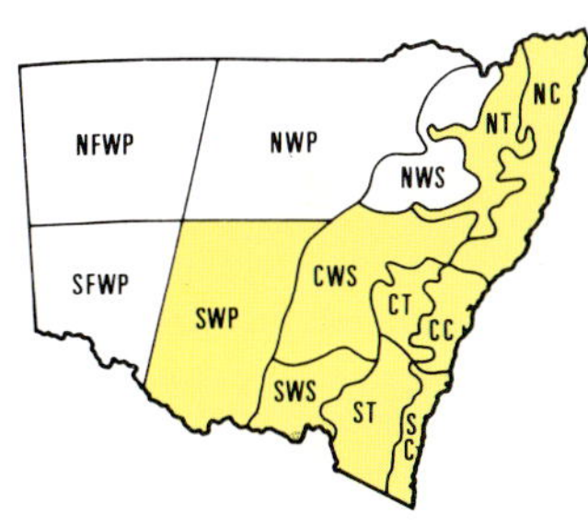

Potamogeton ochreatus in a
supply channel near
Coleambally

Sago Pondweed

Fennel Pondweed

Potamogeton pectinatus

A native submerged rhizomatous perennial with a vigorous rhizome system that penetrates mud to about 1 metre. Stems to about 3 metres long. Leaves to 15 cm long, 2 mm wide, acute, joined to a loose stipular sheath which ± surrounds the stem, upper portion of the sheath forming a stem-clasping ligule. Inflorescence erect or lax, emergent, ± floating or remaining submerged, to 5 cm long, interrupted. Flowers bisexual; perianth 4; stamens 4; carpels 4. Fruiting carpels 2-4 mm long, ± globular, compressed, ± smooth and with a very short beak. *Seed photograph page 444.*

Leaves are similar to *Ruppia* spp. but *P. pectinatus* can be distinguished by the ligule at the leaf-stem junction. Vegetative growth of *Scirpus fluitans* and some species of *Lepilaena* may also be confused with *P. pectinatus.*

A fruiting specimen of *Potamogeton pectinatus* with three tubers on the lower left

Growth Biology

The plant reproduces mainly by specialised fleshy tubers *(see photograph)* that grow in large numbers at the ends of the root system. A 24 square metre culture of *P. pectinatus* produced 36 000 tubers in one growing season. These tubers remain in the soil and develop under suitable conditions. Flowers from spring to autumn; fruits are profuse, mainly in summer. One plant can develop into a large mat, with its new shoots rapidly choking out other plants. In shallow water the leaves are close together; in deep water they are more widely spaced.

Observations made in the Wyangan lakes near Griffith, New South Wales, indicate that *P. pectinatus* will grow in water temperatures varying from 10°C to 25°C and higher. In this situation the plant dies back to the rootstock in winter, making rapid vegetative growth in spring and flowering profusely in summer.

Habitat

Grows generally in static water from 1 to 4 metres deep. Will thrive in saline water containing at least 5000 ppm total dissolved salts.

Economic Significance

This species is at present relatively uncommon in irrigation channels, but is often found in natural lakes, drainage basins and reservoirs, where the strong stem growth may be hazardous to swimmers and an annoyance to power boats and sailing craft. The fruits are readily eaten by waterfowl.

Distribution

NC, CC, ST, SWP, Queensland, Victoria, Tasmania, South Australia, Western Australia.

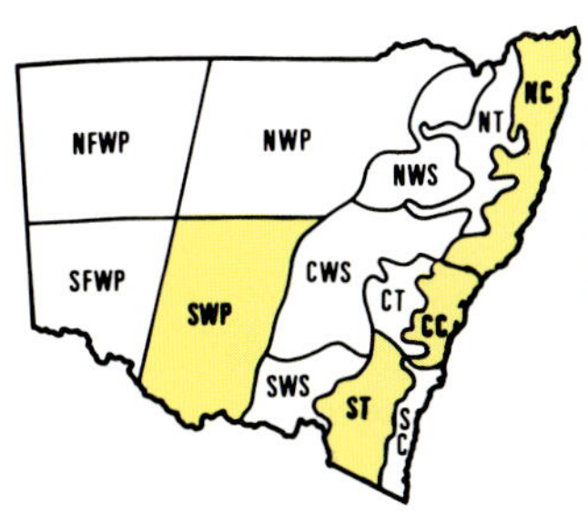

Potamogeton pectinatus in Barren Box Swamp

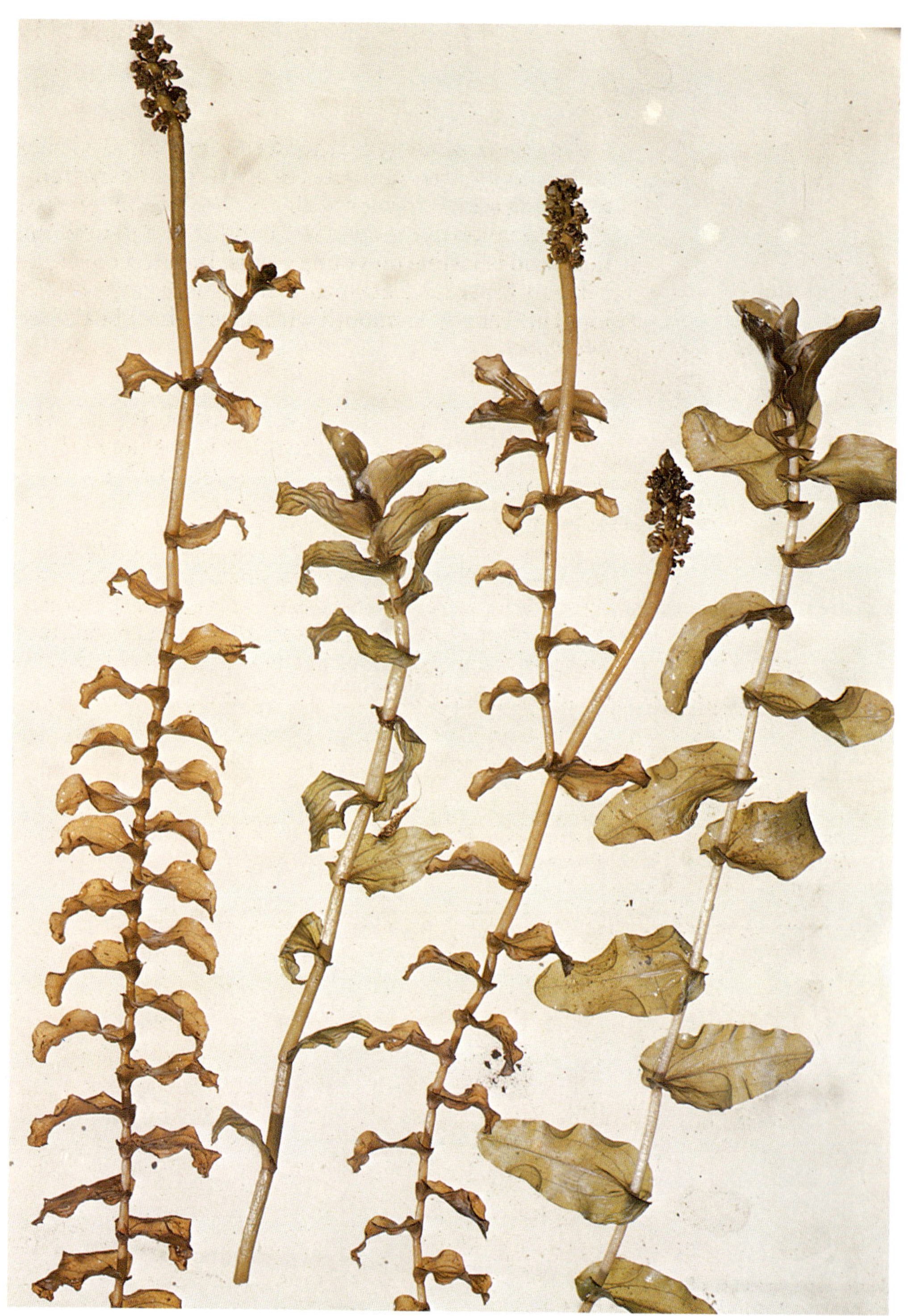

Fruiting and vegetative
stems of *Potamogeton
perfoliatus*

376

Clasped Pondweed

Potamogeton perfoliatus

A native submerged rhizomatous perennial with long, trailing, rounded stems to 3 metres long. Leaves ovate, obviously stem-clasping, cordate, to 6 cm long and 4 cm wide with an undulate but entire margin. A membranous deciduous sheath to 1 cm long surrounds the stem above the node and is visible on young stems. Inflorescence emergent, to 2 cm long. Flowers bisexual. Perianth 4; stamens 4; carpels 4. Fruiting carpels 2-3mm long, flattened, ± smooth with a very short beak. *Seed photograph page 444.*

Growth Biology

Grows vigorously and flowers throughout the summer months. In autumn slender buds are produced in the network of roots. Some of the buds overwinter to grow in subsequent seasons.

Habitat

Tolerant of brackish water, and found growing in Lake Wellington, south-east Victoria, where one measurement indicated 5600 ppm total dissolved salts. Frequently seen as a dark submerged waving mass in fast-moving water downstream of channel regulators or in concrete-lined laterals, where it thrives in a few inches of silt or gravel.

Economic Significance

Occasionally obstructive in constructed waterways. Useful shelter and food for fish and waterfowl.

Distribution

NC, CC, SC, NT, CT, ST, NWS, CWS, SWP, Queensland, Victoria, Tasmania.

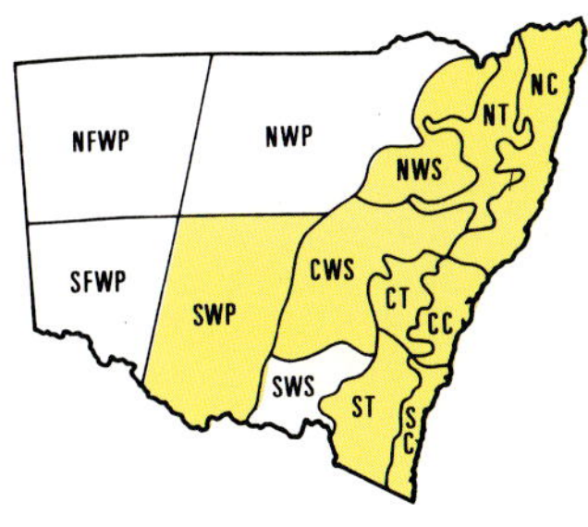

A cultivated specimen of
Potamogeton perfoliatus

Floating Pondweed

Potamogeton tricarinatus

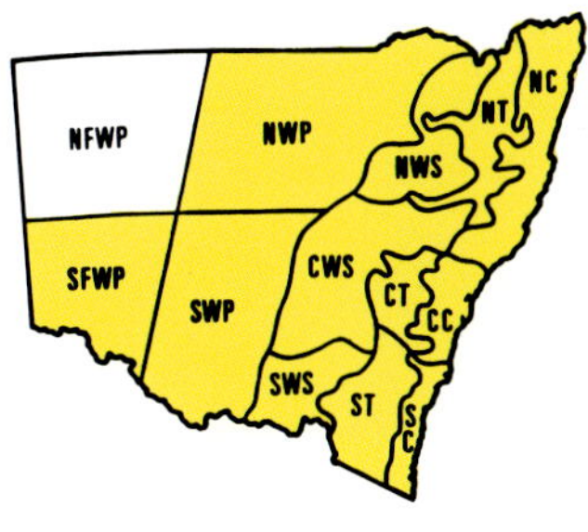

Left: *Potamogeton tricarinatus* in a supply channel near Griffith

Insert: Large-leaved (left) and small-leaved (right) "forms" of *Potamogeton tricarinatus*

A native rhizomatous perennial with both submerged and floating leaves and emergent inflorescences on stems to 4 metres long. Submerged leaves translucent, to 20 cm long and about 1 cm wide, often more or less undulating and with a stipular sheath to about 5 cm long embracing the stem above the axil. Stipular sheath becoming laciniate with age. Floating leaves relatively thick, to about 10 cm long and 7 cm wide with usually 13-17 parallel veins. Inflorescence emergent, to 5 cm long, crowded. Flowers bisexual. Perianth 4; stamens 4; carpels 4. Fruiting carpels 2-4 mm long, laterally compressed, smooth or tuberculate, with a short beak. *Seed photograph page 444.*

This species as presently accepted includes a large range of leaf size and shape and some variation in fruiting carpel shape. There is a wide range of leaf shapes classified as *P. tricarinatus* as shown in the adjacent photograph. It seems unlikely that all these forms are meaningfully included within a single species, but a substantial amount of work is required to investigate the taxonomic status of these forms.

In the early stages of growth the submerged leaves can be mistaken for other species of *Potamogeton,* particularly *P. crispus.*

Growth Biology

Commences rapid growth in the early spring at water temperatures around 15°C, reaching maturity in October-November and usually breaking up or being partly destroyed by insects in mid-summer. A new growth of floating leaves may occur in late summer-early autumn. On maturity the fruits float away, eventually sinking to the mud. In two simple viability tests at Griffith, none of the seeds germinated.

Habitat

The species generally grows in slowly-flowing water in supply channels, rivers and creeks and the margins of lakes, reservoirs and earth dams where it thrives in water 10 cm to 3 metres deep.

Economic Significance

During spring and early summer it causes considerable obstruction to water flow in channel systems, creeks and small rivers.

The waxy coating on the floating leaf, which provides additional buoyancy to the already buoyant submerged leaves of the plant, makes chemical control with herbicides injected into the water more difficult.

Distribution

The most commonly observed *Potamogeton* in the Murray, Lachlan and Macquarie Valleys of New South Wales. NC, CC, SC, NT, CT, ST, NWS, CWS, SWS, NWP, SWP, SFWP, all states.

Ranunculus

Erect or stoloniferous herbs. Leaves either in a basal rosette or alternate along the stem, petiole bases ± inflated and sheath-like; leaf blades usually variously lobed or divided. Flowers bisexual, solitary, axillary and/or terminal. Sepals 4-5; petals 4-12, usually yellow, rarely white, mostly with a characteristic nectary at the base. Stamens numerous. Carpels many, free; style persisting as a beak on the achene.

A genus with only a few truly aquatic species in Australia, viz. *R. inundatus—R. undosus* complex, *R. trichophyllus* and *R. rivularis* but including a large number of species that normally grow in mud or permanently damp ground. A key is given here to the more commonly encountered and/or more economically important species which could, at times, be regarded as aquatics.

1. Petals white. Leaves usually all submerged; much divided into filiform sections. *R. trichophyllus*
1a. Petals yellow. Leaves normally not all submerged but can survive inundation; segments less divided, mostly flattened.
2.

2. Leaves glabrous or with only a few scattered hairs. 3.
2a. Leaves hairy, occasionally only slightly. 8.

3. Achenes with lateral faces smooth, minutely rugulose or ridged. 4.
7.

3a. Achenes tuberculate or warty.

4. Erect much-branched annual; flowers small, numerous; petals 2-3 mm long. *R. sceleratus*
4a. Decumbent perennial herbs rooting at the nodes; petals 3 mm or more long. 5.

5. Nectary without a distinct petaloid lobe. *R. rivularis*
5a. Nectary with a distinct petaloid lobe, often attached at the margins forming a pocket. 6.

6. Mature achenes with a ripple pattern of small rounded ridges; receptacle hispid. *R. undosus*
(see also *R. inundatus*)

6a. Mature achenes with obscure irregular ridges, receptacle glabrous, ± hispid at the base. *R. inundatus*
Note: the two "species" above appear to intergrade and may be difficult to distinguish.

7. Achenes curved, ± twisted. *R. pentandrus*
7a. Achenes flattened, not curved or twisted. *R. muricatus*

8. At least basal leaves pinnate, plant neither stoloniferous nor decumbent. *R. pimpinellifolius*
8a. Leaves either ternate or palmately divided, plants stoloniferous.
9.

9. Leaves ternate, central leaflet stalked. *R. repens*
9a. Leaves palmately divided. *R. papulentus*

Left: A flower of a species of *Ranunculus* showing 4 petals and numerous stamens.

Left: A "form" of *Ranunculus inundatus* with narrow leaf segments

Below: A "form" of *Ranunculus inundatus* with broad leaf segments

River Buttercup

Ranunculus inundatus

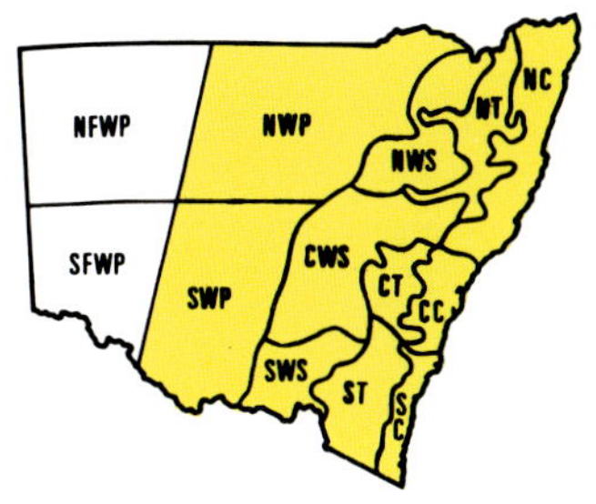

A native rhizomatous and/or stoloniferous perennial herb to 40 cm high. Basal leaves palmately lobed or divided, to about 5 cm wide, with individual lobes to 3 cm long and 2 mm wide, glabrous. Petioles to 15 cm long, mostly glabrous. Flowers yellow, 1-1.5 cm diameter. Petals 5-12, each with a basal nectary consisting of a lobe with the margins adnate to the petal. Achenes 1.5-2 mm long, obscurely wrinkled when dry, sometimes smooth when fresh.

R. undosus is very similar, but intermediates occur and it may be best at present to regard both species as extreme forms of a complex.

R. undosus has leaves to 6 cm in diameter with the segments 1-3 mm wide in submerged leaves and to 1 cm wide in emergent leaves. Flowers 2-2.5 cm in diameter.

Habitat

Either submerged, partially submerged, emergent or terrestrial in or near swamps or slowly-flowing creeks.

Distribution

NC, CC, SC, NT, CT, ST, NWS, CWS, SWS, NWP, SWP, Queensland, Victoria, Tasmania.

Sharp Buttercup

Ranunculus muricatus

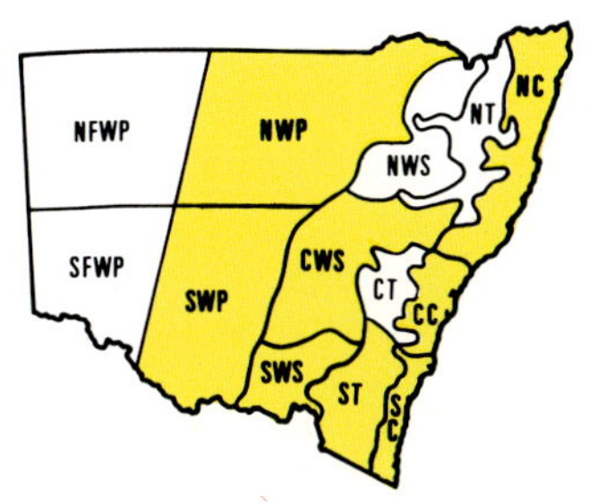

An introduced erect or prostrate annual. Leaves 3-lobed with each lobe deeply to barely crenate. Petioles 10-15 cm long. Flowers 1-1.5 cm diameter. Petals yellow, 6-10 mm long each with a nectary pouch with an ovate lobe. Achenes 3-4 mm long, tuberculate, with a beak more than half as long as the achene. *Seed photograph page 444.*

Habitat

Damp places, usually in disturbed areas.

Economic Significance

A minor weed under some circumstances.

Distribution

NC, CC, SC, ST, CWS, SWS, NWP, SWP, Queensland, Victoria, Tasmania, South Australia.

Ranunculus repens in an
ornamental pond in
Canberra

Ranunculus repens

Creeping Buttercup

Ranunculus repens

An introduced stoloniferous perennial herb to 40 cm high. Leaves alternate, to 7 cm wide. Central leaflet stalked and all leaflets with short or long lobes. Flowers to 3 cm diameter. Achenes smooth with a prominent border, approximately 3 mm long. Beak almost erect.

Habitat

Damp ground, seepage areas, creek banks, swamps, mainly in disturbed areas.

Economic Significance

A minor weed.

Distribution

NC, CC, SC, NT, CT, ST, Victoria, Tasmania, South Australia.

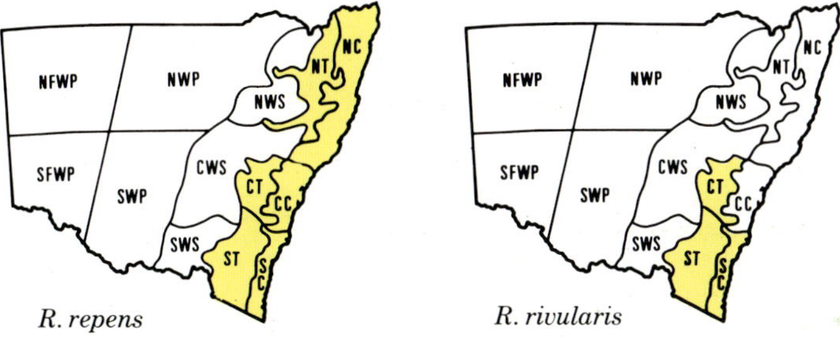

Small River Buttercup

Ranunculus rivularis

A native variable stoloniferous perennial. Leaves to 6 cm across, palmately divided with individual segments to 5 cm long, and these further lobed. The final lobes to 3 or 4 mm across in submerged leaves, 1 cm across in emergent leaves or terrestrial plants. Flowers to 1 cm diameter. Petals 5-9, yellow, 3-4 mm long, each with a swollen nectary at the base which may or may not form a small pocket. Achenes about 2 mm long, obscurely wrinkled when dry, with a slender beak almost 2 mm long.

Habitat

Submerged or emergent in shallow water of swamps or slowly-flowing streams. Also grows on mud or creek banks.

Distribution

SC, CT, ST, Victoria, Tasmania.

Ranunculus sceleratus

                    Ranunculus sceleratus

An erect introduced annual to 50 cm high. Petioles to 4 cm long, leaf blades to 2 cm wide, variously toothed or lobed. Flowers numerous in an open, leafy inflorescence. Petals yellow, subequal to sepals. Achenes numerous, about 1 mm long, apiculate. *Seed photograph page 445.*

The specific name is derived from the Latin "sceleratus", wicked, due to its poisonous properties and not from the similarity of pronunciation between "sceleratus" and celery.

Growth Biology

This species has a long germination season with seedlings appearing in winter, spring and late summer. Flowering mainly spring.

Habitat

Damp places, particularly in disturbed areas. Muddy shallow pools, lake margins and drainage channels.

Economic Significance

Can be poisonous to stock. Infrequently obstructive in drainage channels.

Distribution

NC, CC, ST, SWS, SWP, Victoria.

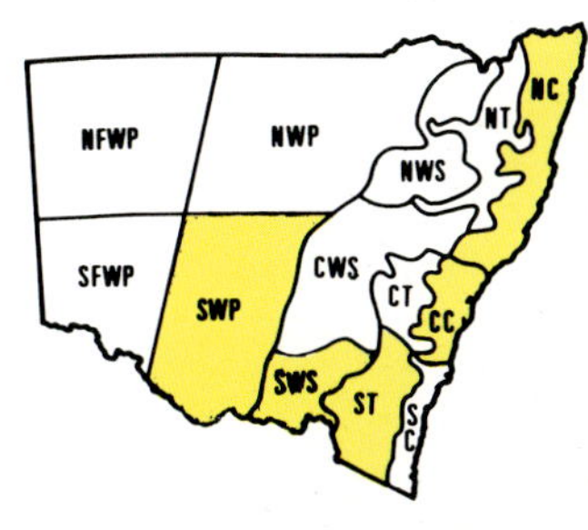

Ranunculus sceleratus in a drainage channel near Lake Wyangan

Ranunculus trichophyllus with submerged dissected leaves and white emergent flowers

RANUNCULACEAE

Ranunculus trichophyllus

A native stoloniferous perennial herb with usually submerged leaves and ± emergent flowers. Stems rooting at the nodes, to 1 metre long. Leaves alternate, to 6 cm long, trichotomously divided in linear or filiform lobes less than 1 mm wide. Flowers to 1.5 cm diameter, but often much less. Sepals 4-5, 2-3 mm long. Petals 4-5, white, 3-6 mm long, each with a swollen nectary near base. Achene about 1.5 mm long with some hairs, strongly wrinkled, with a short beak.

Habitat

More or less permanent creeks and open areas of swamps, mostly in tableland areas.

Distribution

NT, CT, ST, CWS, Victoria, Tasmania, South Australia.

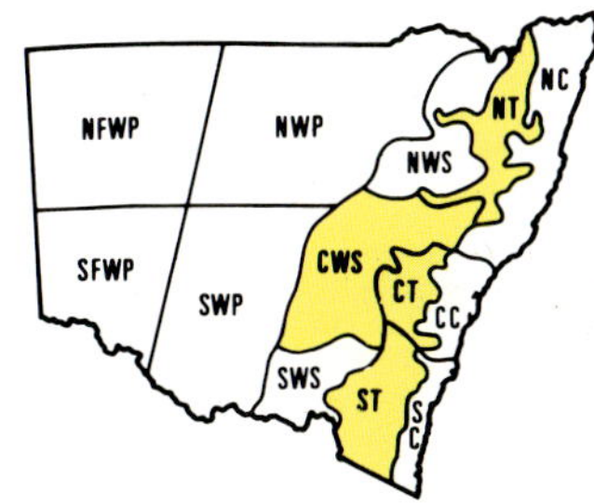

Ranunculus trichophyllus in the Cotter River

(Bryophytes)
Riccia fluitans s. lat.

There are apparently three species in Australia which have been included within *R. fluitans*. The true *R. fluitans* does not seem to occur in New South Wales. The two most common species in New South Wales are *R. duplex* and *R. multifida*. These species can be difficult to separate.

The aquatic forms of these two species consist of an open network of a dichotomously branching thallus up to 5 cm diameter with each individual branch to 2 mm diameter floating at or just below the surface. The aquatic stage is sterile and the species settles and grows on drying mud to produce the spores and sexual stages.

Ricciocarpus natans is also a member of the Ricciaceae. This liverwort consists of a free-floating thallus that lies on the water surface. It is green on top and purple below.

Growth Biology

Plants apparently spread vegetatively, pieces of a thallus breaking off and growing.

Habitat

Slowly flowing water in creeks and swamps.

Economic Significance

None known.

Distribution

CT, CWS, Queensland, Northern Territory.

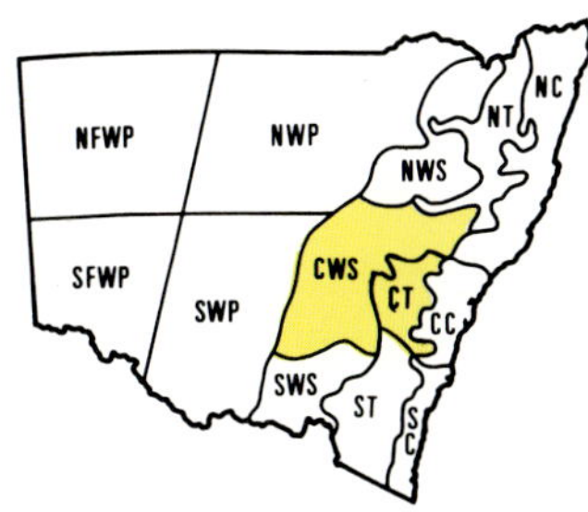

Left: A cultivated specimen of
a species of *Riccia*

A family with a single genus, closely related to the Potamogetonaceae, and possibly best included in that family.

Ruppia

All submerged aquatics, either surface or below-surface flowering. Plants rhizomatous, either producing leaves directly from the rhizome or from stems from the rhizomes. Leaf sheaths loose, inflated, auriculate, non-ligulate. Leaves alternate. Inflorescences apparently axillary, at first enclosed between a leaf and a leaf-like bract. Inflorescence stalk exserting flowers either just above leaves in submerged-flowering species, or to the surface. Flowers 2, small, hermaphrodite, perianth absent. Stamens 2, carpels 2-16 (usually 4). Mature fertilised carpels (fruit) on an individual carpel stalk which may become reflexed. In surface-flowering species the inflorescence becomes loosely to tightly coiled as the carpels mature and the carpel stalks elongate. The length of the inflorescence stalk and the degree of spiralling is partially dependent on water depth.

Most species grow in waters varying from slightly saline to hypersaline.

R. maritima has short (usually 5 cm) inflorescence stalks which do not coil into a spiral at maturity. The fruit is asymmetrical and similar in shape to that of *R. megacarpa,* but *R. maritima* fruit is 2-3 mm long and the fruit of *R. megacarpa* 3-4 mm long. *R. megacarpa* has truncate or notched leaf tips, whereas both *R. polycarpa* and *R. maritima* have obtuse or acute leaf tips.

Ruppia spp. could be confused with *Potamogeton pectinatus*, which may even grow with *Ruppia. P. pectinatus,* has a well-developed, entire ligule, often becoming laciniate with age.

1. Inflorescence stalk short, usually less than 5 cm long, at the most once or twice loosely coiled; flowers pollinated and maturing entirely submerged.

R. maritima group
(including ***R. maritima***
and ***R.*** sp. aff. ***maritima***)

1a. Inflorescence stalk usually more than 5 cm long, often more than 10 cm long, coiled several times; flowers pollinated at the water surface and maturing submerged. **2.**

2. Leaf tips truncate or notched; carpels mostly 4 (2-6); fruit 3-4 mm long.

R. megacarpa

2a. Leaf tips acute or obtuse; carpels usually 6-8 (2-16); fruit 2-3 mm long.

R. polycarpa

Sea Tassel

Ruppia maritima

A native submerged perennial with submerged flowers. Leaves narrow, approximately 0.5 mm wide, less than 5 cm long, with acute or obtuse tips. Inflorescence stalk about 5 cm long. Mature carpel stalks less than 2 cm long (usually about 1 cm). Mature carpels black, strongly asymmetrical, 2-3 mm long.

Growth Biology

Little is known about the growth of this species in Australia, but if it behaves similarly to elsewhere in the world it most likely establishes from seed in summer or autumn. The plants are capable of living almost indefinitely under suitable conditions but can also complete their life cycle, from seed to seed in 60-80 days. Local spread is by the vigorous rhizomes and distance and temporal dispersal by seed.

Habitat

Only recorded in New South Wales from estuarine habitats, but a closely related species has been recorded from Deniliquin.

Economic Significance

Nil.

Distribution

NC, CC, SC, SWP, Europe, America.

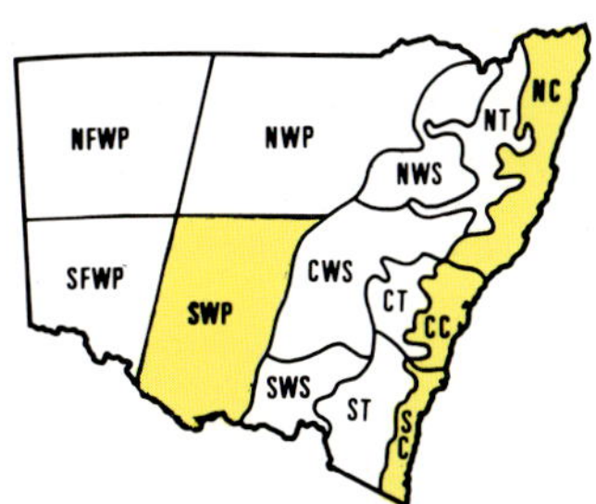

Sea Tassel   *Ruppia megacarpa*

Submerged, rooted, surface-flowering native aquatic perennial. Rhizomes slender, usually producing stems 20-30 cm long (occasionally much longer). Leaf sheath loose, inflated, ligule absent but with two small and variable lateral auricles. Leaves alternate, variable in length, usually 5-15 cm long, approximately 0.5-0.8 mm broad. Leaf tip either truncate or, more usually, with a shallow notch. Length of the inflorescence stalk partially dependent on water depth, but normally at least 15 cm long. Flowers with 2-6 (usually 4) free carpels. Fertilized carpels enlarge, and the carpel stalks elongate greatly to 5 or 6 cm and may become reflexed before the fruit (individual mature carpels) become detached. At maturity the inflorescence stalk coils into a tight or loose spiral. Fruit is asymmetrical, 3-4 mm long, brown, eventually turning black.

Growth Biology

Seeds germinate in summer and autumn, but under suitable conditions mature plants may survive almost indefinitely. Plants flower in summer and autumn. Local vegetative spread is by the vigorous rhizomes; distance and temporal dispersal by seed. The plants have very little food storage and do not seem capable of surviving even brief periods of extreme conditions. They can apparently complete their life cycle, from seed to seed, in 60-80 days.

Habitat

Hypersaline inland lakes, coastal lagoons and estuaries, and inland and coastal lakes of lower salinity than seawater.

Economic Significance

Appears to be a food source for waterbirds.

Distribution

NC, CC, SC, ST, SWP, Victoria, South Australia, New Zealand.

Left: *Ruppia megacarpa* in a lake near Rochester, Victoria Insert: *Ruppia megacarpa* with four carpels per flower, all developed in the stem at left, only two per flower developed on stem at right.

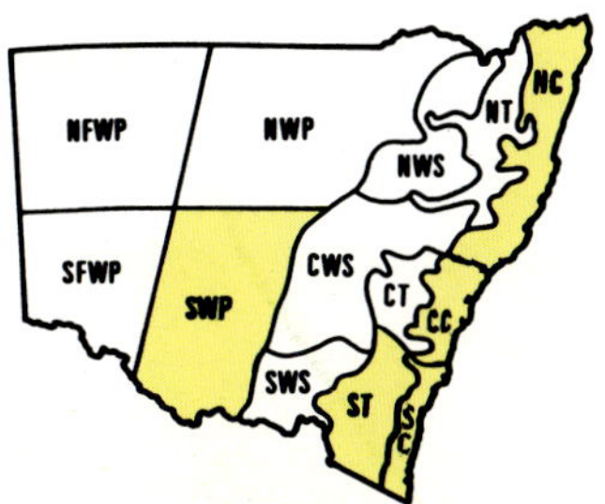

Sea Tassel

Ruppia polycarpa

A rhizomatous, submerged, surface-flowering native species with the leaves mostly produced on short stems, giving a lawn-like appearance. In deeper water longer stems may form. The leaves are variable in length, commonly 5-10 cm long, usually less than 0.5 mm wide. The leaf tip is obtuse or acute. Inflorescence stalk length partially dependent on water depth, but more than 10 cm long. At maturity the inflorescence stalk coils into a tight or loose spiral. Carpels 2-16, but commonly 4-8, per flower. Fruit 2-3 mm long, asymmetrical, but less so than in the other two species. Turions are formed rarely but none recorded from New South Wales specimens to date.

Growth Biology

Seeds presumably germinate in summer or autumn, but mature plants can survive indefinitely under suitable conditions. Plants flower in summer and autumn. Local spread is by the vigorous rhizome, and distance and temporal dispersal by seed. This species commonly behaves more as an annual than do the other two species.

Habitat

Coastal lakes and lagoons at similar or lower salinities than sea water. Usually grows on sandy or alluvial substrates in fairly shallow water, usually less than 1 metre deep.

Economic Significance

Appears to be a food source for waterbirds.

Distribution

CC, SC, Victoria, South Australia, New Zealand.

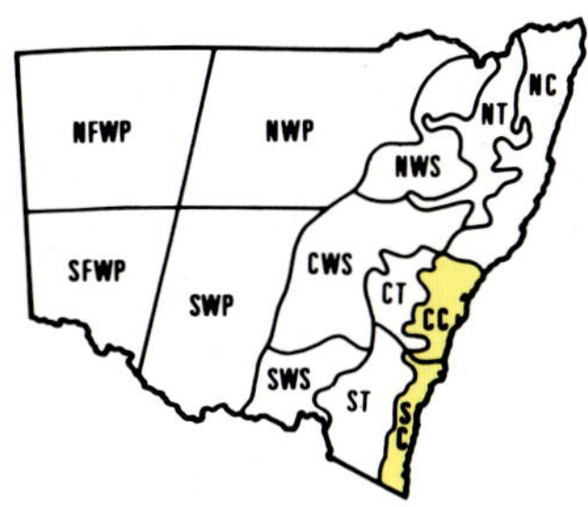

Left: *Ruppia polycarpa* in Narrabeen Lakes

Willows

Salix

Many species, hybrids and cultivars have been introduced both as horticultural subjects and to help control soil erosion. Most of these introductions have not become naturalised and only a few species have become widespread. The key below may help in identifying the species introduced but hybrids, especially between *S. alba* and *S. fragilis,* can be difficult to identify. In some areas, one or more extra introductions may have become established but have not spread and have not yet been recorded. These plants will obviously make it difficult to produce a reliable key. A study of the introduced taxa of *Salix* is required before reliable identification is possible.

1. Leaves hairy (under 10x magnification). **2.**
1a. Leaves glabrous. **3.**

2. Leaves about 5-7 times as long as wide. *S. alba*
2a. Leaves about 1-3 times as long as wide. *S. caprea*

3. Trees with penultimate or ultimate branches pendant or drooping. *S. babylonica*
3a. Trees without pendant branches; penultimate or ultimate branches fragile and readily disarticulating. *S. fragilis*

Left: *Salix alba* (foreground) and *Salix babylonica* in a river near Wonthaggi, Victoria

Right: A flowering branch of *Salix alba*

Weeping Willow

Salix babylonica

Introduced spreading deciduous dioecious tree to 12 metres tall with pendulous penultimate branches. Trunk up to 75 cm diameter, rarely straight. Primary branches brittle, commonly eventually breaking from wind or water action. Leaves alternate on short ultimate branches borne alternately on the pendulous simple or divided penultimate branches which are up to 6 metres long. Leaves 2-12 cm long, 0.5-2 cm wide with $\pm$ serrate margins. Inflorescences or catkins-unisexual, a compact spike about 4 cm long and less than 1 cm wide, terminal on short ultimate branches. Male flowers rare in New South Wales. Female flowers with a solitary sessile ovary and 2-fid style. *Fruiting flowers photograph page 445.*

Growth Biology

Spread by vegetative means, branches are brittle and break in floods or strong winds, dispersing stem portions capable of taking root along the banks. Plants are winter deciduous, producing the inflorescences about the same time as the first leaves in early spring.

Habitat

Creek and river banks, drainage and channel banks, seepage areas.

Economic Significance

Planted to reduce erosion after removal of native vegetation. *Salix* spp. can actively compete against native species. Broken branches may jam and block smaller streams. The vigorous root systems of *Salix* damage drains, channel banks and concrete structures, and their pendulous branches may reduce water flow.

Distribution

Most divisions of the State, but difficult to differentiate between naturalised and cultivated plants. Victoria, Tasmania, South Australia.

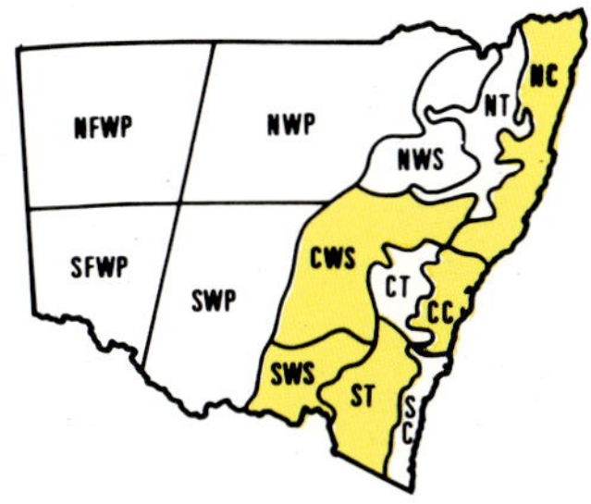

Left: *Salix babylonica* on the banks of a creek near Merriwa

Right: Inflorescences of *Salix babylonica*

Salvinia

Salvinia molesta
[S. auriculata]

An introduced free-floating aquatic fern. The main axis fragments easily, but plants are commonly up to 30 cm long. Leaves in whorls of 3, the upper 2 with a ± folded lamina, the lower much divided and rootlike, submerged. Floating leaves shortly stalked, 1-4 cm long and 1-5 cm wide, flat when in open water, folded when plants crowded in dense mat. Upper surface covered with "basket" hairs. Mature plants may produce sporangia in sporocarps on segments of the submerged leaves. Up to 55 sporocarps produced in 2 rows along a segment, the first 2 or 3 macrosporangia, the remainder microsporangia. Sporocarps hairy, usually less than 1 mm long, mostly empty. *Sporocarp photograph page 445.*

Growth Biology

This species is apparently of hybrid origin and fertile spores have yet to be reported. The plant spreads vegetatively by fragmentation and can grow from a single node. The plants look very different in the early colonising stages, the leaves being barely keeled and well spaced, than in the closely packed, dense mat forming stage. Plant fragments invariably manage to survive any treatment covering large areas and, coupled with its rapid growth rate (doubling time 5-10 days), can soon re-cover very large areas. It generally grows best in high nutrient levels and water temperatures around 20-30°C, but can survive 10-35°C. Although frost sensitive, it can survive these conditions.

Left: *Salvinia molesta*
showing the mature leaves
with sterile sporocarps (top)
and juvenile leaves (bottom)

Salvinia molesta in a small dam near Sydney

Habitat

Permanent still or slowly-flowing water or on land frequently inundated. Will survive lying on mud. This species will not thrive free-floating in water containing low nutrient levels.

Economic Significance

Taints water storages and seriously affects other aquatic plants and animals, blocks pump inlets or other structures. A very expensive weed, almost impossible to eradicate from large areas.

Control

Limit inflow of nutrients from sewerage, urban runoff and intensive agriculture. *S. molesta* has been controlled by flooding with saline water. It is quickly killed by sea water and 3000-4000 ppm total dissolved salts will severely limit growth after a few months exposure. Also see section on biological control, page 488.

Distribution

NC, CC, SC, CWS, SWP, Queensland, Victoria, Northern Territory, South Australia, Western Australia.

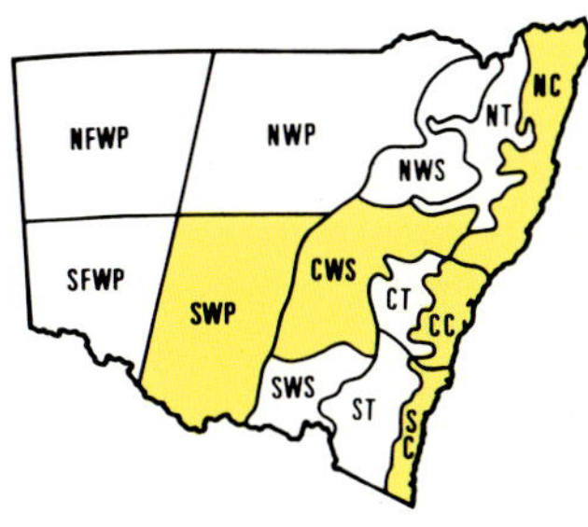

Salvinia molesta
on Lake Moondarra, Queensland

Bacopa monniera

A prostrate native herb to 30 cm high. Stems much-branched and rooting at the nodes. Leaves opposite, to about 1.5 cm long and less than 1 cm wide, almost sessile. Flowers bisexual, solitary, axillary, on pedicels about as long as or longer than the leaves. Sepals 5, about 5 mm long. Corolla tube 5-lobed with the lobes about as long as the entire part (approximately 4 mm), white or light blue. Stamens 4 (2 long, 2 short). Ovary 2-locular; capsule shorter than the calyx.

Habitat

A species characteristic of damp ground rather than true aquatic situations, but grown as an aquatic in aquaria.

Economic Significance

Used in aquarium plant trade.

Distribution

NC, CC, Queensland.

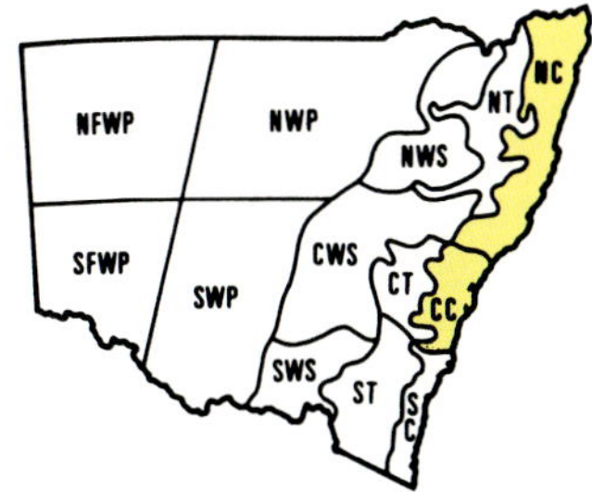

Left: The white flowers and opposite, slightly fleshy leaves of *Bacopa monniera*

Right: Bacopa monniera in a swamp on the North Coast

Top: The minute plants of *Glossostigma diandrum* with small blue flowers and flattened fleshy leaves

Left: *Glossostigma diandrum* in a supply channel near Widgelli

Glossostigma

Prostrate native annuals or short-lived perennials usually less than 20 cm high. Leaves opposite, usually with relatively long petioles, entire. Flowers bisexual, solitary, axillary, small. Calyx 3- or 4-lobed; corolla with 5 subequal lobes. Stamens 2 or 4. Ovary 2-celled. Capsule enclosed in calyx; seeds numerous.

1. Stamens 2, calyx 3-lobed	***G. diandrum***
1a. Stamens 4, calyx 4-lobed	***G. elatinoides***

Glossostigma diandrum

A native prostrate herb, leaves to 1.5 cm long, including the petiole, which is longer than the blade. Pedicels usually shorter than the leaves. Calyx 1.5 mm long, doubling its size in the fruiting stage, unequally 3-lobed. Corolla slightly longer than calyx, blue. Stamens 2. *Seed photograph page 445.*

Habitat

Swamps and periodically inundated areas. Occasionally found in rice crops.

Economic Significance

Nil.

Distribution

NT, ST, SWS, NWP, Queensland, South Australia, Western Australia, Asia, Africa.

Glossostigma elatinoides

Similar but generally smaller than *G. diandrum*. Calyx 4-lobed (often unequal); stamens 4.

Distribution

NC, NT, CT, ST, SWS, Victoria, Tasmania.

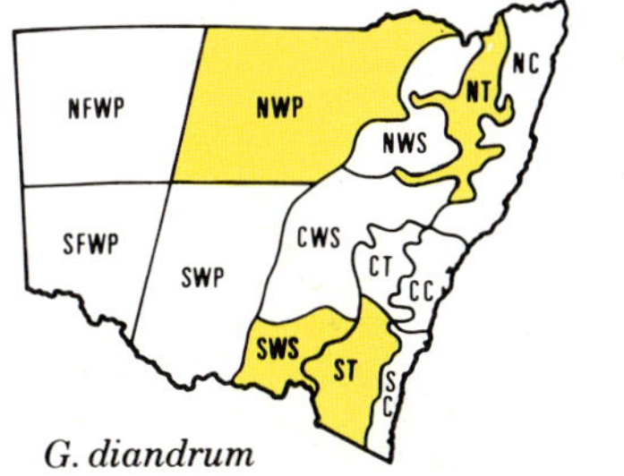

G. diandrum

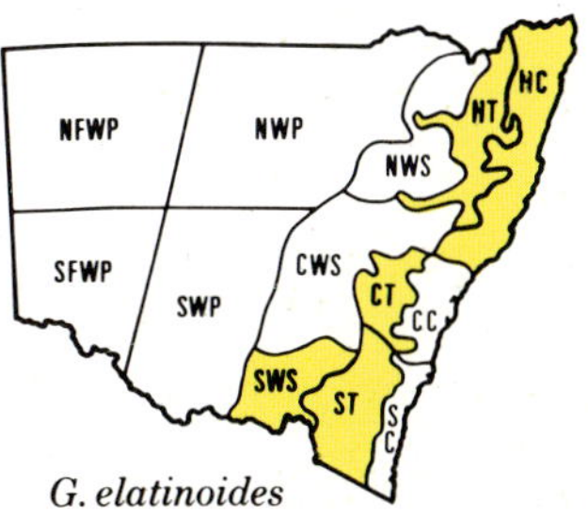

G. elatinoides

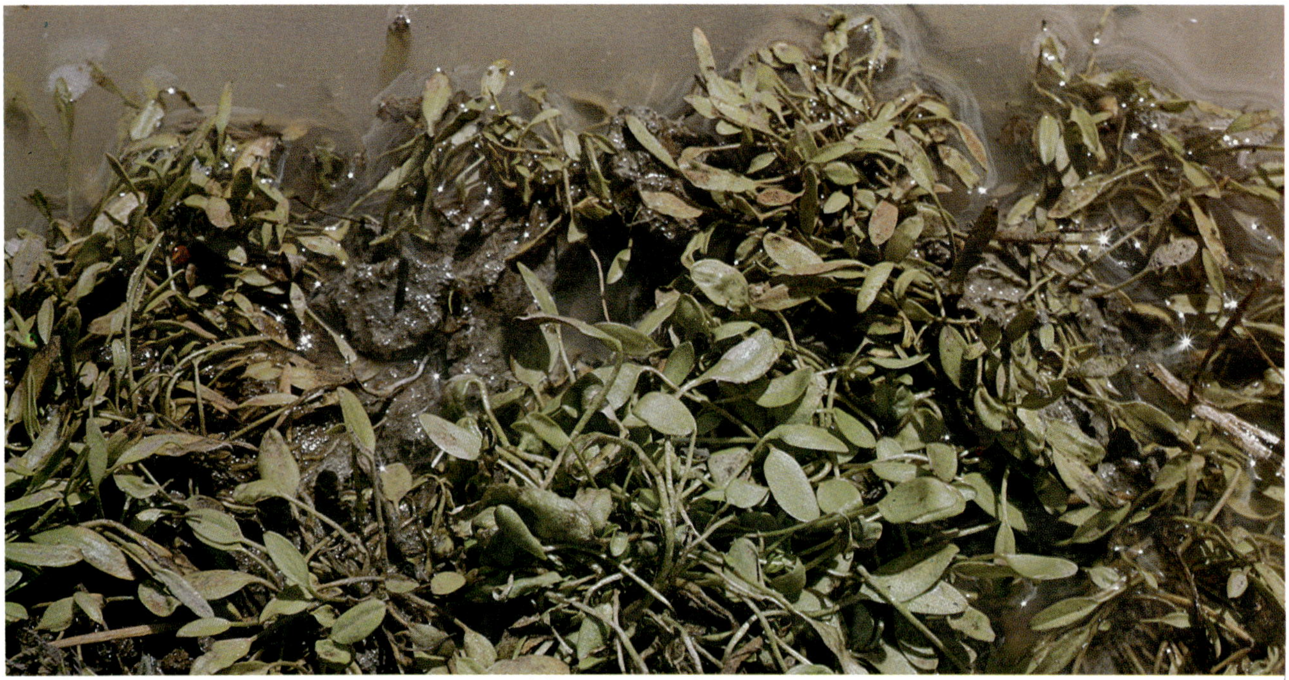

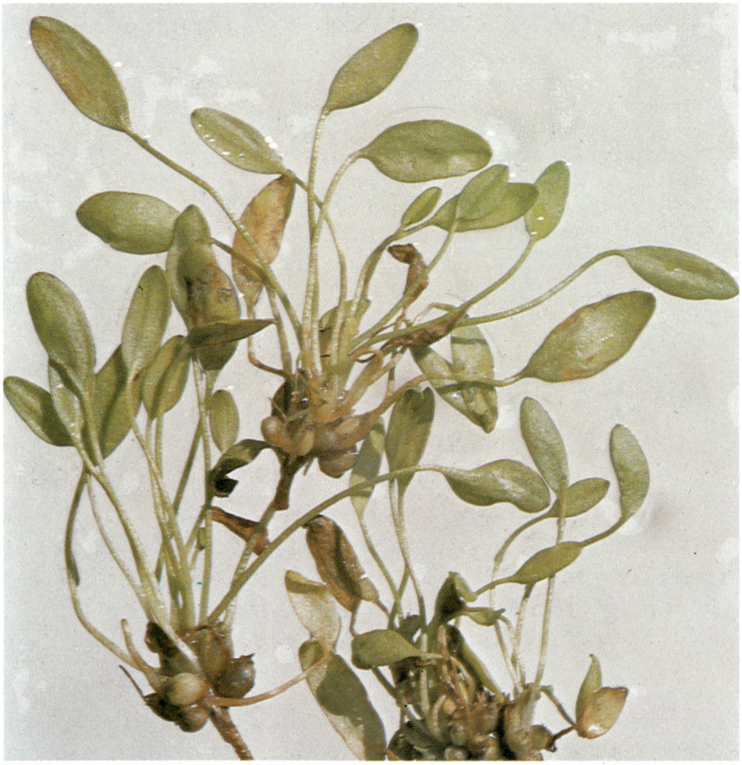

Top: *Limosella curdiana* on the margins of a rice crop

Left: *Limosella curdiana* showing the long petioled leaves and sessile flowers

L. curdiana

L. australis

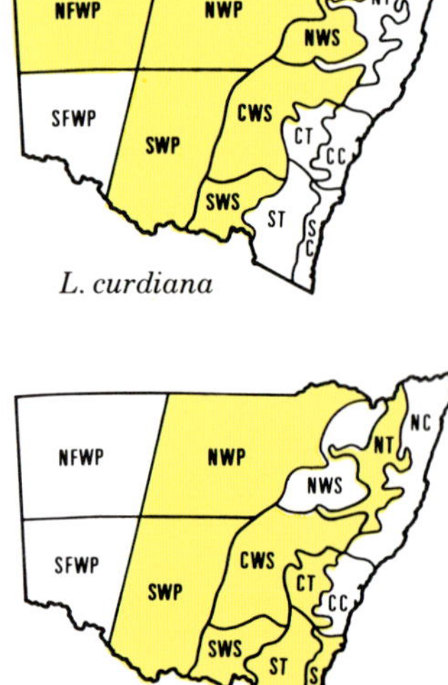

Limosella

Prostrate native annuals or short-lived perennials up to 10 cm high. Leaves mostly basal. Flowers bisexual, solitary, axillary, pedicellate or sessile, but always shorter than the leaves. Calyx with 5 small lobes. Corolla with 5 almost equal lobes. Stamens 4. Ovary 2-celled. Capsule about same length as calyx; seeds numerous.

1. Flowers sessile, leaf blades usually expanded.   *L. curdiana*
1a. Flowers pedicellate, leaf blade usually not expanded. *L. australis*

Mudwort

Limosella curdiana

Stoloniferous prostrate to tufted native annual or short-lived perennial. Leaf blade up to 4 cm long and 1.5 cm wide on petioles of different length in different habitats, 1-2 cm long on drying mud, to 15 cm long in water. Flowers sessile. Calyx about 3 mm long. Corolla about as long as calyx, white or light blue. *Seed photograph page 445.*

Growth Biology

Plants can grow in water, but will successfully flower only out of water, usually on drying mud.

Habitat

Damp ground or in ephemeral pools. Occasionally found in rice crops.

Economic Significance

None known.

Distribution

NWS, CWS, SWS, NWP, SWP, NFWP, Victoria, South Australia.

Limosella australis

Tufted native annual with linear leaves up to 5 cm long. Leaves may or may not have an expanded blade, but only poorly developed at best. Flowers pedicellate. Pedicels shorter than leaves. Calyx about 2 mm long, 5-lobed. Corolla white, blue or pink, 5-lobed, about 3 mm long.

Distribution

SC, NT, CT, ST, CWS, SWS, NWP, SWP, Victoria, Tasmania, South Australia, Africa, America.

A large plant of *Veronica anagallis-aquatica*

Blue Water Speedwell

Veronica anagallis-aquatica

An introduced decumbent or erect short-lived perennial herb to 1.5 metres high. Stems often creeping and rooting from the lower nodes. Leaves opposite, to 10 cm long, 1-2 cm wide, the margins mostly serrulate, base stem-clasping. Inflorescences 1-several per plant, axillary racemes to 25 cm long, each few to about 50-flowered. Flowers bisexual, pedicellate. Calyx with 4 nearly free lobes, each 2-3 mm long. Corolla pale-blue to white, 5-6 mm across, 4-lobed, the two lateral lobes narrower than the upper and broader than the lower lobe. Stamens 2. Ovary 2-celled with a small 2-lobed stigma. Fruit a capsule 2-3.5 mm long, seeds numerous. *Seed photograph page 445.*

Habitat

In stationary or slowly-flowing water to about 1 metre deep (but usually less). Margins of lakes and beds of intermittently flowing channels. Will also grow submerged in clear spring water.

Economic Significance

Infrequently reported obstructive in drainage channels.

Distribution

CC, NT, CT, ST, NWS, CWS, SWS, Victoria, Tasmania.

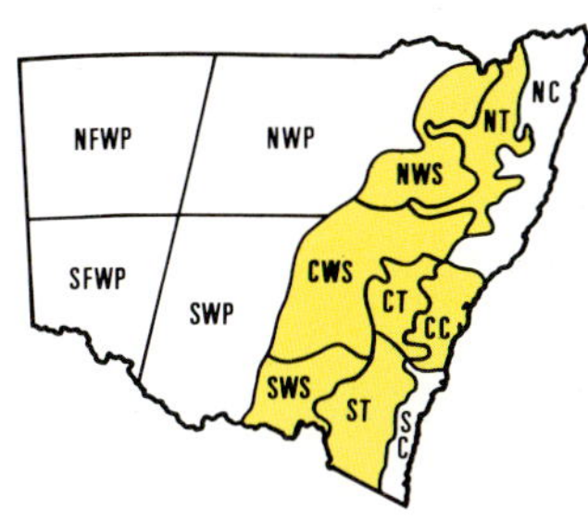

Veronica anagallis-aquatica in a drainage channel near Lake Wyangan

Floating Bur-reed

Sparganium antipodum

A native erect rhizomatous emergent monoecious perennial to 1 metre high. Leaves emergent, erect, occasionally floating, to 1 metre long and 2 cm wide, mostly basal, but progressively smaller leaves produced up the stem grading into the leaflike inflorescence bracts. Inflorescence with 1-several alternate branches, the lowest to 15 cm long, becoming progressively shorter upwards. Each branch with usually 2-20 dense globose heads of unisexual flowers, the lowest 1-4 female, the upper progressively smaller clusters of male flowers. Female flowers with 3-6 perianth segments and a solitary, sessile ovary. Fruiting clusters crowded, 1-2 cm diameter. Male cluster about 1 cm diameter at anthesis, flowers with 3-6 perianth segments and 3-6 stamens. *Seed photograph page 445.* Reports of *S. erectum* in New South Wales seem to be based on larger specimens of this species.

Habitat

Still or slowy-flowing water less than 1 metre deep and on the margins or dry beds of creeks, rivers, lagoons or swamps. Apparently can grow equally well on sand or mud.

Economic Significance

A significant food source for waterbirds. May help reduce bank erosion in some cases.

Distribution

NC, CC, SC, NT, CT, ST, Queensland, Victoria, New Zealand.

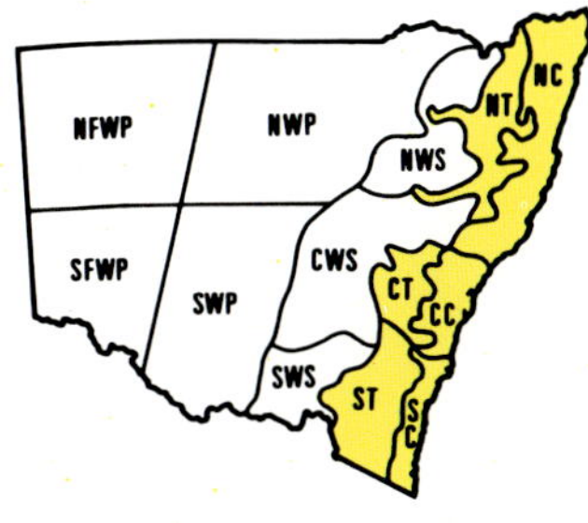

Left: *Sparganium antipodum* showing the inflorescence with clusters of white male flowers above and brown female flowers below

Right: *Sparganium antipodum* on the banks of the Coldstream River

Typha orientalis (left) and
T. domingensis (right)

416

Cumbungi

Typha

Erect rhizomatous native perennials to 4 metres high. Rhizomes extensive, branched, up to about 2 cm diameter. Stems cylindrical, pithy, up to 2 cm diameter. Leaf sheaths overlapping, in 2 rows; leaf blades often more than 1 metre long, up to 3 cm wide. Inflorescence a densely compact spike. Flowers unisexual, males above, females below ± separated by a length of stem. Male flowers with 1-7 stamens. Female flowers borne on very short crowded lateral branches. Usually both sterile and fertile female flowers present, subtended by several hairs on an elongated axis. Fertile female flower with 1-ovuled ovary with a filiform style ± terminally expanded into a stigma. Sterile female either imperfectly developed or modified into a club-shaped carpodium.

The two common species in New South Wales can be difficult to distinguish without experience, mainly because of the variation shown in individual characters. When the two species are growing together (which often happens), it is usually quite easy to separate them. The following key is a guide to the differences between them.

1. Stigmas linear; bracts in female inflorescence numerous, spathulate (4-8 cells across). Mature female spikes usually less than 2 cm diameter; male and female spikes separated by 0.5-5 cm.

T. domingensis
Narrowleaf Cumbungi

1a. Stigmas narrow-obovate; bracts in female inflorescence few, narrow-spathulate (3-4 cells across). Mature female spikes 1-4 cm diameter; male and female spikes separated by 0-5cm.

Narrowleaf Cumbungi
Broadleaf Cumbungi

T. orientalis
Broadleaf Cumbungi

Growth Biology

Seeds of both species are wind-dispersed, apparently over large distances. One spike can produce up to 200,000 seeds with a high percentage of viability.[1,2] The plants grow in fresh to brackish water up to 2 metres deep, forming an extensive underground network of fleshy white rhizomes that produce aerial shoots at intervals. A plant of *T. latifolia* has

[1] PRUNSTER, R. W., 1940. The control of cumbungi (*Typha* spp.) in irrigation channels. *Journal Scientific and Industrial Research,* 13:1-6.
[2] YEO, R. R., 1964. Life history of common cattail. *Weeds,* 12:284.

been measured to extend over an area of 8 square metres in 6 months. Plants flower most of the year, but less commonly in the colder months. The inflorescence is protandrous and the male flowers drop off soon after anthesis. The female flower will remain intact for a long time, apparently until dried out, either by heat or frost.

Habitat

Stationary or slowly-flowing water up to 2 metres deep.

Economic Significance

Provides food and shelter for animals and prevents erosion of gullies and creeks. Also one of the major weeds of irrigation systems and a weed of rice crops. In shallow creeks dense stands slow water movement and cause flooding of surrounding land in periods of high flow. Mosquitoes breed rapidly in the protection given by dense stands of *Typha* spp.

Evaporation of water from swamps or shallow lakes may be reduced by scattered growth of *Typha* spp.[3] This is because the tall plants shelter the water surface from wind and there is an increase in humidity near the water surface. Species of *Typha* may be useful to extract pollutants from waste or drainage water.

Control

Grazing and cutting are useful alternates to the use of herbicides. Cutting below the water line in autumn will effectively reduce the size of stands.

Distribution

T. domingensis: NC, CC, CT, ST, NWS, CWS, SWS, NWP, SWP, NFWP, SFWP, Queensland, Victoria, Tasmania, Northern Territory, South Australia, tropical and temperate.

T. orientalis: NC, CC, SC, NT, CT, ST, NWS, SWS, NWP, SWP, Victoria, South Australia, Western Australia, Malesia.

[3] LINACRE, E. T., HICKS, B. B., SAINTY, G. R., GRAUZE, G., 1970. The evaporation from a swamp. *Agricultural Meteorology,* 7:375-386.

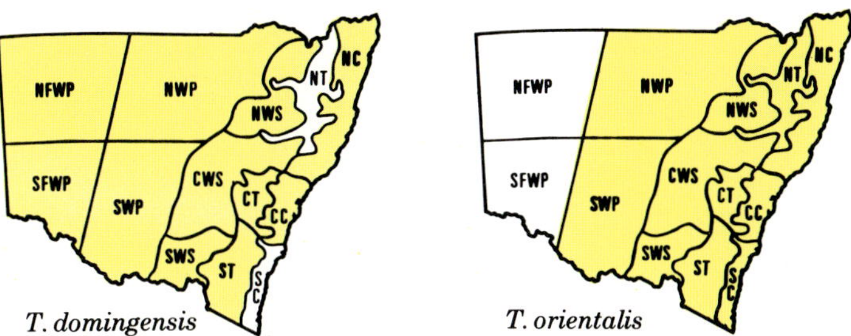

T. domingensis

T. orientalis

Left: *Typha orientalis* in a supply channel near Wumbulgal

Two genera from this family, *Lepilaena* and *Zannichellia,* have been recorded in New South Wales. There are four species of *Lepilaena* and one Australian species of *Zannichellia. Lepilaena australis* and *Z. palustris* have each been recorded once and *L. bilocularis* three times from New South Wales. Only one of the five species normally occurs in New South Wales, but any of the other four could reappear briefly and it seems improbable that only those species previously recorded from New South Wales are likely to reappear. Consequently we have included a brief generic description of *Lepilaena* with a key to the species, descriptions of two of the species, and a description of *Zannichellia palustris.*

1. Leaves mostly opposite, without a sheath but with ± sheathing stipules; female flower with 2-8 carpels enclosed within a cup-like perianth; mature carpels dorsally curved, usually tuberculate on dorsal surface. ***Zannichellia palustris***
1a. Leaves mostly alternate, but leaves subtending flowers ± opposite; leaves with basal sheaths; female flowers with 3 carpels and 3 distinct perianth segments. ***Lepilaena* 2.**

2. Stigma deeply divided (laciniate); male flower with a single 2-celled anther; leaf tip more or less truncate, sometimes mucronate.
L. bilocularis
2a. Stigmas not laciniate; male flower with a 6-celled anther mass; leaf tip acute to obtuse. **3.**

3. Leaves with minutely serrulate margins towards tip; style shorter than fruiting carpel; fruiting carpel ± tuberculate, half as broad as long.
L. australis
3a. Leaves with entire margins; style subequal to fruiting carpel; fruiting carpel smooth, usually less than half as broad as long. **4.**

4. Pedicels of fruiting flowers elongated beyond leaf sheath; fruiting carpel about one-third as broad as long; anther mass 1-3 mm long.
L. cylindrocarpa
4a. Pedicels of fruiting flowers not elongating beyond leaf sheaths; fruiting carpel about half as broad as long; anther mass 1-2 mm long.
L. preissii

ZANNICHELLIACEAE

Lepilaena

Native submerged dioecious or monoecious annuals (in New South Wales) or short-lived perennials with slender rhizomes. Plants often much-branched and, in flowing water, may be 1 metre long. Leaves alternate or ± opposite, 1-nerved, mostly 3-7 mm long, less than 1 mm wide, with basal sheath. Flowers unisexual, axillary, solitary. Female flower with 3 membranous perianth segments and 3 free carpels, each with an enlarged stigma. Male flower with a reduced perianth with a single anther mass, either 2-celled or 6-celled.

Lepilaena australis

Small much-branched monoecious native annual with stems up to about 20 cm long. Leaves alternate, 1-nerved, acute up to 6 cm long and less than 1 mm wide; sheath with small auricles. Female flowers on variable length pedicels (up to 10 cm long) which are reflexed in fruit; perianth segments 3, 1-2 mm long; style much shorter than the fruiting carpel; stigma asymmetrically funnel-shaped. Male flower with a vestigial 3-lobed perianth subtending a 6-celled anther mass. Fruiting carpel 1.5-2.5 mm long with 1 ventral and 3 dorsal keels, scattered blunt tubercles and persistent style base. Pedicels of fruiting carpels elongating at maturity.

Habitat

Ephemeral fresh or slightly brackish water.

Distribution

SWS, SWP, Victoria, Tasmania, South Australia, Western Australia.

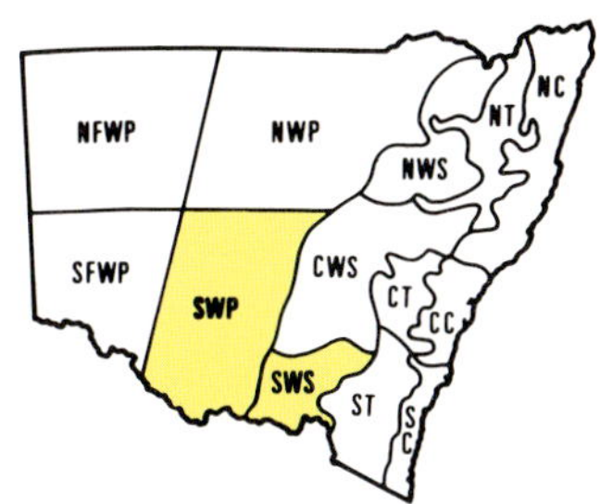

Lepilaena bilocularis in a lake south of Cooma

Lepilaena bilocularis

Delicate dioecious native annuals or short-lived perennials from less than 1 cm tall to about 1 metre long with alternate 1-nerved leaves. Leaves from very short to about 4 cm long, auriculate on female plants, ligulate on male plants, usually less than 1 mm wide with a truncate tip and usually 3 ± distinct mucros. Female flowers shortly pedicellate with 3 perianth segments each about 2 mm long; carpels 3, free; style thin, about as long as the mature ovary; stigma broad, deeply divided, nearly as long as the style. Male flower with a cup-shaped perianth subtending a single 2-celled, mucronate anther. Fruiting carpels usually retained in the persistent perianth.

Growth Biology

A summer growing species that mostly seems to die in winter. This species is capable of forming extensive mats of minute plants, even in quite deep water, and can be easily overlooked. At one of its localities in New South Wales it has been known for about 30 years.

Habitat

In temporary or permanent saline or brackish lakes; not common in New South Wales.

Economic Significance

Readily eaten by many waterbirds.

Distribution

ST, NFWP, Queensland, Victoria, Tasmania, South Australia, Western Australia.

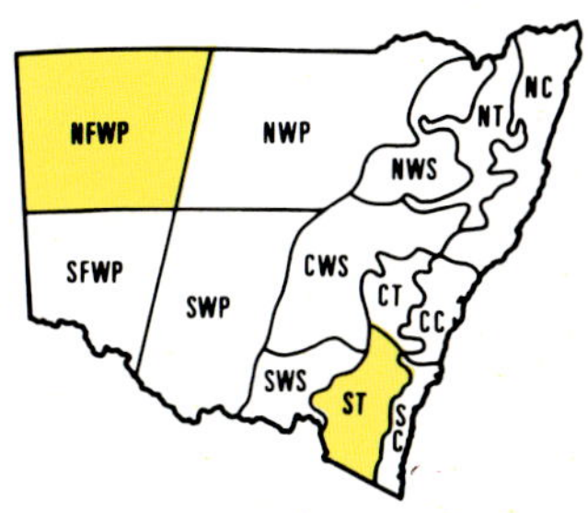

Submerged growth of *Lepilaena bilocularis* and floating plants uprooted by feeding swans

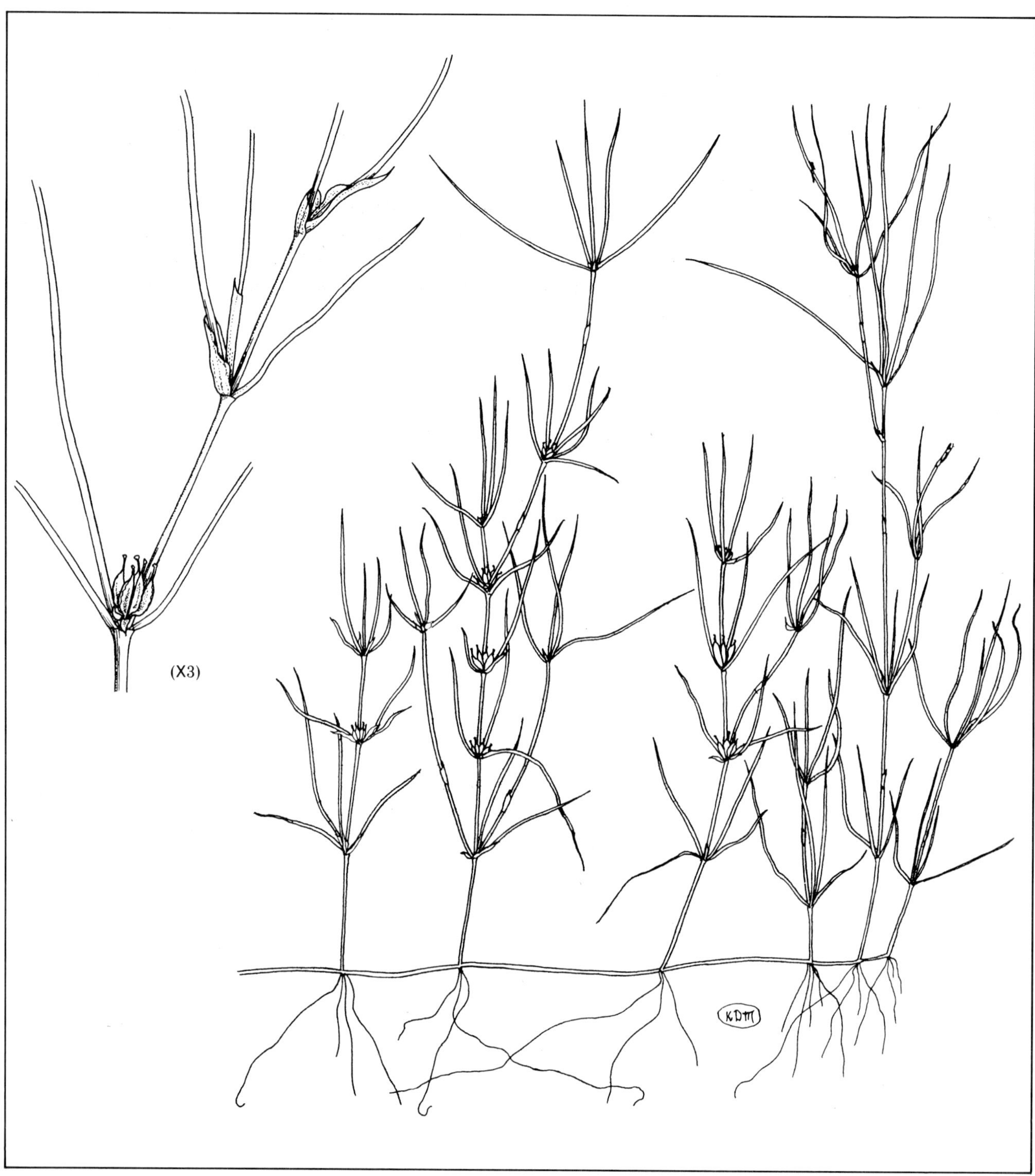

(X3)
KDM

ZANNICHELLIACEAE

Zannichellia palustris

A native submerged monoecious annual (in New South Wales) or perennial. Leaves mostly opposite, without sheaths but with ± clasping membranous stipules, 2-7 cm long, less than 1 mm wide. Leaf margins entire. Flowers unisexual, solitary, axillary. Female flower with a cuplike perianth and mainly 3 (2-8) carpels, each with a much-enlarged stigma; fruiting carpels dorsally curved and mostly with tubercles on dorsal surface, 2-4 mm long. Male flower with a perianth and with a single 2-celled anther.

Habitat

Saline or brackish ephemeral pools in water up to 50 cm deep.

Economic Significance

Nil in New South Wales. A rare plant.

Distribution

NC, South Australia, cosmopolitan.

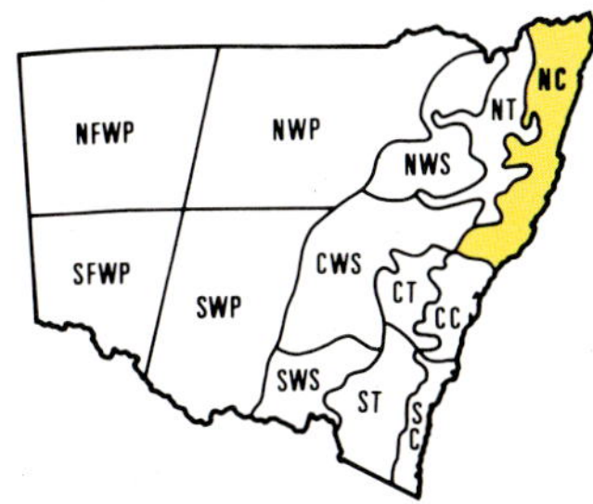

Left: Fruiting inflorescence (X3) and plant of *Zannichellia palustris* (X0.7)

CM 1 2 3 4 5 6 7 8 9 10 11 12

Zostera. Heterozostera

Fully submerged native marine or estuarine rhizomatous species. The roots are fibrous and arise from the nodes of the extensive rhizome system. The stem-clasping leaf sheath has both auricles and ligules. The leaves are linear, up to 5 mm wide and about 30 cm long, with either entire or notched tips. The elongate inflorescence consists of several spatheate spadices, each resembling vegetative leaves. The retinaculae (bractlike appendages on spadix) are narrowly or broadly triangular. There are about 6 or more male flowers and each consists of 2 unilocular anthers. The female flowers consist of a solitary laterally attached carpel with a single pendulous ovule, short style and bifid stigma, and are approximately equal in number to the male flowers. Seeds usually about 5 mm long, striate.

Eelgrass　　　　　　　　　　　　　*Zostera capricorni*

Zostera muelleri

These two *Zostera* species normally only produce erect stems when flowering. Both species have only 2 accessory vascular bundles in the internodes of the stems and rhizomes. Both species have broadly triangular obtuse retinaculae. *Z. capricorni,* with its entire, truncate leaf tip, can be distinguished, sometimes only with difficulty, from the usually notched leaf of *Z. muelleri. Heterozostera tasmanica* has notched leaves but produces both erect vegetative and erect flowering stems. *H. tasmanica* has narrow triangular, acute retinaculae and 4-8 accessory bundles in the internodes of stems and rhizomes.

Growth Biology

All three species have similar life cycles. The seeds germinate on a sand or mud substrate in spring. The young plant grows slowly, apparently not flowering for several (?) years. Plants either frequently exposed by the tide or growing on nutritionally poor substrate may not flower until they spread vegetatively into more favourable conditions.

The three species appear capable of extensive vegetative spread. The inflorescences are completely submerged and are normally produced between December and February, although flowering specimens may be found from November to March. The threadlike pollen is released and drifts around suspended in the currents until it lodges on the exserted stigmas. As the seed is maturing the rhizome becomes progressively more brittle, and in late summer or autumn the whole inflorescence breaks off at the base and the seed is released as the inflorescence floats around. This dispersal seems to be effective over large distances.

Left: An inflorescence (left) and plant of *Zostera capricorni*

Zostera capricorni in Narrabeen Lakes

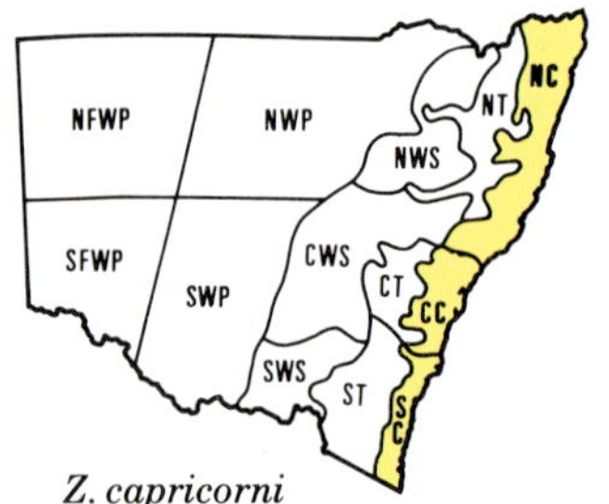

Z. capricorni

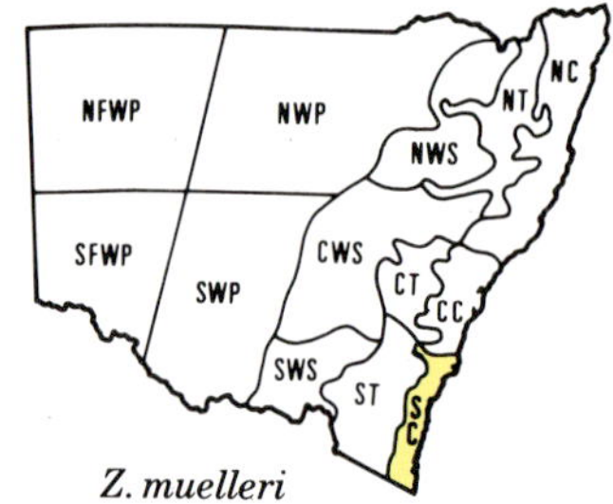

Z. muelleri

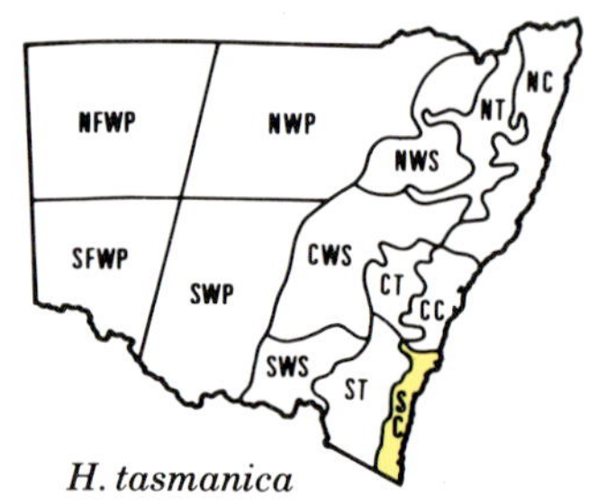

H. tasmanica

Habitat

Z. capricorni grows in estuaries, bays or protected ocean habitats, from exposed at low tides to water 4-5 metres deep. In southern New South Wales it becomes restricted more to estuarine entrances where the larger volume of water prevents the water temperature from dropping too low. Consistent water temperatures of less than 10°C limit growth of *Z. capricorni.*

Distribution   *Z. capricorni*

NC, CC, SC, Queensland, Victoria, (?) Northern Territory, New Guinea, Lord Howe Island and New Zealand.

Z. muelleri grows in very similar situations to those outlined above, but is tolerant of colder water temperatures. *Z. muelleri* occurs much further up-river on the New South Wales Coast and is more common along the southern continental shoreline.

Distribution   *Z. muelleri*

SC, Victoria, Tasmania and South Australia. It may also occur in New Zealand.

Tasman Grasswrack

Heterozostera tasmanica

Eelgrass

H. tasmanica occurs in rather more exposed conditions than the two *Zostera* spp. The plant grows in water up to 5-6 metres deep on either a sandy or muddy bottom.

Economic Significance

All three species are extremely important and valuable for both commercial and recreational fishing. Areas of Eelgrass provide extensive fish nurseries and suitable habitats for the young of many fish species. Few fish actually eat the Eelgrass; they mostly eat the other plants (epiphytes) and animals that grow on the leaves, or are carnivorous fish preying on the herbivorous species.

Occasionally *Z. capricorni* or *Z. muelleri* produce sufficient growth to be regarded as undesirable in swimming areas. This usually reflects either a change in sedimentation patterns or an unsuitable siting of swimming facilities. *Zostera* spp. rarely form extensive stands on clean sandy bottoms.

Distribution

Rarely found in New South Wales, SC, Victoria, Tasmania, South Australia and Western Australia.

WATERPLANTS IN RICEFIELDS

The photographs in this section include the most common waterplants found amongst rice in south-western New South Wales. A major factor causing an increase of these plants is the trend in recent years to plant rice crops in short rotation. The aquatic plant propagules are able to survive the short interval between crops.

Left-right seedlings of Starfruit
Damasonium minus; (page 49)
Dirty Dora
Cyperus difformis; (page 115)
Barnyard Grass *Echinochloa crus-galli;* (page 174)
Rice *Oryza sativa*

430

Arrowhead
Sagittaria montevidensis (page 53)

Dirty Dora
Cyperus difformis (page 115)

Starfruit
Damsonium minus (page 49)

Young plants of Starfruit
Damasonium minus (page 49)

Barnyard Grass
Echinochloa crus-galli (page 174)

Brown Beetle Grass
Diplachne fusca (page 171)

Waterwort
Elatine gratioloides (page 155)

Common Nardoo
Marsilea drummondii (page 297)

Curled Dock
Rumex crispus (page 355)

Narrowleaf Cumbungi
Typha domingensis (page 417)

Umbrella Sedge
Cyperus eragrostis (page 119)

SEEDS OF WATERPLANTS

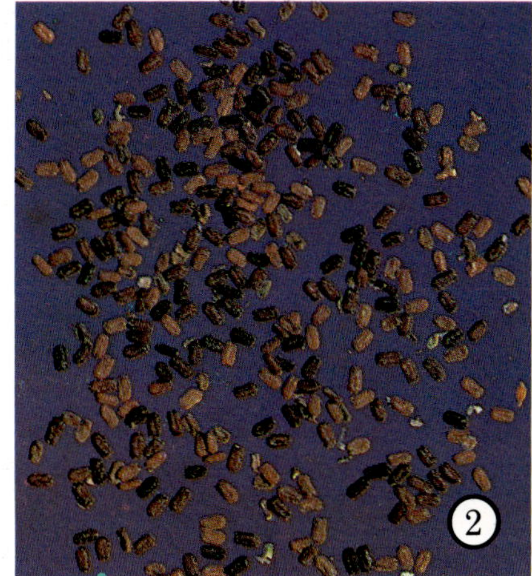

1. *Alisma plantago-aquatica*

2. *Damasonium minus*

3. *Sagittaria graminea* var. *platyphylla*

4. *Sagittaria montevidensis*

5. *Lilaeopsis polyantha*

6. *Sium latifolium*

7. *Aster subulatus*

8. *Eclipta platyglossa*

All X3.5

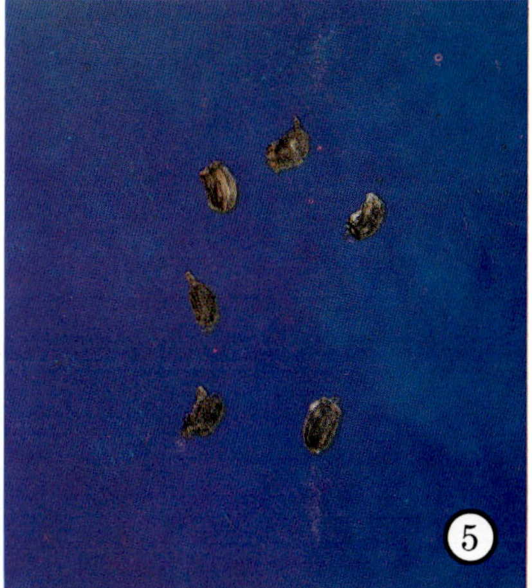

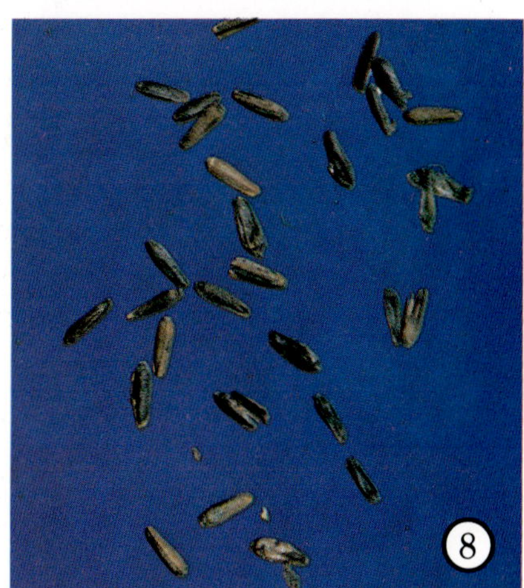

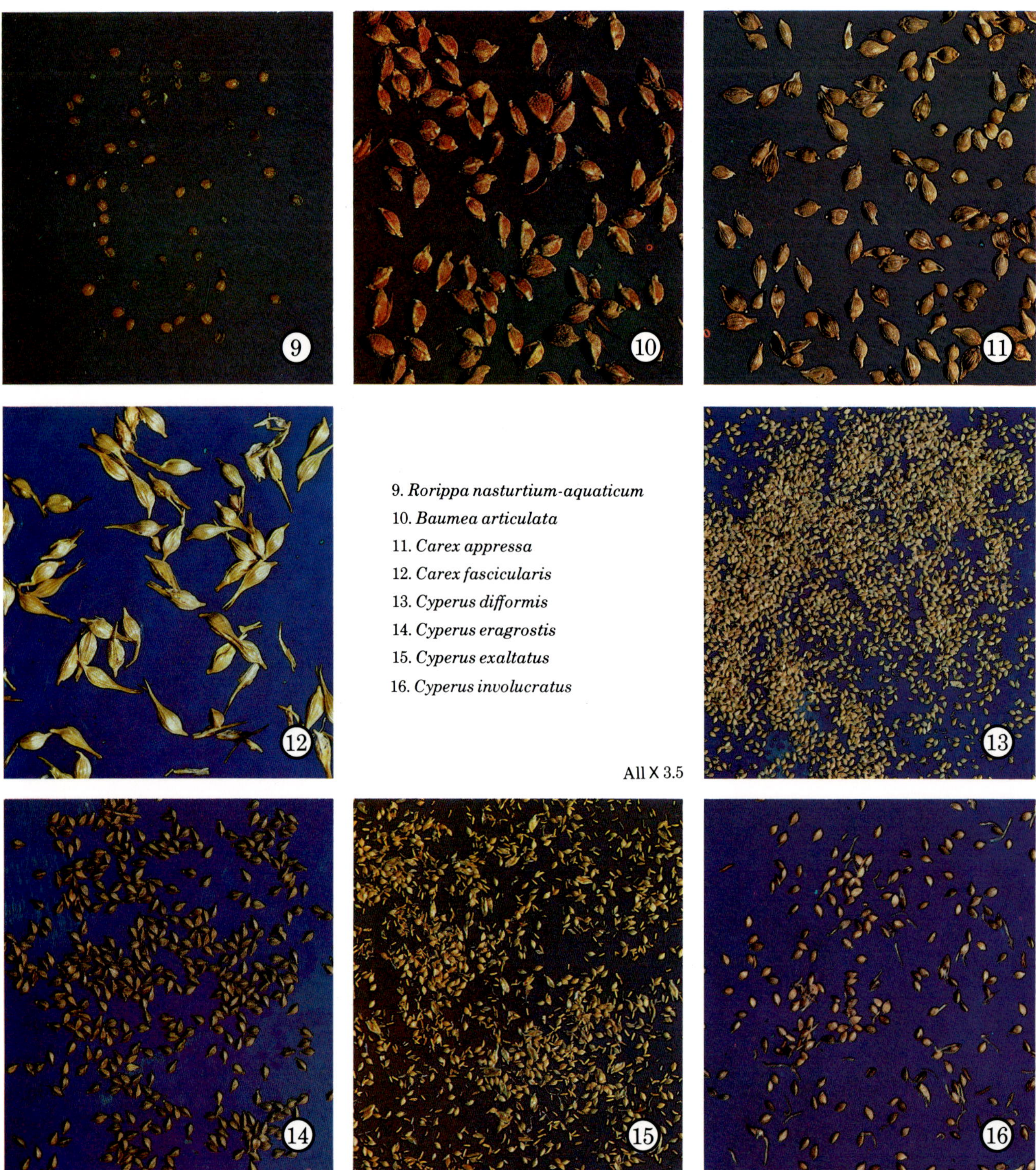

9. *Rorippa nasturtium-aquaticum*

10. *Baumea articulata*

11. *Carex appressa*

12. *Carex fascicularis*

13. *Cyperus difformis*

14. *Cyperus eragrostis*

15. *Cyperus exaltatus*

16. *Cyperus involucratus*

All X 3.5

438

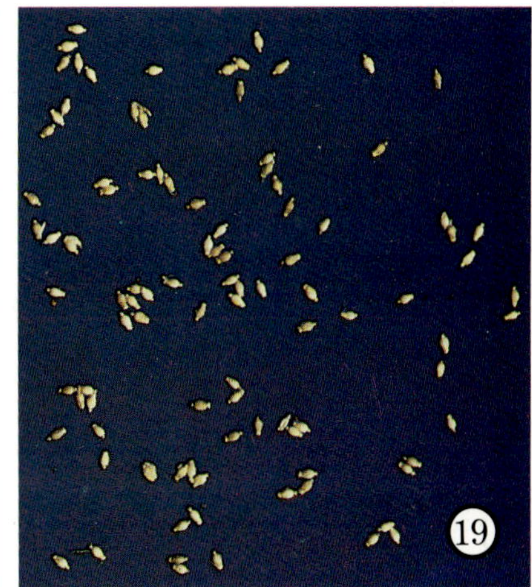

17. *Eleocharis acuta*
18. Spikelets of *Eleocharis pusilla*
19. *Eleocharis* sp.
20. *Gahnia sieberana*
21. *Scirpus fluviatilis*
22. *Scirpus mucronatus*
23. *Scirpus prolifer*
24. *Scirpus validus*

All X 3.5

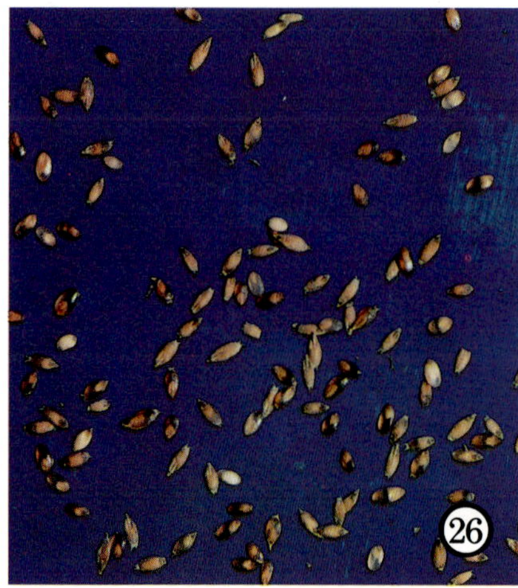

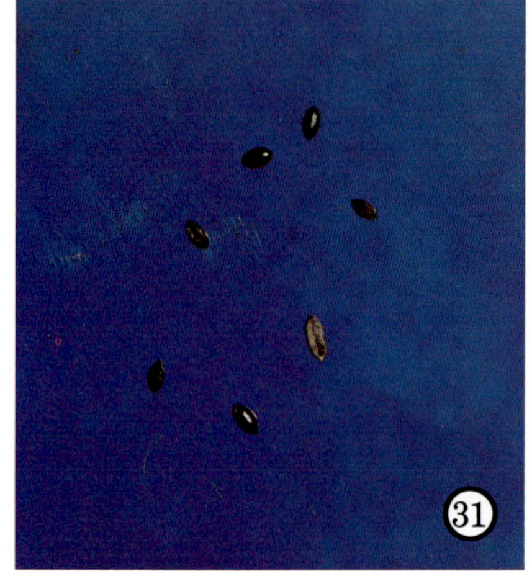

25. Spikelets of *Brachiaria mutica*

26. *Diplachne fusca*

27. Spikelets and florets of *Echinochloa colona*

28. Spikelets and florets of *Echinochloa crus-galli*

29. Spikelets and florets of *Echinochloa oryzoides*

30. Spikelets and florets of *Echinochloa telmatophila*

31. *Glyceria declinata*

32. Spikelets and florets of *Glyceria maxima*

All X 3.5

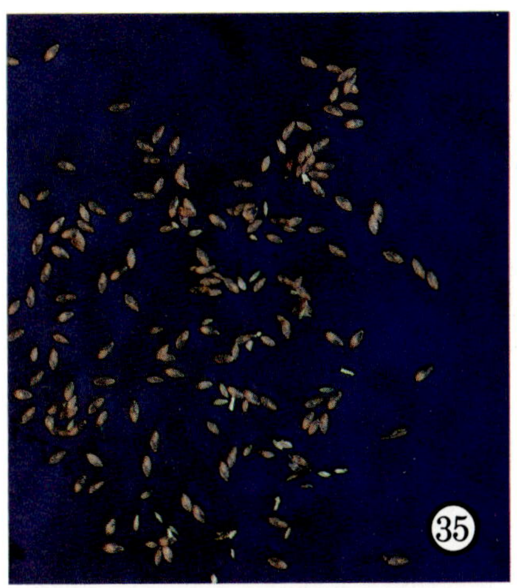

33. Spikelets and florets of *Isachne globosa*

34. Spikelets of *Leersia hexandra*

35. *Leptochloa digitata*

36. Spikelets of *Panicum bisulcatum*

37. Spikelets and florets of *Paspalum dilatatum*

38. Florets of *Panicum paludosum*

39. Florets of *Phalaris arundinacea*

40. *Phragmites australis*

All X 3.5

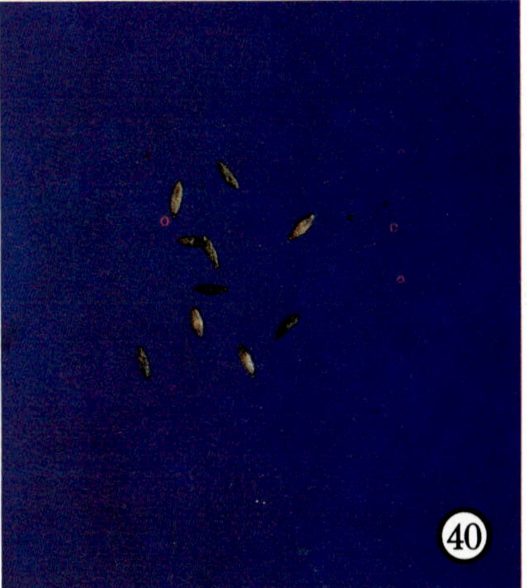

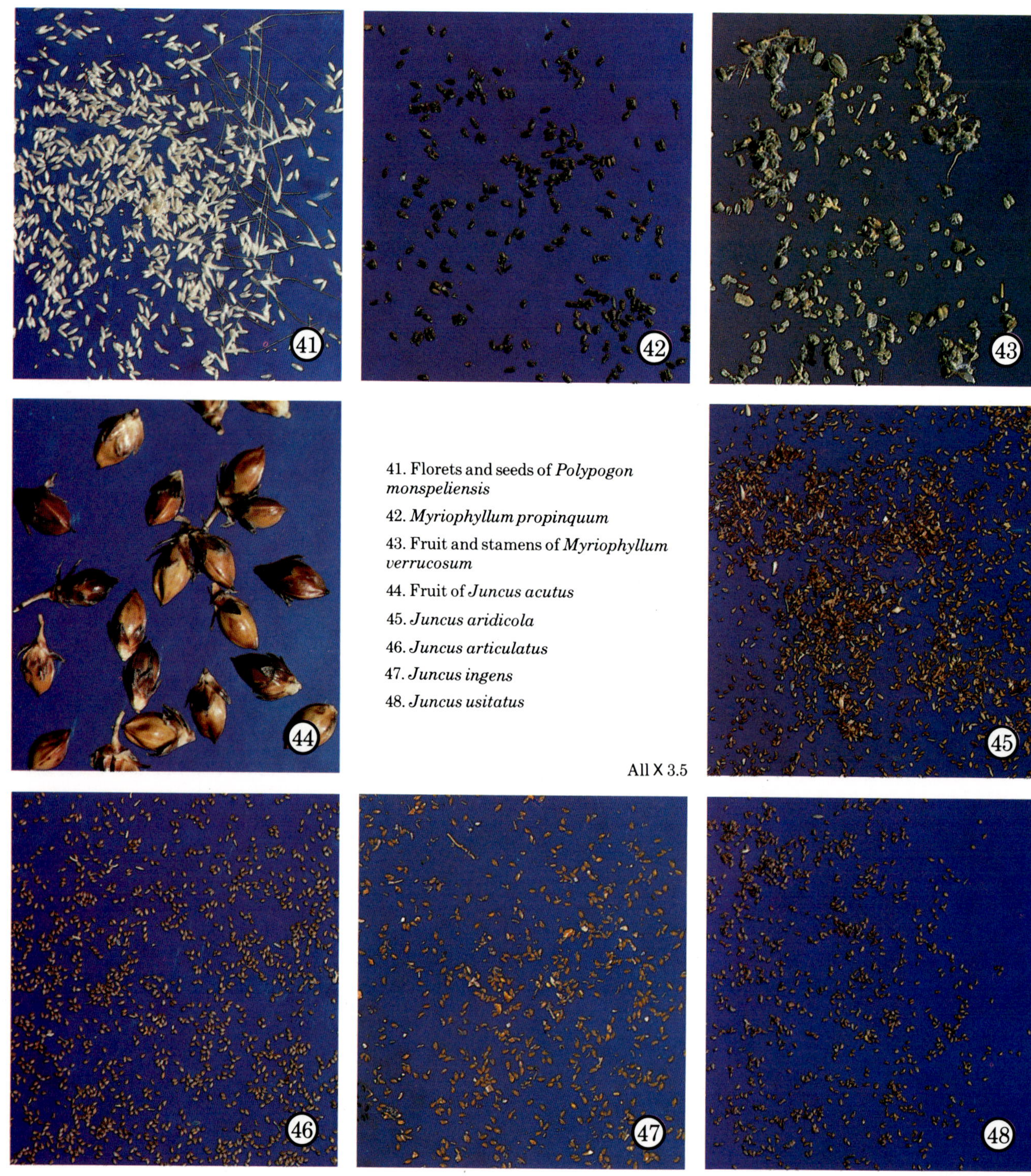

41. Florets and seeds of *Polypogon monspeliensis*

42. *Myriophyllum propinquum*

43. Fruit and stamens of *Myriophyllum verrucosum*

44. Fruit of *Juncus acutus*

45. *Juncus aridicola*

46. *Juncus articulatus*

47. *Juncus ingens*

48. *Juncus usitatus*

All X 3.5

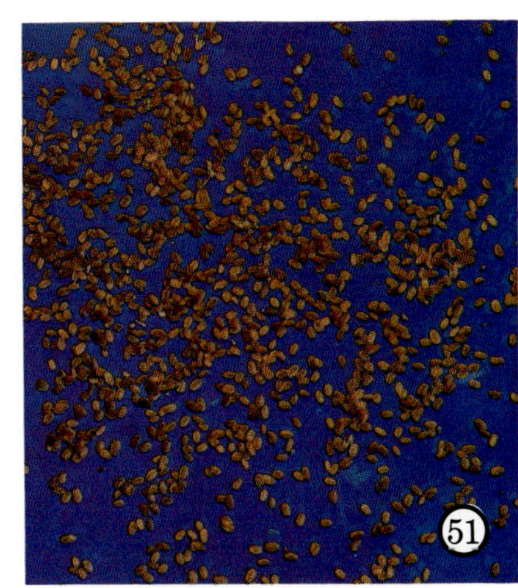

49. Fruit of *Triglochin procera*

50. Fruit of *Triglochin striata*

51. *Lythrum hyssopifolia*

52. Fruit of *Thalia dealbata*

53. *Melaleuca quinquenervia*

54. *Philydrum lanuginosum*

55. *Muehlenbeckia cunninghamii*

56. *Polygonum decipiens*

All X 3.5

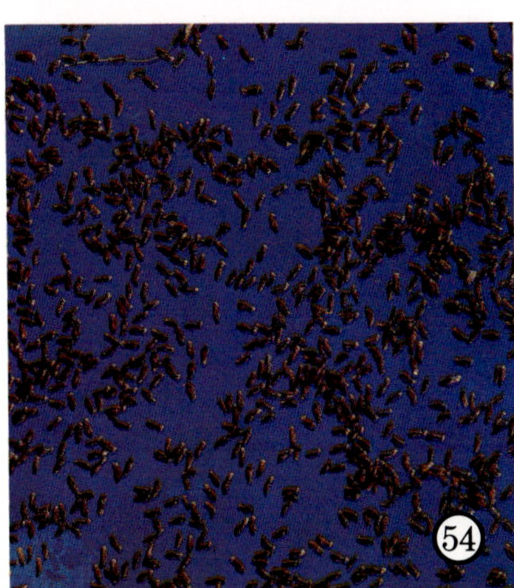

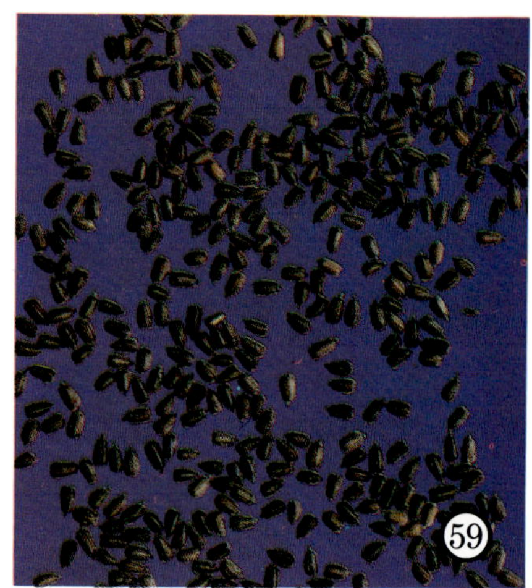

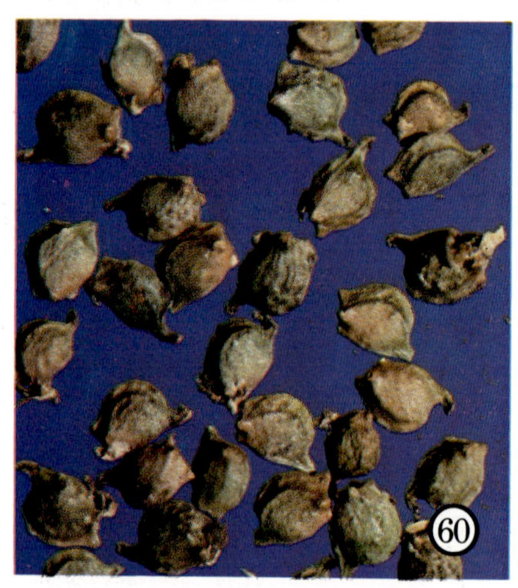

57. *Polygonum lapathifolium*

58. *Rumex crispus*

59. *Eichhornia crassipes*

60. *Potamogeton ochreatus*

61. *Potamogeton pectinatus*

62. *Potamogeton perfoliatus*

63. *Potamogeton tricarinatus*

64. *Ranunculus muricatus*

All X3.5

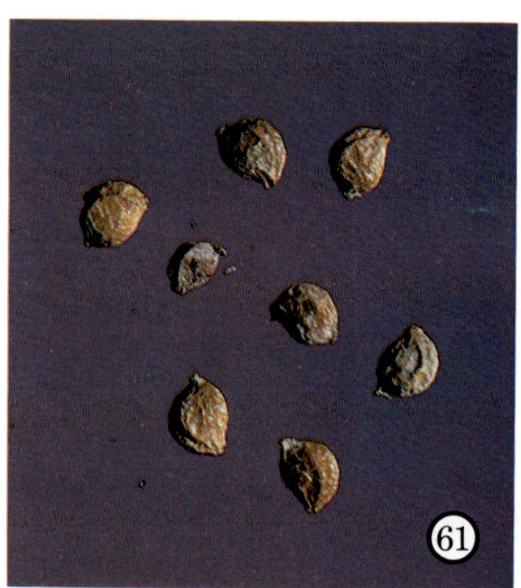

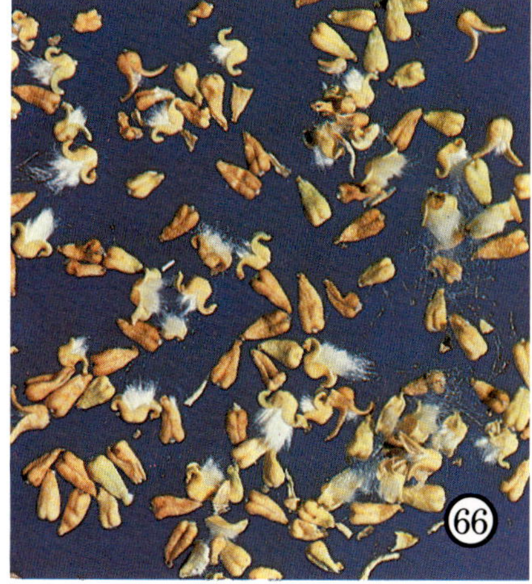

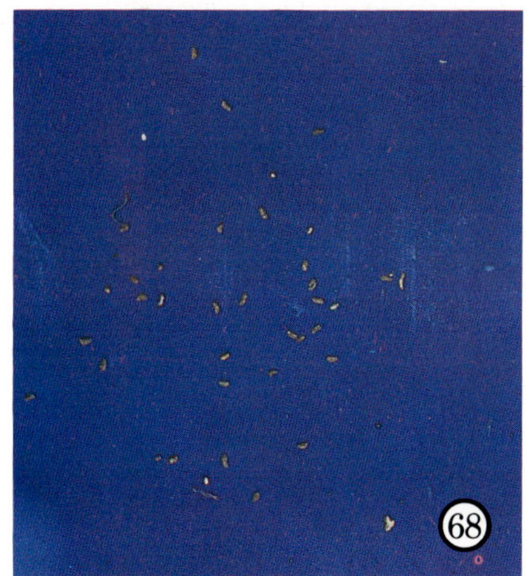

65. *Ranunculus sceleratus*

66. Fruiting flowers of *Salix babylonica*

67. Sterile sporocarps of *Salvinia molesta*

68. *Glossostigma diandrum*

69. *Limosella curdiana*

70. *Veronica anagallis-aquatica*

71. *Sparganium antipodum*

72. Sporocarps of *Marsilea drummondii*

All X3.5

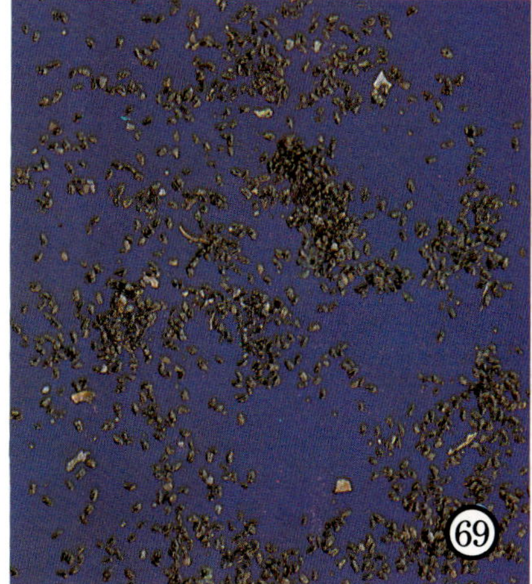
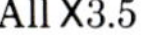
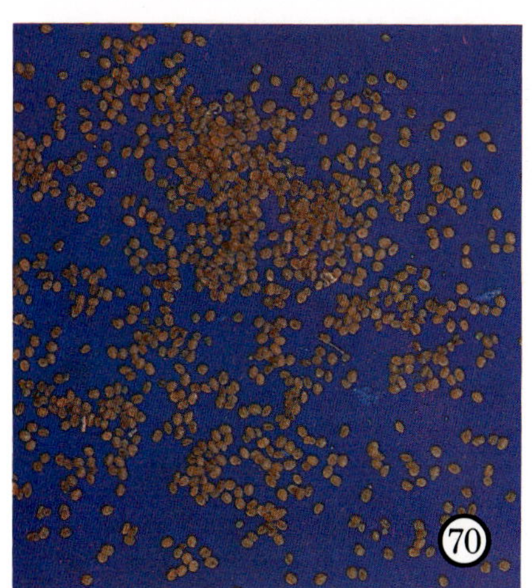

FRESHWATER ALGAE

by Valerie May

Range and Habitat

Algae are simple plants, usually found in water and are extremely varied in size, shape, colour and habitat. They are classified primarily according to pigments, storage products and reproductive structures. Included in the group are small single-celled organisms less than 0.001 mm in diameter, and at the other end of the scale the brown seaweed, ocean kelp, *Macrocystis* known to reach 45 metres in length.

The range of form and structure shown by algae is enormous. Some species are so variable or plastic that forms growing under different conditions have been described as belonging to different genera (*Schizothrix* sp.). Other species, uniform under one set of environmental conditions, separate into distinct forms when grown under a different set of environmental conditions (*Scenedesmus* spp.).

Algae occur in vast numbers floating near the surface of oceans, lakes or rivers, or attached to aquatic plants, stones, shells, logs or concrete structures, etc., in the shallows of these waters.

A wide range of small algae occur on or in soils. Some algal species grow at the edge of hot springs at temperatures rising to over 80°C, others grow in snow or under ice.

Algae also occur on rocks, trees or even on or in certain plants or animals, such as *Anabaena* inside the frond of the fern *Azolla,* and *Basicladia* on the back of freshwater turtles. All lichens are associations of various algae and fungi.

Some algae can break down rocks or penetrate them, while others can secrete lime and so actually form solid deposits. The walls of diatoms, which contain silica, accumulate to form diatomaceous earth.

This section deals with freshwater algae, which are mostly microscopic; only a few genera become large enough for their structure to be apparent to the naked eye. Many unicellular species are mobile throughout most of their life.

There is usually a much larger range of algal species in dams, lakes and ponds than in moving water; this is particularly so when the stored water contains low levels of nitrogen and phosphorus. Where these nutrient levels are high and eutrophication occurs, the range of species decreases, but the number of individuals of certain species may increase. At times, usually during summer, this leads to the development of excessively high numbers of individuals, often with a resultant colouring of the water, and such an increase is known as an algal bloom. During an algal bloom underwater visibility can be reduced to about 10 cm, resembling the effect caused by suspended solids. Storms and flooding can reduce or destroy these blooms, the changes sometimes being dramatic.

Although there is generally a mixture of species of algae in a given situation, one species is often dominant, and in a lake, for example, a series of dominant species can replace each other as the season progresses.

The algae can be classified into about 15 groups, but of these there are four widely distributed groups often found in fresh water. These four groups are:

1. Green Algae (Chlorophyceae or Isokontae)

These plants are mainly grass-green and usually store starch. They include a very large number of diverse aquatic and terrestrial forms. Different species may be motile or sedentary, single-celled or colonial, filamentous in form or sufficiently complex in structure that they appear more like flowering plants.

Common small green algae include species of the genera *Scenedesmus, Staurastrum, Cosmarium, Oocystis* and *Chlorella*.

The larger green algae in natural watercourses and in irrigation systems of New South Wales include the following:

Filamentous Green Algae

These threadlike algae often occur in huge masses. The plants occur either attached or free-floating and can form dense tangled mats in static water or long ropelike strands in flowing water. They may multiply vegetatively by fragmentation and in many cases produce thick-walled microscopic spores, or resting bodies, which survive during unfavourable growing conditions. Examples of this group of algae are:

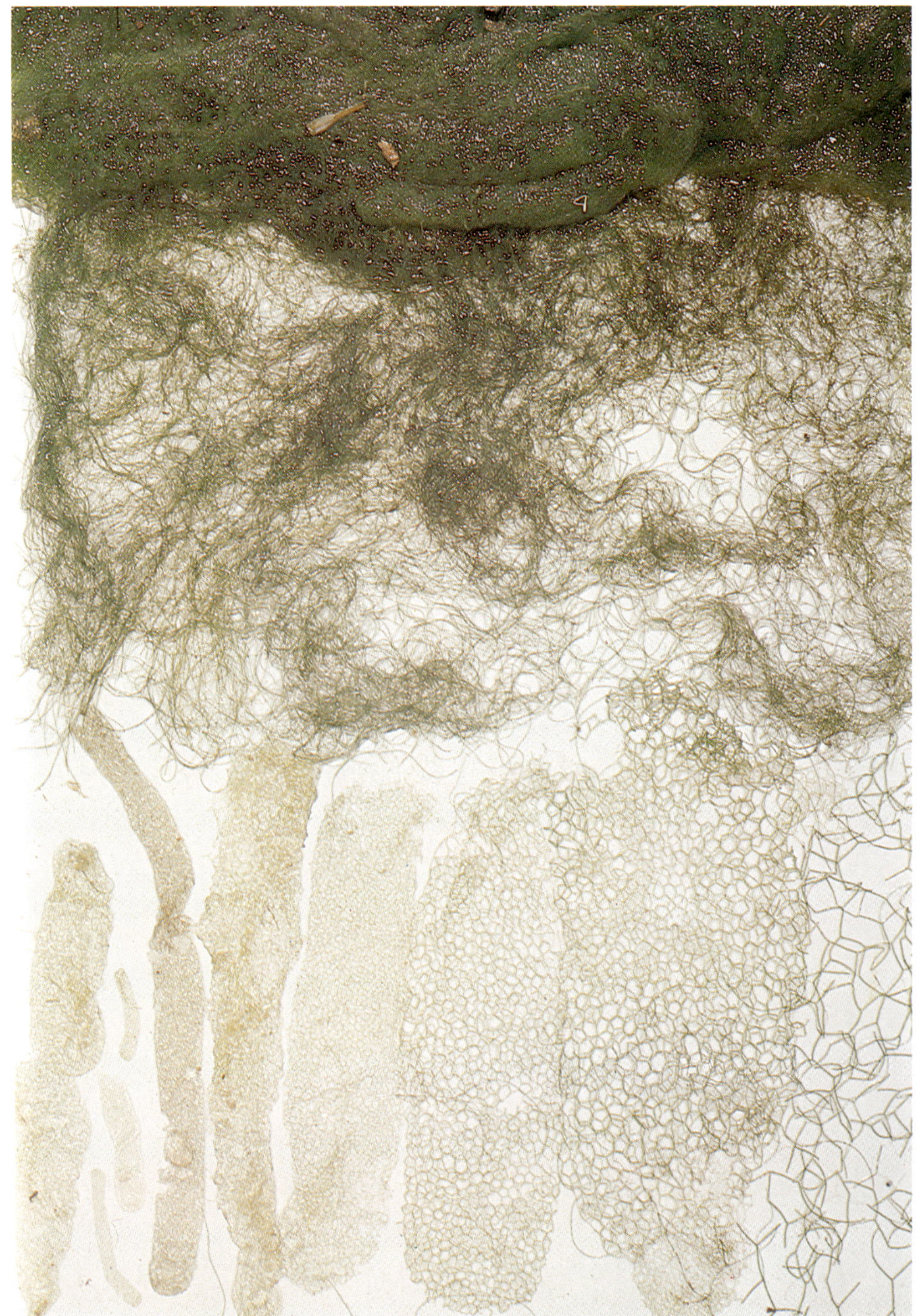

Species of the filamentous algae *Cladophora spiragyra* and Water Net *Hydrodictyon reticulatum* are often found growing together in static or slowly flowing water. The illustration shows a dense mass of *Cladophora,* top and the developing stages of *H. reticulatum,* bottom. Young growth of *H. reticulatum* is on the left, maturing progressively to the right.

Cladophora. The more common species of this genus are coarse and green to dull brownish green, depending on growth stage and the muddiness of the water. The filaments are usually about the thickness of a human hair and are composed of long, cylindrical, thick-walled cells. These filaments are richly branched, at least during times of active growth.

Hydrodictyon (Water Net). The species of this genus usually form greenish-yellow bubbly masses in still or nearly still water. The plant cells are arranged like a small hair net.

Mats of the filamentous alga *Cladophora* in a drainage channel near Hanwood

Nitella sp. (left) and *Chara* sp. (right)

Spirogyra. Species of *Spirogyra* are very common green algae which feel like wet soapy hair, bright green and often found free-floating in static water near the surface, or in masses on the sediment. The cells of the small unbranched filaments have their green chloroplasts arranged in a characteristic spiral.

Stoneworts

These plants are often regarded as quite separate from the green algae. Stoneworts are superficially similar to flowering plants, such as species of *Ceratophyllum* or *Myriophyllum,* so that they are often mistaken for one of these. However, Stoneworts are not flowering plants and their structure is very distinctive, with whorled branches consisting of large individual cells, sometimes covered by a small-celled cortex.

The following genera belong to the stonewort group: *Chara*, sometimes referred to as Muskgrass, is commonly found in lakes and dams and slow-flowing streams where calcium is abundant. Some common species become encrusted with lime so that when the plant dies calcareous accumulation remains intact—hence the name Stonewort. Species of *Chara* are often dark grey-green with "fruits" seen as orange or green pinpoints on the branches. *Chara* often grows on the bottom of lakes in water from one to six metres deep. It is usually noticed more in time of drought, when the water level has dropped.

Nitella also occurs in still or slow-moving water. However, species of this genus thrive under mildly acidic rather than alkaline conditions. The plants are usually greener and less often encrusted with lime than *Chara* spp. Unlike *Chara*, *Nitella* has branches which are themselves repeatedly branched. Both genera may be found in ricefields when the crop is thin.

2. Blue-green Algae (Myxophyceae or Cyanophyceae)

These plants are mainly blue-green and lack the cell nucleus found in all other living organisms except bacteria. Food is stored as sugar or glycogen. Blue-green algae occur as individual cells, filaments, or in colonies. Often they are embedded in mucilage and may be closely associated with bacteria.

Anacystis cyanea on the margins of a lake near Griffith

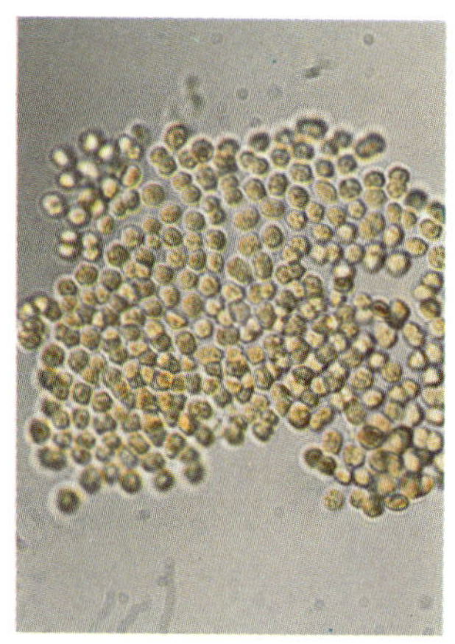

A colony of *Anacystis cyanea* photographed through a microscope *(X1000)*

Blue-green algae can grow and multiply under a wide range of conditions, in arid deserts, lakes, the icy waters of the Antarctic and hot thermal springs. Some species show great resistance to desiccation, being known to survive in air-dry conditions for over 100 years. This characteristic allows many species to be disseminated in airborne dust, or by insects or birds. The effectiveness of distribution of the blue-green algae is further illustrated by the fact that they occur in the pioneer flora of newly formed volcanic islands.

Gas vacuoles, in the cells of many species of these algae, probably allow the plant to move vertically within stored water, thus giving the algae the advantage of access to both high mineral concentrations (near the sediment) and high light concentrations (near the air-water surface).

A species included in this group of algae is the widely distributed *Anacystis cyanea* (also known under its synonym of *Microcystis aeruginosa*). This species has numerous small cells crowded within a gelatinous matrix, forming a colony which may be ovate or an open meshwork *(see photograph)*. Another blue-green species, *Anabaena circinalis,* grows as spirally coiled filaments that are obvious under a microscope, but can also be seen when examined carefully with a hand lens. Both of these species often occur as water blooms, which can be concentrated by wind action.

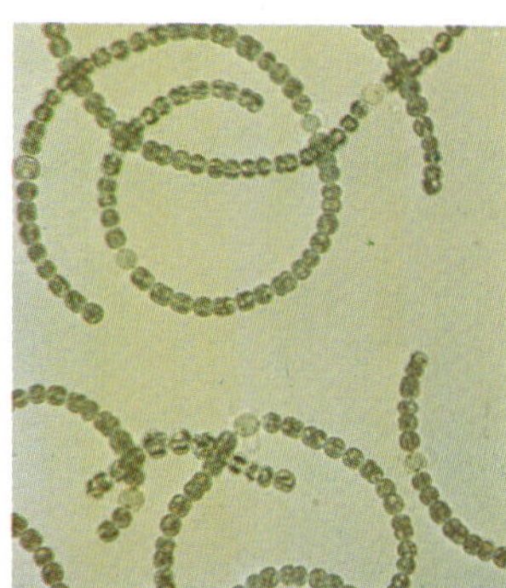

A microscopic view of *Anabaena circinalis (X200)*

Right: *Anabaena circinalis* in the Georges River

3. Diatoms (Bacillariophyceae)

These microscopic algae, often brown or yellow in colour, always have cell walls containing silica. The wall is commonly richly ornamented and is used to classify and identify the many species. Food is stored as fat or volutin. Diatoms are extremely widely distributed and occur as single cells, colonies, free-floating in water, on sediment or attached to aquatic plants, sometimes coating them with a brown slime. They may also be dispersed by wind.

Common diatoms in fresh water in New South Wales are *Navicula* spp., *Cyclotella* spp., and the filamentous colonies of *Melosira* spp.

4. Euglenoids (Euglenineae)

These microscopic plants are usually clear green and store paramylon as food. The euglenoids mostly consist of single cells without cell walls and are nearly all motile, swimming by means of one or two flagella which propel the cell through the water. Many species change shape as they swim about. This specialised group of algae includes some species which are colourless and closely resemble simple animals. Some species are coloured under some conditions but colourless under others. A common and widespread species of the genus *Euglena* can change colour in a few hours, creating either a red or a green scum on the water surface *(see photographs)*.

The Euglenineae occur mainly in fresh water which has a high organic content. Many species can form resting spores to enable them to survive periods when the pool dries out.

Red-coloured *Euglena* in a dam near Narrandera

Right: A close-up of green-coloured *Euglena* in a dam near Yenda

Benefits Derived from Freshwater Algae

Almost all algae are photosynthetic organisms, i.e., they can use light energy to combine carbon dioxide and water, along with added minerals to form organic foodstuffs and oxygen as shown by the equation:

Carbon Dioxide (CO_2)
or   + water (H_2O) + light energy $\rightarrow$ sugars + Oxygen (O_2).
Bicarbonate (HCO_3^-)

These simple sugars are a major energy source for all further chemicals synthesised by the plant.

Hence algae, and other water plants, are the basis of food chains in aquatic ecosystems. Many algae can also be used as food for both stock and people. Algae provide a rich assortment of amino acids for protein production, and also have a high vitamin and mineral content. On the whole they appear to be poor in available carbohydrate, but this is readily and cheaply supplied. Some blue-green algae have proved to be such a rich food source for humans that massive cultures of one (*Spirulina* sp.) are being grown in many parts of the world. The small green algae, *Chlorella* sp., has been studied as a possible foodstuff for use in spaceships. Algae growing in water intended for stock can actually provide nutrients. Hence using an algicide to clear a dense algal growth may remove a source of animal food.

Oxygen given off during photosynthesis is vitally important to aquatic animals and tends to minimise the re-solution of iron and manganese, the production of ammonia and sulphides, and the release of other nutrients, such as phosphorus, from the sediment beneath stored water.

Algal growth assists in purifying polluted water, such as sewerage, by removing nutrients. Such algal growth from sewerage ponds may in time be used as a source of stock food. At present the cost of collecting the faster-growing, smaller algae makes the scheme uneconomic. In some cases algae can also be used to extract toxic substances from waste water.

Some blue-green algae can fix atmospheric nitrogen, producing an effect in soil or water similar to that of the symbiotic nitrogen-fixing bacteria in legumes. The fixed nitrogen has been shown to be highly advantageous to rice production in some areas, especially in less developed countries which use little or no nitrate fertiliser or algicide. In Vietnam ploughing in *Azolla* (a fern which contains the nitrogen-fixing blue-green alga *Anabaena*) has led to an increase in rice yield.

Soil algae improve the organic content and aeration of the soil, and help to prevent erosion. Certain algae are pioneer species, colonising newly exposed rocks or stabilising river banks. There is also some evidence of antibiotic activity by algae.

Colonies of some algae (*Botryococcus*) become embedded in a tough mucous envelope; the smaller colonies can become united into larger ones and a water bloom can develop. The whole structure can yield an oil which, it has been claimed, could prove of use commercially.

Problems Caused by Freshwater Algae

Excessive growth of algae can reduce the aesthetic and recreational values of a water body; it can also trap litter and sediment, thus reducing the holding capacity of a dam. Decaying algae often form unsightly brown to white scums and may cause odours. Cases of nausea from swallowing algal debris have been recorded, as has contact dermatitis and allergic conjunctivitis. Respiration of excessive growth of algae (as can happen at night), or rapid rotting of algae (whether due to natural causes or to algicides), can lower the amount of oxygen in the water and this in turn can cause the death of fish and other aquatic animals. Floating algal mats can reduce the light and so inhibit the photosynthesis of underlying vegetation, leading to further oxygen depletion.

Blue-green algae are often called the "nuisance algae" because they include species which can become dominant and form unwanted blooms in waters where nutrients are plentiful. One probable reason for these algae forming blooms is that they can grow under conditions of low oxygen and low light that often occur in stored water.

A gelatinous mass of a species of blue-green alga, *Nostoc,* floating in a rice field near Benerembah

Certain blue-green algal blooms are toxic. Stock, including cattle, sheep, horses and turkeys, as well as test mice and some wildlife, have been killed by the blue-green alga *Anacystis cyanea*. This is probably the most common toxic alga, occurring in farm dams of many districts in New South Wales. Acute toxicity has also been demonstrated with *Anabaena circinalis* from Burrinjuck Dam, New South Wales, while *Nodularia* sp. has caused stock deaths in other States. Toxins produced by these algae have been blamed for many undetermined stock deaths in Australia. It is likely that the toxins could be dangerous to man and research in the United States of America strongly suggests that this is so. *Anacystis cyanea* and *Anabaena circinalis* occur commonly in New South Wales in water with a pH greater than 6 and with a high phosphorus level, i.e., in polluted water. They develop mostly in late summer or autumn, but may be found at any time.

Taste and odour problems can be caused by certain algae in storage water. Fishy and other tastes or odours have been attributed to species of *Synura, Cyclotella, Asteroniella, Euglena, Dinobryon, Aphanizomenon* and *Oscillatoria*. Species of the latter genus have been reported as imparting a muddy taste to fish. In 1973 a massive bloom of *Anabaena circinalis* in Burrinjuck Dam created an odour similar to that of the gamma isomer of benzene hexachloride (BHC) commonly used in grasshopper control. The odour could be detected 5 kilometres downwind from the dam water and was noticeable both at the height of the bloom and during its decomposition.

A rice crop near Warrawidgee with an algae slime containing species of *Cladophora, Melosira, Microcoleus, Nostoc, Oedogonium, Oscillatoria, Spirogyra, Zygnema,* and *Zygnemopsis*

Water flow can be severely impeded by algae when they grow as slimes or encrustations on rocks, pipes, concrete, or more commonly when they form floating mats often entangled in other aquatic plants. They may reduce or block flow in pipes, irrigation channels, siphon tubes, pump inlets and orifices of trickle irrigation channels and diatoms such as a species of *Melosira* may block sand filters in water works.

Some species of blue-green algae form a mat on soil, clogging the soil-air surface with a resultant soil-oxygen depletion. Such mats (e.g. of *Microcoleus* sp.) also retain moisture around the stems of young potted nursery plants providing a suitable habitat for the pathogenic "collar-rot" fungi. Another clogging effect of mats formed by algae occurs in ricefields, where excessive algal growth can weigh down the leaves of young rice plants, preventing them from emerging above water.

Water Management

Algal blooms have been recorded since the time of Exodus and have always been considered by man as an ill omen. However, it is only since agricultural practices and pollution have raised the nutrient levels (particularly of phosphorus) of so much water that blooms of toxic species have become a widespread problem.

Preventing the excessive growth of nuisance algae is preferable to treating an existing problem.

Enclosing storage tanks to eliminate light will stop algal growth. Roofing has been effective in some cases. Cheap methods suggested for reducing light to stored water include the use of black plastic covers, activated carbon or even masses of floating black table tennis balls. Where light exclusion is not practical, algal growth can be controlled by limiting the nutrient supply. To do this, large amounts of organic matter, phosphorus and nitrogen should be kept out of any water storage. It is more feasible to control phosphorus than nitrogen entering a given system, since nitrogen can be fixed and so added to the available nutrients by blue-green algae such as *Anabaena*. Before a reservoir is filled, all vegetation should be cleared from below the high water mark and either removed or burnt and buried. If weeds in a storage are cut, they should also be removed.

Additional sources of nutrients which could be controlled include run-off or direct contamination from grazing animals, dairies, piggeries, stockyards, abattoirs, feed lots, bird rookeries or colonies, urban houses, gardens and streets, sewerage outfalls and septic systems, agricultural fertilizers, disturbed soil and decaying vegetable matter and perhaps recreational activity. Bushfires in the catchment area may lead to increased nutrient levels in the reservoir, especially if heavy rain occurs soon after the burn. Sediments in water storages release soluble nutrients (particularly phosphorus), especially when there are low levels of oxygen in the water.

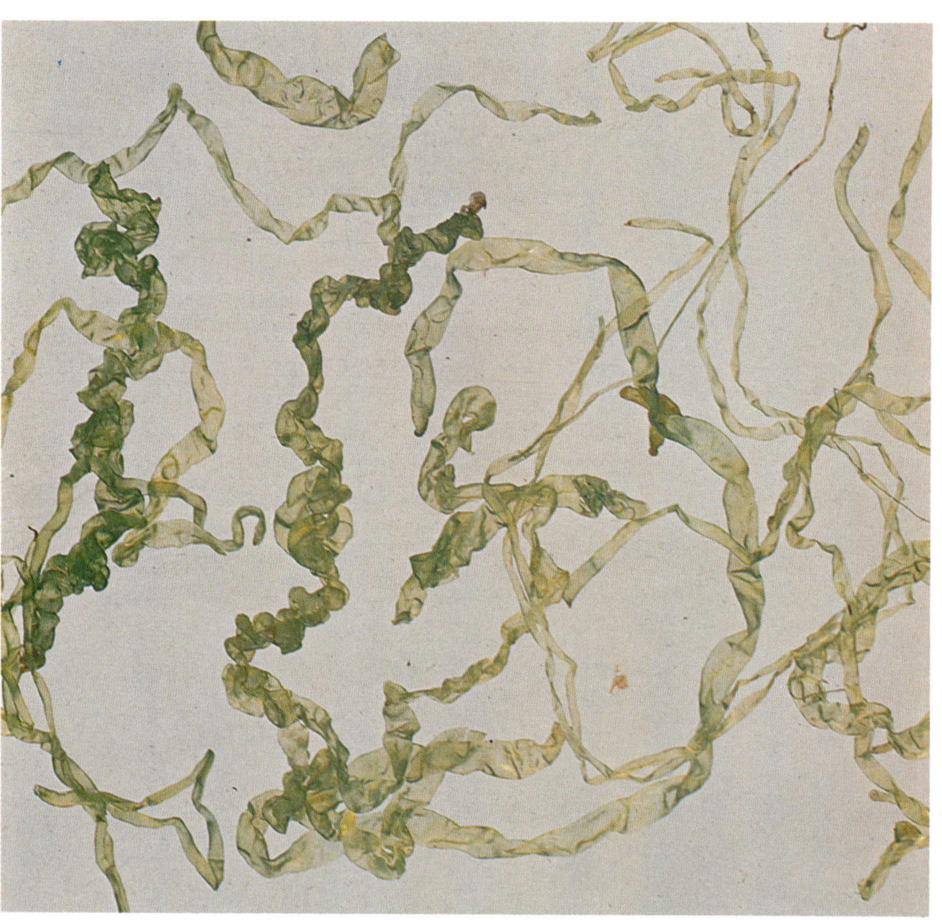

Enteromorpha is a green alga capable of tolerating a wide range of conditions and is often indicative of high levels of nitrogen

It may be feasible to divert nutrient inflows into a lake, to a locality where the nutrients will do less harm. This is usually expensive, but effective after a variable time lag, allowing the water body to recover from the nutrients already present in the sediment. The sediment could, of course, be sealed off, cleaned, or removed.

Since a high concentration of phosphorus is a major factor in the development of blooms of algae, it is advisable to store superphosphate away from the catchment area of any farm dam. It is also wise to keep a verge of unstocked land adjoining any large dam or reservoir, and, where possible, to use troughs for stock water supplies in order to reduce the contamination from animal waste.

Infestations of toxic algae in small farm dams have been either prevented or reduced by the addition of block ferric alum (registered trade name alumina ferric: $100mg/L = 1kg/10^4L$) prior to the summer growing season.

The alum has been shown to reduce the peak seasonal production of phosphorus in the dam water. Alum treatment for the removal of nutrients is suitable for small dams. The corresponding treatment for large reservoirs is aeration of the deeper waters. This increases the oxygen content of the underlying layers of water and so decreases the re-solution of nutrients (such as phosphorus) from the sediment, thus reducing the stimulus towards excessive algal growth. The initial cost of this equipment and its installation usually prohibits its use for farm dams.

Periodic removal of excess plants or animals growing in the storage is a long-term method of removing nutrients from the system. If there is an adequate renewable supply of fish, their removal by public fishing is likely to be popular! In any case, physical removal of massive plant growths (either algal or angiosperm) is effective and not likely to lead to other ill effects. The technique of using plants to remove nutrients from a water supply is often the basis of sewerage treatment pondage recycling (*see page 512*). If, despite precautions, an unwanted algal growth occurs, mechanical clearing is often sufficient treatment. The use of a mesh guard or filter at the pump intake pipe may prevent algae entering the pump and storage tank.

In some cases drawing off water from a different level of the stored water is all that is needed.

Algicide treatments (i.e., using herbicides for algal control) should not be attempted unless absolutely necessary as most algal growth is advantageous. Many algicides have some side effects, so they should be used only when the alga present is known to be a significant problem and immediate control is necessary. There is always the risk of killing other organisms either *in situ* or downstream.

Blooms of blue-green algae, diatoms and other algae can generally be controlled by copper sulphate (bluestone) at a dose rate of about one part per million. Larger filamentous green algae are usually controlled by a dose of up to 19 ppm, preferably applied in early stages of growth. In flowing water the rate needed varies from 1 to 26 ppm depending on the duration of the injection, the state of the water and its rate of flow. As copper sulphate is corrosive to galvanised iron pipes and tanks, caution is needed in its use. A level of 1 ppm should not be exceeded where water is to be used for stock or domestic purposes. When applying an algicide the volume of the water to be treated should be calculated carefully and the dose measured accurately. In dams, copper sulphate is best applied by spraying a concentrated solution evenly over the water surface. An alternative method is to place the copper sulphate in hessian bags and drag these through the water, ensuring adequate distribution over the infected part of the dam.

Copper sulphate rapidly kills the algae and the residual copper normally is precipated to the bottom. Precipitation is rapid in hard water and slower in soft, hence there is a variable time lag before regeneration of the bloom is likely to occur.

Other algicides are available, but these are usually less satisfactory and cost more. Expert advice should be sought on their use and label recommendations followed carefully. Many instances have been recorded of algicides damaging other plants, fish, etc. The addition of chlorine to domestic water supplies with excessive algae may worsen a problem due to the formation of chloro-organic taste-bearing compounds.

Work is still proceeding to test the effectiveness of viral, fungal or bacterial control of algal blooms.

After any massive killing of algae, resultant decay and bacteriological breakdown may cause taste, odour and colour problems, while the resultant shortage of oxygen may lead to fish-kill or other troubles, and the concurrent release of mineral nutrients can stimulate further growth of aquatic plants. The usual procedure to reduce these problems—e.g. killing limited areas of unwanted plants at intervals of about 10 days—is not applicable when the plant being treated is free-floating.

Further Reading

DESIKACHARY, T.V. (1959) "Cyanophyta." (Indian Council of Agricultural Research: New Delhi) 686 pp.

FRITSCH, F.E. (1935) "The Structure and Reproduction of the Algae." Vol. 1. (Cambridge University Press: Cambridge) 792 pp.

PRESCOTT, G.W. (1962) "Algae of the Western Great Lakes Area." (Wm C. Brown: Dubuque, Iowa) 977 pp.

SMITH, G.M. (1950) "The Fresh-water Algae of the United States." 2nd ed (McGraw-Hill Book: New York) 559 pp.

A simpler reference is:

PRESCOTT, G.W. (1964) "How to Know the Fresh-water Algae." (Wm C. Brown: Dubuque, Iowa) 272 pp.

About the Author

Valerie May is an algologist with long and varied experience in the study of Australian algae, both marine and freshwater.

She trained at Sydney University, where she was awarded a B.Sc.(Hons 1) and later a M.Sc. degree. After some years on research scholarships and a Linnean Macleay Fellowship she joined the staff of CSIRO, working at the Division of Fisheries and Oceanography.

Valerie now works at the National Herbarium of New South Wales, lately specialising in the distribution and control of toxic algae in inland waters.

CARP, CRAYFISH AND WATER RAT

by Karl Shearer

Four inland animals commonly seen in irrigation systems are regarded as pests; two fishes, the Common Carp *Cyprinus carpio,* and the Goldfish *Carassius auratus;* a crustacean, the Yabby *Cherax destructor;* and a small mammal, the Water Rat *Hydromys chrysogaster.* These animals are regarded as pests because they reputedly either damage irrigation channels and fixtures or reduce the quality of the water these channels carry.

Goldfish and Carp

The two fishes are members of the cyprinid family, of which there are no native representatives. They are typically found in shallow, warm, stagnant or slow-moving water and prefer areas with moderate cover, which is usually provided by aquatic plant growth or overhanging vegetation. The two introduced fishes are typical members of the family.

The Goldfish was introduced by European migrants late last century, probably as an ornamental fish, as it has poor eating qualities. Larger specimens can often be captured by rod and reel but they are not regarded as an angling species. The Goldfish exhibits a wide range of colour and red, black, olive, yellow and grey individuals can often be found in a single school. Variegated specimens are not uncommon. Goldfish can reach 30 cm or more in length and weigh more than a kilogram, but seldom exceed 15 cm.

Although Goldfish have been in the irrigation system for many years, concern about the possible damaging effects of this species on irrigation channels and rice crops was minimal until its larger relative, the Common Carp, began appearing in large numbers during the early 1970's.

Records of the Common or European Carp in New South Wales date from over 100 years ago, but until recently they occurred only in small numbers. Two varieties of Common Carp are present in the irrigation system, the red or orange Carp and the more recently introduced silver-grey variety. It is the second variety that has caused concern. Evidence indicates that it was introduced from Europe into south-eastern Victoria in about 1960, but it is now quite common throughout the Murray-Darling drainage system.

FISHES

Carp   *Cyprinus carpio*   "Yanco"

Carp   *Cyprinus carpio*   "Prospect"

Goldfish   *Carassius auratus*

Top: *Cyprinus carpio* "Yanco", the red variety of the Common Carp
Centre: *Cyprinus carpio* "Prospect" the silver-grey variety of the Common Carp
Below: *Carassius auratus* the Common Goldfish

Each fish photographed approx. 15 cm long

The two pairs of barbels or whiskers on the lower jaw of the Common or European Carp readily distinguish it from its relative the Goldfish. Carp can also grow to a much larger size and fish longer than 1 metre have been reported.

Carp normally breed once each spring but during some seasons they have been known to breed two or three times. The eggs, up to several million, are scattered and rapidly adhere to aquatic vegetation. Hatching takes place in about 5 days. A possible method of Carp control is to temporarily reduce the water level in a lake or channel, as the eggs die when exposed.

The feeding habits of the Carp and the Goldfish are the cause of any reputed damage. Carp feed by taking up bottom sediment in their mouth and expelling the mud back into the water. Heavy materials such as stones and sticks settle and the Carp can then select small organisms which are left suspended in the water. This feeding habit has two effects; first, the turbidity or muddiness of the water increases and second, silt on the bottom of irrigation channels tends to be carried downstream after having been dislodged. Increased turbidity also reduces the light that can reach aquatic plants, and in many cases reduces their growth or even eliminates certain species. In some regions Carp have reduced water quality and also the ability of an area to support other aquatic life and waterfowl, but generally these effects have been exaggerated. Several government agencies have programs in progress to assess the impact of Carp, and it may be proved that in certain areas some form of control is justified.

Although Carp are a nuisance in many situations they are of some value. Hundreds of tonnes are caught commercially each year and processed into pet food or used as bait in crayfish pots. The fish also provides good angling on light tackle and they are good eating if properly prepared. The muddy taste can be removed if fish are bled after being caught and then soaked overnight in fresh water to which salt has been added. They can then be baked or poached, and are also excellent smoked or dried. The numerous small bones present a problem, but larger specimens can be filleted.

EUROPEAN CARP

Further Reading

LAKE, J. S. (1978) "Australian Freshwater Fishes". (Thomas Nelson: Melbourne) 160 pp.

SHEARER, K. D. and MULLEY, J. C. (1978) The Introduction and Distribution of the Carp, *Cyprinus carpio* Linnaeus, in Australia. *Australian Journal of Marine and Freshwater Research* 29:551-563.

VICTORIA. FISHERIES AND WILDLIFE DIVISION (1976) "A Proposal to Assess the Impact of European Carp on Fish and Waterfowl". *(Victoria. Fisheries and Wildlife Division, Melbourne).*

WHARTON, J. C. F. (1971) European Carp in Victoria. *Fur, Feathers and Fins* No. 130:3-11.

WHARTON, J. C. F. (1977) Impact of Exotic Animals, especially European Carp, *Cyprinus carpio,* on Native Fauna. *Fisheries and Wildlife Paper, Victoria* No. 20.

WHITLEY, G. P. (1951) Introduced Fishes. II. *Australian Museum Magazine* 10:234-238.

Left: The silver-grey variety of *Cyprinus carpio* (Common or European Carp) in a supply channel near Murrami

THE CRAYFISH OR YABBY

Male

Female carrying young

Dorsal view of male

Cherax destructor
Top left, ventral view of male; top right, ventral view of female; bottom, dorsal view of male.

Crayfish or Yabby

The Yabby has been accused of causing erosion of levee banks and channel stops. Large animals can dig burrows up to 10 cm in diameter and these may be several metres in length, but are usually less than one metre long. Under normal circumstances a burrow this size should not damage a well constructed bank. The Yabby is ideally suited to the irrigation system because it can tolerate high water temperatures and low oxygen and can actually walk away from areas where conditions are unfavourable.

Yabbies reproduce in the spring and summer and during some seasons may breed more than once. Up to 600 eggs are produced by a female during each breeding. Yabbies can live for more than six years and reach a size large enough to be eaten at about two years. They can be easily sexed by noting if the genital pore is located at the base of the last walking leg (male) or the third last leg (female). Yabbies often lose a leg, which may regrow in a few weeks.

Although many people in inland areas are familiar with the eating qualities of the Yabby, it has not yet received the commercial interest it deserves. Export of Yabbies to overseas markets has now exceeded 100 tonnes, and demand is still increasing.

When preparing Yabbies they should be placed into boiling water containing a generous amount of salt. They should boil for at least 15 minutes and then be placed directly into cold water, preferably ice water. With the price of prawns and marine crayfish increasing, the irrigation areas of inland Australia offer a cheaper, readily available alternative.

Further Reading

HALE, H. M. (1927) The Crustaceans of South Australia. (South Australian Museum: Adelaide)

MORRISSY, N. M. (1974) The Ecology of Marron. *Cherax tenuimanus.* Fisheries Research Bulletin of Western Australia 12:5-55

RIEK, R. F. (1969) The Australian Freshwater Crayfish (Crustaceae: Decapoda: Parastacidae) with Description of New Species. *Australian Journal of Zoology* 17:855-918.

WILLIAMS, W. C. (1968) "Australian Freshwater Life. The Invertebrates of Australian Inland Waters". (Sun Books: Melbourne) pp. 147-150.

WATER RAT

Hydromys chrysogaster
("Water Rat")

468

Water Rat

The Water Rat is one of few native Australian mammals that have adapted to a semi-aquatic life. They can co-exist with man, adapting well to habitat changes. The animal reaches a maximum size of about 60 cm and one kilogram. They are now protected, but large numbers have been killed in the past for the excellent quality of their fur.

The diet of the Water Rat consists mainly of fish and insects, and in the irrigation system they feed mainly on Carp and English Perch (Redfin). They will also eat Yabbies and Shrimps and have been known to place Mussels in the sun until they open and can be eaten.

Water Rats can construct large burrows in levee banks and in some cases have caused damage. In situations where they are suspected of causing damage, the local wildlife authorities should be contacted.

Further Reading

McNALLY, J. (1960) The Biology of the Water Rat *Hydromys chrysogaster* Geoffroy (Muridae: Hydromynae) in Victoria. *Australian Journal of Zoology* 8:170-180.

TROUGHTON, Ellis (1948) "Furred Animals of Australia". (Angus & Robertson: Sydney) pp. 266-273.

WOOLLARD, Penny, VESTJENS, W. J. M. and MACLEAN, L. (1978) The Ecology of the Eastern Water Rat, *Hydromys chrysogaster,* at Griffith, N.S.W. Food and Feeding Habits. *Australian Wildlife Research* 5:59-73.

About the Author

Karl Shearer is a native of Seattle, Washington, U.S.A., and a graduate of the University of Washington College of Fisheries. He came to Australia on a brief working holiday in 1971 and stayed until 1978.

During the whole of his stay in Australia, Karl worked with the New South Wales State Fisheries, mainly at the Inland Fisheries Research Station, Narrandera. Two of his most significant projects at Narrandera were a study of carp and the development of a successful method of boosting production of native fish for the stocking of farm dams and other impoundments.

Corbiculina australis, the Little Basket Shell. Immature forms can block sprinklers

LITTLE BASKET SHELL
Corbiculina australis

This bivalve has been reported blocking sprinkler nozzles in the Murray Valley. A build-up of these animals in pipelines also increases frictional resistance to water flow and consequently reduces output and increases pumping times and hence pumping costs.

Information on the biology of *Corbiculina australis* is limited. It is known that the adult measures up to 2 cm long while young bivalves may be smaller than 1 mm. The young forms readily pass through normal mesh filters and settle and grow in mud deposits inside the pipes; they may also reproduce there.

Overhead irrigation systems normally use nozzles with apertures in the 3 mm to 6 mm range and mesh filters with about 2 mm apertures.

Unfortunately, a filter fine enough to exclude the young bivalves would become excessively clogged due to the quantity of fine debris which passes through the standard filter and sprinkler outlets.

Flushing the pipeline with water at high velocity will remove many of the bivalves, but this offers only short-term relief from blocking problems.

Copper sulphate at 50 ppm (though lower levels may have been equally effective) for 24 hours has been used to control the bivalves. Copper sulphate will also corrode iron pipes and fittings and will contaminate water used for domestic and stock purposes. Potable water should not exceed 1 ppm copper. If domestic water supply pipelines are treated, it is necessary to isolate consumers from treated lines. After treatment is completed, flush the treated lines with fresh water before reconnecting consumers. Consult with local authorities to determine if it is permissible to use copper sulphate.

Further Reading

McMICHAEL, D. F. (1976) Australian Freshwater Mollusca and their Probable Evolutionary Relationships: A Summary of Present Knowledge. In "Australian Inland Waters and their Fauna". (Weatherley, A. H., Ed.) (Australian National University Press: Canberra) pp. 126-127.

SMITH, B. J. and KERSHAW, R. C. (1979) "Field Guide to the Non-Marine Molluscs of South-Eastern Australia". (Australian National University Press: Canberra).

WILLIAMS, W. D. (1968) "Australian Freshwater Life. The Invertebrates of Australian Inland Waters". (Sun Books: Melbourne) pp. 65-70.

Species of *Gallionella* and *Leptothrix* appearing as red sludge in a small spring on the edge of a lake in Centennial Park, Sydney

IRON AND MANGANESE BACTERIA

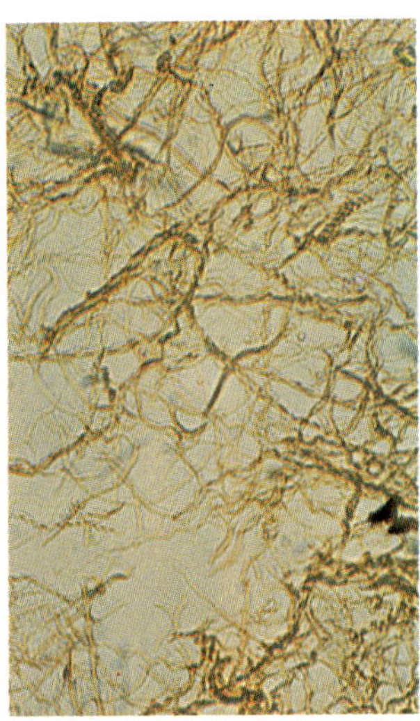

Iron bacteria photographed through a microscope (X2600)

Iron and manganese bacteria usually appear as a brown sludge or ochre in clay, steel or plastic pipes, streams or in the soil. These bacteria oxidise ferrous or manganous ions and the resultant insoluble compound is precipitated on or in their mucilaginous secretions. The sludge is made up of numerous microscopic tubules which remain after the bacteria have died. These bacteria grow attached to the lining of pipes or other surfaces in contact with a water supply containing low concentrations of iron or manganese. Under certain water conditions iron compounds may also be oxidised chemically without the aid of bacteria.

Under special circumstances these bacteria can cause problems by:

(i) depositing sludge in subsoil drainage pipes;
(ii) blocking water pipes and filters;
(iii) discolouring and tainting water supplies.

Control measures using chemical and mechanical cleaning techniques have been tried with only limited success.

Further Reading

CULLIMORE, R. D. and McCANN, A. E. (1977) The Identification, Cultivation and Control of Iron Bacteria in Groundwater. In "Aquatic Microbiology." (Skinner, F. A. and Shewan, J. M. Eds) (Academic Press: London) pp. 219-257.

HUTCHINSON, M. and RIDGWAY, J. W. (1977) Microbiological Aspects of Drinking Water Supplies. In "Aquatic Microbiology" (Skinner, F. A. and Shewan, J. M. Eds) (Academic Press: London) pp. 179-218.

IVARSON, K. C. and SOJAK, M. (1978) Microorganisms and Ochre Deposits in Field Drains of Ontario. *Canadian Journal of Soil Science* 58:1-17.

TYLER, P. A. (1970) Hyphomicrobia and the Oxidation of Manganese in Aquatic Ecosystems. *Antonie van Leeuwenhoek* 36:567-578.

BIOLOGICAL CONTROL

by K. L. S. Harley

Weeds very often are plants that are not a problem in their native range, but which have increased to abnormally high proportions following their introduction into another geographic region. An important reason for plants proliferating and becoming weeds in this way is that in being moved from one region to another they have been freed from the controlling effects of their natural predators. Biological control programs seek to re-establish the balance between a plant and its natural predators.

Safe and effective biological control needs extensive and thorough research. The major steps in developing a program may be summarised as follows:

1. Thorough studies of the taxonomy and ecological characteristics of the target weed both where it is a pest and in its native range.

2. Exploration for natural enemies in the native range of the weed.

3. Study of the ecology, biology, and host specificity of natural enemies which show potential as control agents.

4. Final selection of natural enemies for liberation as control agents.

5. Mass propagation, liberation, and establishment of control agents.

6. Study of the effects of these agents on the target weed.

Prior to liberation, potential control agents are thoroughly studied to ensure that they are safe and that they meet the stringent requirements of Australian quarantine regulations.

Weeds and their natural enemies both may occur as a number of biotypes, extending over a broad geographic range and embracing a variety of climate types.

To maximise the effectiveness of a biological control program it may be necessary (i) to match the biotype of the control agent with the biotype of the target weed and (ii) to collect potential control agents in a region climatically similar to that in which the weed is to be controlled. However, close matching of biotype and climatic region is not always possible and it may be more useful to collect a particular control agent from localities scattered throughout its natural range. Widespread collection of a control agent may result in the importation of a number of biotypes with potentially wide genetic variability. Natural selection operating on such a gene pool, together with possible recombination of genetic material, may produce a biotype particularly suited to conditions in the region of introduction.

In the process of introduction, control agents are freed from the effects of their own controlling agents. This results in increased survival and, with an initially unlimited food supply, the population of control agents may explode and exert a high level of control over the target weed. As the abundance of the weed is reduced, the population of a control agent falls in proportion to the available food supply.

There are of course other factors influencing the weed/control agent interaction which may affect the degree of weed control achieved. Some of these factors may be amenable to manipulation but the majority are not and the most effective control is achieved by selecting agents best adapted to the overall environment.

Biological control cannot solve all weed problems. However, it may be applied in a vast array of situations where it is by far the most satisfactory and economical form of control. Biological control can reduce the infestation of a weed to an acceptable level and produce a permanent balance between the target weed and its introduced natural enemies. It has no undesirable side effects.

Though biological control programs may take a number of years to implement, their costs are low when account is taken of the fact that the control achieved is usually permanent. Hence, benefits accrue indefinitely from successful biological control programs.

Aquatic weeds

In Australia the majority of the more serious aquatic weeds are introduced plants. Released from their natural enemies, these weeds have grown virtually unchecked and are drastically altering many of our relatively scarce freshwater aquatic habitats. They overwhelm native aquatic plants which are often essential parts of the environment for native wildlife, limit the use of water and water bodies, reduce water quality, increase health hazards, increase water loss through evapotranspiration, and increase silting and the severity of flooding.

Some control is being effected by application of herbicides, but the cost is high (around $6 million per annum), the problems are ever-recurring, and the continual pollution of the aquatic environment is highly undesirable. Biological control is a much more desirable method.

Insects were first used for the biological control of terrestrial weeds over 100 years ago. However, it was not until 1964, when the Alligatorweed Flea-beetle was liberated in the United States, that insects were used against an aquatic weed. To date the United States Department of Agriculture (USDA) has been very active in research into the biological control of aquatic weeds, with further work being undertaken by the Commonwealth Institute of Biological Control (CIBC) and relatively minor amounts by other countries.

The USDA has concentrated on Alligatorweed, *Alternanthera philoxeroides,* and Water Hyacinth, *Eichhornia crassipes,* and CIBC on Water Hyacinth and Salvinia, *Salvinia molesta.*

We have been able to avail ourselves of the results of these programs as a first step towards biological control of aquatic weeds in Australia. The Division of Entomology, CSIRO, commenced work on the biological control of Water Hyacinth in 1975, Alligatorweed in 1976 and Salvinia in 1978. The Division's Biological Control Unit based at Curitiba, Brazil, has undertaken a vigorous exploratory program seeking control agents for aquatic weeds originating in South America. Initial emphasis is on organisms specially suited to control of Salvinia, Water Hyacinth and Alligatorweed in the Australian environment.

It is necessary to attack aquatic weed problems on a broad, multi-species front. Unless this is done, control of one weed species will very likely result in its replacement by another. For example, in Australia at the present time a major program has been mounted for the biological control of Salvinia because, not only is Salvinia a problem in its own right, but also the expected progressive control of Water Hyacinth will open additional areas to Salvinia invasion.

A multi-species approach is particularly important with submerged weeds where the problem is often the result of excessive growth of several species. A possible alternative to mounting a number of biological control programs each directed at a different submerged weed may be the use of an organism with suitably restricted but less specific feeding behaviour than most biological control agents.

Experience in many parts of the world indicates that a herbivorous fish, the White Amur (or Chinese Grass Carp) *Ctenopharyngodon idella* can achieve satisfactory control of many submersed weeds. The suitability of this fish to Australian conditions should be evaluated.

Water Hyacinth showing brown areas of damage near the margins caused by the weevil *Neochetina eichhorniae*. The two lighter-green patches in the centre foreground are Para Grass *(Brachiaria mutica)*

A dense growth of Water Hyacinth choking a coastal stream

Pontederiaceae
Eichhornia crassipes

Distribution in Australia

Water Hyacinth is generally regarded as originating in tropical South America. From there it has been spread to most other tropical and sub-tropical regions of the world to become an extremely serious weed.

Water Hyacinth was first recorded from the Australian States of Queensland and New South Wales about 1895 but the exact date of introduction is uncertain. By the early 1900's it had spread along the east coast of Queensland and the north-eastern region of New South Wales. It now occurs along the east coast from Sydney to southern Cape York Peninsula, in the Mitchell River on western Cape York Peninsula, inland near Moree, Mt Isa and Georgetown, near Darwin and near Perth. It has been recorded from southern New South Wales, Victoria and eastern South Australia but is believed to have been eradicated in those regions.

In Australia, Water Hyacinth causes serious problems in natural watercourses, natural and man-made lakes, irrigation and flood mitigation channels, and dams. There is grave concern that Water Hyacinth may spread into regions that are presently uninfested.

Adult Water Hyacinth Weevil, *Neochetina eichhorniae* and adult feeding scars. Larvae tunnel in plant tissues, chiefly in the petioles

Water Hyacinth damaged by larvae of the moth *Sameodes albiguttalis*

Biological Control

Work on the biological control of Water Hyacinth in Australia commenced in January, 1975, with the importation of a South American weevil, *Neochetina eichhorniae,* from the United States Department of Agriculture laboratories in Florida. Studies conducted in CSIRO quarantine facilities in Brisbane confirmed the host specificity of this insect and indicated that it may be a useful control agent under Australian conditions. It was approved for liberation in June, 1975, and is now established on numerous weed infestations along the eastern coast.

The adult weevils feed on the leaf blades and petioles, making small sub-circular scars. Eggs are usually deposited in the petioles and larval tunnelling is followed by rotting of plant tissues, by waterlogging and, if heavily attacked, by death of the plants. During their development, larvae can transfer to fresh petioles. At maturity, larvae form a pupal cell from rootlets beneath the water surface. The adult weevils are nocturnal, very susceptible to humidity change and strictly tied to the aquatic environment. Completion of the life cycle requires about three months. The insect is relatively inactive during winter.

Damage to Water Hyacinth by larvae of the weevil *Neochetina eichhorniae*

Adult male (left) and larva (right) of the Water Hyacinth Moth *Sameodes albiguttalis*. The larvae prefer to tunnel in bulbous petioles

It will be several years before the effectiveness of this insect as a control agent can be assessed. In central coastal Queensland, some plants have been killed only two years after the first liberation of *N. eichhorniae* and it is probable that this weevil will be a valuable control agent. However, it is anticipated that a complex of natural enemies will be required for satisfactory control of Water Hyacinth over the whole of its range in Australia.

A pyralid moth, *Sameodes albiguttalis,* was selected as the second control agent in the complex to be liberated against Water Hyacinth in Australia. This South American insect was first liberated in October, 1977, and has now successfully completed several field generations in south-eastern and central coastal Queensland, and is well established.

The eggs of this moth are laid in cracks or scars on the pseudolamina of the leaf and on petioles. Larvae tunnel in the petioles and associated rotting of the tissues causes extensive damage to the Water Hyacinth plants. *S. albiguttalis* appears to have a preference for plants with bulbous petioles. Female moths each lay about 200 eggs and the life cycle is completed in about six weeks.

Other potential control agents are currently being studied in Brazil and it is anticipated that a progressive increase in the biological control of Water Hyacinth will be achieved during the next few years.

Damage to Alligatorweed by adults and larvae of the flea-beetle, *Agasicles hygrophila*

Amaranthaceae
Alternanthera philoxeroides

Distribution in Australia

Alligatorweed originated in South America where it has an extensive distribution through tropical and sub-tropical regions. Following introduction into North America it created serious problems in freshwater bodies in the more southerly States.

Alligatorweed has been in Australia for at least 34 years and occurs in Georges River and some of its tributaries, and Duck River, a tributary of Parramatta River. Drainage canals and adjacent low-lying land in the Fullerton Cove/Williamtown area north of Newcastle, and a small area at Woomargama near Albury, have also been infested by the weed.

This weed has the potential to occupy and seriously affect most Australian freshwater bodies. Infestations are generated by plants rooted in the banks of lakes, streams and canals. Under favourable conditions of temperature and nutrition, the plant grows outwards from the banks and forms a floating mat, which is replaced rapidly if swept away by floods or killed by herbicides. The thick, dense mats of floating, hollow stems and upright, leafy stems limit the recreational and industrial use of water, clog watercourses and affect water quality in general.

A healthy infestation of Alligatorweed near Liverpool, March 1977

The same infestation of Alligatorweed 12 months (top) and 14 months (lower) after release of the flea-beetle, *Agasicles hygrophila*

Biological Control

Biological control of Alligatorweed in Australia commenced in June, 1976, with the importation of a South American insect, the Alligatorweed Flea-beetle, *Agasicles hygrophila,* from the United States Department of Agriculture laboratories in Florida.

Following studies to confirm its host specificity and suitability to Australian conditions, liberation was approved in January, 1977. During the remainder of that summer 4,390 adult beetles were liberated on Alligatorweed infestations in Georges River, Duck River and Fullerton Cove.

At Chipping Norton, on Georges River, and at Duck River these insects bred slowly during the ensuing winter and, together with a further 25,100 adult beetles which were liberated, provided a nucleus from which a large, damaging population developed during the summer of 1977-78. Thereafter, beetle populations rose rapidly, spreading over all the floating material within and downstream of the liberation areas, and damaging a high proportion of the foliage and emergent stems within five months. Subsequently, the much weakened floating mats were washed away by flood waters or sank. During 1977-78, 4,500 beetles were liberated on weed growing in drainage canals in low-lying areas at Fullerton Cove, however, comparatively little damage has resulted so far as the micro-habitat appears less suitable for the development of *A. hygrophila.*

Enormous numbers of beetles died following collapse of the floating mats, but sufficient numbers survived the winter period to build up a large population during the ensuing warmer months. The beetles caused heavy damage by mid-summer and this was quickly followed by the collapse of emergent stem material.

Adult (left) and larvae (right) of the Alligatorweed flea-beetle, *Agasicles hygrophila*

A. hygrophila lays its eggs in groups on the underside of leaves. Larvae and adults both feed on foliage and to a lesser extent on stems. Final instar larvae pupate inside the hollow stems. During summer the total life cycle is about five weeks.

A second insect introduced for Alligatorweed control is a pyralid moth, *Vogtia malloi.* Liberations of this insect began during the summer of 1977-78. Breeding was observed in the field and establishment was confirmed in late 1978. Massive damage caused by *A. hygrophila* generally obscured damage resulting from *V. malloi* attack. However, in several small areas total emergent stem destruction was attributed entirely to the effects of *V. malloi.*

The massive and spectacular knock-down of the floating mats of Alligatorweed during the summers of 1977-78 and 1978-79 demonstrated the damage which introduced biological control agents may inflict on an introduced weed. This damage was especially impressive because of the rapidity of the action of the control agent. However, it must be emphasised that such rapid results are exceptional and that the majority of biological control agents take much longer to reach damagingly high population levels.

It appears likely that *A. hygrophila,* and possibly *V. malloi,* will keep the floating mat of Alligatorweed at an acceptably low level, but these insects are not expected to control the weed growing on dry land.

In the Fullerton Cove area in particular, weed growing on dry land constitutes a large reservoir from which it may be spread to uninfested regions. For example, removal and sale of lawn turf originating in this area and infested with Alligatorweed has been reported. Movement of livestock from this area to uninfested properties is another cause for concern. Terrestrial Alligatorweed is now seen as the principal problem and research at the CSIRO Biological Control Unit in Brazil is directed towards finding agents which may be effective in that situation.

Top: Larvae and (right) Adult of the Alligatorweed Moth, *Vogtia malloi.* Larvae tunnel in the stems of the plant

Salviniaceae
Salvinia molesta

Distribution in Australia

Salvinia molesta is one of four described species in the *S. auriculata* complex, a species group indigenous to South America. *S. auriculata* is widespread, occurring from the Caribbean to Argentina, but the other species occur mainly in south-eastern Brazil. *S. molesta* has become a very serious weed in regions of South East Asia, India, Sri Lanka, Africa, Australia, Papua New Guinea and Fiji.

In Australia, *S. molesta* occurs in South Australia, Western Australia, Northern Territory, Queensland and New South Wales. It infests natural and man-made lakes, streams and swamps and can very rapidly form dense, thick floating mats that can, in turn, be colonised by other plants and lead to severe disruption of the ecology of an area. *S. molesta* is rapidly increasing in distribution and abundance and is posing a serious threat to Australian freshwater bodies.

A heavy infestation of Salvinia, Lake Moondarra, Mt Isa, Queensland

Biological control

A research program on the factors influencing the abundance and distribution of *S. molesta* in South America, its biology and the potential of its natural enemies as biological control agents in Australia was mounted early in 1978. This program is operating from the CSIRO Biological Control Unit in Brazil. The native range of *S. molesta* was previously unknown, but extensive surveys have led to the discovery of *S. molesta* in south-eastern Brazil and a weevil, *Cyrtobagous singularis,* has been imported into CSIRO quarantine facilities in Australia for detailed evaluation as a control agent. Progress is encouraging but it will be several years before the long-term results of the program can be assessed.

Top: A weevil, *Cyrtobagus singularis* which attacks Salvinia in South America

Right: Damage to Salvinia by natural enemies in its native range in south-eastern Brazil

About the Author

Ken Harley is Officer-in-Charge of the Division of Entomology at the CSIRO Long Pocket Laboratories in Queensland. He is also responsible for the Organisation's biological weed control laboratory in Curitaba, Brazil.

Dr Harley has a B.Sc and M.Sc from the University of Queensland and a Ph.D from the University of Manitoba, Canada. His 23 years' specialised scientific experience includes six years as Officer-in-Charge of the CSIRO Cattle Tick Research Station at Ingham in Queensland and two years in charge of the Biological Control of Weeds Research Station in Hawaii.

Nutrient Enrichment. A white scum of dead algae on the margins of a water storage. The algae (mostly *Anacystis cyanea)* have died as a natural conclusion to an algal bloom caused by nutrient enrichment

NUTRIENT ENRICHMENT EUTROPHICATION

Cause of Problem Plant Growth—Possibilities for Regulation

by K. H. Bowmer

Nutrient enrichment of water stimulates the growth of aquatic plants. This increase in plant growth may become a problem directly *(see page 501)* or the increase in activity of organisms responsible for decay and breakdown of plant material may deplete the water of oxygen causing algal blooms or fish kills. Throughout the world the process of nutrient enrichment has been accelerated dramatically with more intensive land use and fertiliser application in agriculture, and increasing discharge into waterways of stormwater, sewerage, detergents and industrial effluents.

Information is required on the current status of inland waters so that deterioration in water quality can be monitored and remedial measures taken where appropriate. Where information on nutrient levels is adequate there has been limited success in predicting the potential for phytoplankton growth in lakes and reservoirs, but the prediction of macrophyte growth, and extrapolation to other systems and to flowing waters is far more complicated. Also there are several features of Australian inland waters which invalidate many of the observations made overseas. Some of the problems of predicting the extent of plant growth from a knowledge of nutrient levels and inputs will be discussed at greater length in this chapter.

Phosphorus and, to a lesser extent, nitrogen, appear to be the most important nutrients involved.

The principal forms of phosphorus in natural waters are soluble inorganic phosphates, phosphorus as a component of dissolved organic materials, and phosphorus associated with suspended particles. All these forms of phosphorus also occur in sediments. The particulate phosphorus comprises organic forms of phosphorus in algal and bacterial cells, and also inorganic phosphates which are adsorbed or precipitated on suspended matter. Precipitation occurs by reaction of phosphate with calcium, aluminium and iron compounds to form complex insoluble substances. Adsorption occurs at the broken edges of clay particles where aluminium is exposed, and on amorphous (non-crystalline) gel-like complexes of hydrated iron oxides which have large and reactive surfaces.[34]

The phosphorus content of water is usually reported as phosphate-phosphorus and/or total phosphorus. Phosphate-phosphorus gives an estimate of the amount of phosphorus which is readily available to plants. Total phosphorus is a measure of all the principal forms of phosphorus and represents all the phosphorus which might eventually become available. The phosphates associated with suspended materials are made available by dissolution and desorption, and the availability of the organic component is increased with time by the activities of bacterial enzymes which release phosphates from organic materials.

The balance between dissolved and particulate phosphate is controlled largely by the pH and redox (the reduction-oxidation state) of the water since these factors control the solubility of phosphate compounds and the adsorption-desorption reactions on the iron-gel complexes.[19, 20] As anaerobic conditions develop iron is reduced from the ferric to the ferrous state, with release of the more soluble phosphate to the water. The ferrous oxides have a greater surface area and affinity for phosphorus than ferric oxides, so that some of the phosphate released into the solution may be re-adsorbed.[38] However, the overall effect of development of anaerobic conditions is usually to increase the availability of phosphorus to plants.

The situation is further complicated by diurnal and/or seasonal fluctuations in all the factors which determine the rate of supply of phosphates to plants: namely, nutrient supply from "biological turnover" (decay of organisms), and from external sources to the water; transport processes (sedimentation, destratification, erosion of sediments, convection and diffusion); and exchange with the sediment. *So it is impossible to assess the nutrient status of a water body without a great deal of information, and certainly not from a few values of average concentration.*

Phosphorus concentrations in water are increased by discharge of sewerage effluents including detergents, urban stormwater, abattoir wastes, etc. Inputs from agriculture of phosphorus fertiliser in run-off have generally been regarded as less important than urban sources, because of the relative immobility of phosphorus compounds through rapid fixation in the soil. However, there is some evidence that a large proportion of the particulate phosphate in streams originates from soil erosion when selective removal of fine clay-sized particles occurs in surface run-off.[28] This process is especially relevant in Australia because of the susceptibility of the dry landscape to erosion during storms. In fact, very high agricultural inputs of phosphorus have been recorded in Australian waters, in the River Murray downstream of pasture and cereal areas,[11] and in farm dams in orchard and grazing catchments in South Australia.[24]

Nitrogen contamination is caused by run-off of fertiliser, as well as discharge of sewerage into water. Nitrogen occurs in gaseous forms (nitrous oxides and ammonia), in inorganic forms (nitrates, nitrites and ammonium compounds) and as organic complexes. Since nitrate-nitrogen is the predominant form of inorganic nitrogen in aerated waters, it is usually a good approximation to total inorganic nitrogen. It is readily absorbed by plants and is widely used in assessing water quality criteria.

In contrast to phosphorus compounds, which mainly come from well-defined or "point" sources, nitrogen compounds are readily mobile and often derived from "diffuse" sources, so they are not easily controlled. Also, in quite mildly reducing conditions, there is a rapid conversion of nitrate-nitrogen to gaseous nitrous oxides and nitrogen which are lost to the atmosphere. Conversely, some of the blue-green algae which occur in "bloom" proportions in eutrophic water are able to fix atmospheric nitrogen when nitrogen concentrations in water are low. This ready exchange of nitrogen between the water and the air means that the impact of any given nitrogen input is difficult to assess. For these reasons more emphasis is usually given to phosphorus than nitrogen in assessing the effects of nutrient inputs on water quality, and in attempting to regulate nuisance growth by removing nutrients from effluents.

Some critical levels of nutrients for eutrophication were first derived from observations of lakes in Wisconsin, U.S.A.[29] which suggested that concentrations of 10 mg/m^3 phosphate-phosphorus and 300 mg/m^3 of nitrate-nitrogen, measured in the spring, are critical. Subsequently a relationship between the phosphorus loading per unit area of water surface and the degree of eutrophication was proposed.[35] The critical rates of total phosphorus and total nitrogen were 0.2 g/m^2/year and 5.0 g/m^2/year respectively for a lake 10 metres deep, with modification according to depth. For example, a lake of 50 metres mean depth would be eutrophic if its total phosphorus and total nitrogen inputs exceeded 5.0 g/m^2/year and 8.0 g/m^2/year respectively. Other factors, including the residence time of the water, seasonal variations in nutrient input, stratification, and the proportion of nutrients retained in the lake sediment have also been considered.[6, 7, 18, 36]

The role of sediments in maintaining nutrient supply is the subject of dispute. One view[34] is that it would be of minor importance, at least in deeper lakes, since the release of phosphorus from undisturbed sediments to the water can reach a maximum of only 0.27 mg/m^2/day (0.1 g/m^2/year). In fact there are several examples of successful and prompt reversal of eutrophication in lakes by diversion of sewerage effluents, also suggesting that sediment phosphorus was not an effective and continuing source of nutrients.[9, 31]

On the other hand there is experimental evidence that sediments do supply substantial quantities of nutrients to water and that this continues for many years after the addition of nutrients has ceased. It has been suggested that to restore seriously-enriched lakes, it will be necessary to remove much of the sediment or to lock accumulated nutrients into it, for example, with weighted polythene sheet.[14] In

Australia it has been shown that the addition of alum to inhibit the release of phosphate from bottom muds of farm dams can prevent the development of algal blooms.[21] A great deal of effort has been expended in minimising the influx of nutrients, especially phosphorus, to lakes. Several hundred lake treatments have been catalogued,[8] ranging from wastewater treatment and diversion to lake bottom sealing, precipitation of nutrients, and dredging.

Others[30, 31] have questioned the usefulness of the regulation of nutrient input because of the immense costs often involved, and because large numbers of eutrophic lakes receive nutrients from diffuse or non-controllable sources. They also feel that the role of phosphorus in eutrophication has been over-emphasized and over-simplified; that the prediction of plant growth from phosphorus loadings is only useful on a global or average basis, and not in individual lakes where biological processes can have an overriding influence.

Some of the more important modifying biological processes include the excretion of nutrients by bottom-feeding fish, the "pumping" of nutrients by macrophytes from sediments to water, and the grazing of algae by zooplankton, which is in turn regulated by populations of plankton-eating fish. As an alternative to regulation of nutrient input, it is suggested, for example, that algae could be kept in check by regulation of fish populations. It is admitted by the exponents of this kind of approach that biological regulation depends on ecological relationships which are "slippery and not under our control", so they are not likely to be attractive to government and regulating authorities.

Town planners and engineers require information on the loadings of nutrients which can be tolerated by waters receiving urban effluents, so that they can plan the development of inland towns and water-treatment facilities to avoid the development of eutrophication. Unfortunately, for reasons already discussed, there are many problems in identifying this critical level or loading. There is also a danger that, in the general application of a given critical level, the important ecological and biological characteristics of individual situations may be neglected.

These difficulties, and the problems of assessing the monetary value of water quality, have sometimes tempted regulating authorities either to do nothing, or to require the maximum feasible treatment using the latest technology available, regardless of the ecological benefits and costs.

The Lower Molonglo Water Quality Control Centre is designed to treat the sewerage effluent from Canberra, Australia's largest inland city, and is an example of sophisticated technology in wastewater treatment. Here, an advanced degree of treatment involves precipitation of phosphorus with lime, and removal of nitrogen by biological processes requiring the addition of methanol.[33] Thereby, approximately 99 per cent of the phosphorus and 95 per cent of the nitrogen will be removed from wastewater at a capital cost of the order of 48 million dollars, and with considerable maintenance costs. However, other nutrient sources—urban runoff, rural runoff and erosion—could contribute approximately 20 per cent of the nutrient loading in Burrinjuck Reservoir, downstream of Canberra, on which all the tributaries of the Upper Murrumbidgee Basin converge. Because of their scattered origin, large volume, and variability in flow, these nutrient sources are not amenable to control by conventional water treatment. So in this system, self-purification during retention in urban lakes and settling basins plays an important role in improving water quality.[13]

Several people are interested in extending the processes of natural cleansing of water[10] by retaining or developing wetlands for economical water treatment in Australia. It is proposed that swampy areas of Cumbungi *(Typha* spp.), Common Reed *(Phragmites australis)* or Tall Spikerush *(Eleocharis sphacelata)* might be used as "wet filters" to remove suspended matter, nutrients, bacteria and viruses, with associated benefits in aesthetics and wildlife conservation. The usefulness of wet filters is being investigated at the University of New England at Armidale, the Canberra College of Advanced Education,[4] and the CSIRO Division of Irrigation Research.[23]

The whole subject of eutrophication has been reviewed recently,[26] and a co-operative project has been initiated with several countries including Australia for measuring and monitoring the rates of change, so that different remedial measures can be compared. Two intensive studies have been undertaken, of Mount Bold and Prospect Reservoirs, important water storages for Adelaide and Sydney respectively. Lake Burley Griffin,[4] Lake Burragorang,[32] Eildon Reservoir, the North Pine River Dam,[15] Dartmouth Dam, Lake Hume, Lake Mulwala,[12] and the Murray,[3, 11] Murrumbidgee,[25] and Onkaparinga Rivers,[2] as well as various farm dams and coastal lakes and estuaries,[1] have also been investigated in detail with respect to their current or potential eutrophic state. Many of these reports will be published soon in a comprehensive book.[40] Reviews of the extent of eutrophication[41] and of the impact of aquatic weeds[22] in Australian inland waters are also available.

In some of these and other studies, attempts are made to predict the probability of excessive growth of plants in lakes and reservoirs. Providing that phosphorus is the limiting nutrient and with good data on phosphorus loadings and water retention time in relatively stable conditions, predictions of the growth of green algae have been partly successful.

However, there are difficulties in extending the predictions of plant growth from green algae to blue-green algae, diatoms and macrophytes, and from lakes and reservoirs to streams, rivers and flowing waters of irrigation systems.

For flowing waters the natural capacity for self-purification is greater than for storages, particularly where turbulence can reoxygenate the water. Also, for a given concentration, the effective supply of nutrients to a plant is increased with flow for two reasons: the volume of water which can be scavenged for nutrients increases, and the diffusion barriers for uptake adjacent to plant surfaces are reduced by turbulence. Consequently much lower concentrations of nutrients might be expected to promote growth in rivers and streams than in relatively static water. In fact, in a recent review of the concept of critical concentrations of nutrients,[37] it was noted that for many European rivers and streams the annual throughput of nutrients is usually over five, and sometimes over forty times the plant demand. Evidently in flowing water plant growth must be limited by other factors, such as excessive flow, high turbidity or the suitability of the substrate or sediment for anchorage.

In predicting the probability of macrophyte growth there is also the problem of assessing the relative importance of sediment and water in respect to nutrient uptake by roots or shoots of water plants. It has been shown that in some rooted species both pathways of uptake are possible.[5] The relative importance of the nutrient uptake pathways, as well as competition for light, will affect the relative success of rooted species of macrophytes compared with floating species and phytoplankton.

In fact, nutrient-enrichment of lakes has often been associated with a decline rather than a stimulation in submerged macrophyte growth, and this has been attributed to shading by increasing populations of phytoplankton, though recent evidence[27] suggests that the competitive mechanism is more complicated, and associated with the shading caused by algal epiphytes (colonising algae which may form a thick layer over the leaves).

In addition the critical standards of nutrient concentrations and loadings developed in Europe and North America may not be applicable to many Australian waters, where there are higher salinities, generally higher temperatures and larger fluctuations in water levels.[39] There are often very large and unpredictable fluctuations in water flow, especially at flood times,[2, 4, 13] with very rapid changes in sediment load, associated movement of nutrients, and turbidity.

High turbidity, together with the yellow colour originating from dissolved organic compounds, severely restricts the depth to which light can penetrate in quantities sufficient for photosynthesis. Changes in light quality are also significant in that the absorption of blue light in the coloured water disadvantages the green chlorophyll-containing species, which absorb primarily in the blue wavelength region. Diatoms, containing the carotenoid pigment fucoxanthin, are able to exploit the non-absorbed, longer-wavelength light to some extent, whilst the undesirable blue-green algae and cryptomonads with phycobiliprotein pigments are best equipped to take advantage of the underwater light environment. This aspect of water quality has recently been studied in considerable detail.[16, 17] Again the situation is complicated, by the motility of some species of green algae, and by the ability of the blue-green algae to move vertically in the water by expansion and contraction of their gas vacuoles.

Although many Australian waters are already eutrophic with respect to classical standards of nutrient status, they often do not show the expected nuisance symptoms. Evidently there are many features which distinguish Australian waters from those elsewhere. In particular, fluctuating water flows and levels, the yellow colour of the water and high turbidities may counteract the effects of high nutrient levels and regulate plant growth. Nevertheless, perhaps further contamination ought to be avoided until the biology of the aquatic ecosystem and the unique features of Australian inland waters are better understood.

References

[*1]BAYLY, I. A. E. and WILLIAMS, W. D. (1973) "Inland Waters and their Ecology". (Longman: Hawthorn, Vic.) 316 pp.

[2]BUCKNEY, R. T. (1979) Chemical Loadings in a Small River, with Observations on the Role of Suspended Matter in the Nutrient Flux. *Australian Water Resources Council Technical Paper* no. 40. 72 pp.

[3]CROOME, R. L., TYLER, P. A., WALKER, K. F. and WILLIAMS, W. D. (1976) A Limnological Survey of the River Murray in the Albury-Wodonga Area. *Search* 7:14-17.

[4]CULLEN, P., ROSICH, R. and BEK, P. (1978) A Phosphorus Budget for Lake Burley Griffin and Management Implications for Urban Lakes. *Australian Water Resources Council Technical Paper* no. 31. 220 pp.

[5]DENNY, P. (1972) Sites of Nutrient Absorption in Aquatic Macrophytes. *Journal of Ecology* 60:819-829.

[6]DILLON, P. J. (1974) A Critical Review of Vollenweider's Nutrient Budget Model and Other Related Models. *Water Resources Bulletin* 10:969-989.

[7]DILLON, P. J. (1975) The Phosphorus Budget of Cameron Lake, Ontario: The Importance of Flushing Rate to Degree of Eutrophy of Lakes. *Limnology and Oceanography* 20:28-39.

[*8]DUNST, R. C. *et al.* (1974) Survey of Lake Rehabilitation Techniques and Experiences. *Wisconsin Department of Natural Resources Technical Bulletin* no. 75. 179 pp.

[9]EDMONDSON, W. T. (1970) Phosphorus, Nitrogen and Algae in Lake Washington after Diversion of Sewage. *Science* 169:690-691.

[10]GOOD, R. E., WHIGHAM, D. F., SIMPSON, R. L. and JACKSON, C. G. (Eds) (1978) "Freshwater Wetlands". (Academic Press: New York) 378 pp.

[11]GUTTERIDGE, HASKINS and DAVEY (1974) River Murray in Relation to Albury-Wodonga. (Cities Commission: Canberra) Volume 3:17, 66-68, 197, 201.

[12]HART, B. T., McGREGOR, R. J. and PERRIMAN, W. S. (1976) Nutrient Status of the Sediments in Lake Mulwala. I. Total Phosphorus. *Australian Journal of Marine and Freshwater Research* 27:129-136.

[13]HIGGINS, W. L. and LAWRENCE, A. I. (1977) Water Quality Management in the Canberra Metropolitan Region. *Australian Water and Wastewater Association Proceedings, Canberra, 1977* pp. 95-115.

[14]HYNES, H. B. N. and GREIB, B. J. (1970) Movement of Phosphate and Other Ions From and Through Lake Muds. *Journal of the Fisheries Research Board of Canada* 27:653-658.

[15]KING, C. R. (1979) Secondary Productivity of the North Pine Dam. *Australian Water Resources Council Technical Paper* no. 39. 75 pp.

[16]KIRK, J. T. O. (1976) Yellow Substance (Gelbstoff) and its Contribution to the Attenuation of Photosynthetically Active Radiation in Some Inland and Coastal South-eastern Australian Waters. *Australian Journal of Marine and Freshwater Research* 27:61-71.

[17]KIRK, J. T. O. (1979) Spectral Distribution of Photosynthetically Active Radiation in Some South-eastern Australian Waters. *Australian Journal of Marine and Freshwater Research* 30:81-91.

[18]LERMAN, A. (1974) Eutrophication and Water Quality of Lakes: Control by Water Residence Time and Transport to Sediments. *Hydrological Sciences Bulletin* 19:25-34.

[19]LI, W. C., ARMSTRONG, D. E., WILLIAMS, J. D. H., HARRIS, R. F. and SYERS, J. K. (1972) Rate and Extent of Inorganic Phosphate Exchange in Lake Sediments. *Soil Science Society of America Proceedings* 36:279-285.

[20]McCALLISTER, D. L. and LOGAN, T. J. (1978) Phosphate Adsorption-Desorption Characteristics of Soils and Bottom Sediments in the Maumee River Basin of Ohio. *Journal of Environmental Quality* 7:87-92.

[21]MAY, V. (1974) Suppression of Blue-Green Algal Blooms in Braidwood Lagoon with Alum. *Journal of the Australian Institute of Agricultural Science* 40:54-57.

[*22]MITCHELL, D. S. (1978) "Aquatic Weeds in Australian Inland Waters". (Australian Government Publishing Service: Canberra) 189 pp.

[23]MITCHELL, D. S. (1979) The Potential for Reclamation of Water and Other Resources from Waste Waters in Australia. *Victorian Agricultural Year Book* 49(1):45-50.

[24]MOORE, S. D. (1974) Changing Land Use Increases Nitrate-Nitrogen & Phosphate Concentrations in Farm Dams. *South Australia Soil Conservation Branch Report* S14/74. 7 pp.

[25]NEW SOUTH WALES STATE POLLUTION CONTROL COMMISSION (1976) "An Investigation into Pollution of the Lower Murrumbidgee River (1972-1974)". 52 pp.

[26]ORGANISATION FOR ECONOMIC CO-OPERATION AND DEVELOPMENT (In preparation). Co-operative Programme for Monitoring of Inland Waters (Eutrophication Control). Regional Project Shallow Lakes and Reservoirs. (Clasen, J. Ed.) (O.E.C.D.: Paris)

[27]PHILLIPS, G. L., EMINSON, D. and MOSS, B. (1978) A Mechanism to Account for Macrophyte Decline in Progressively Eutrophicated Freshwaters. *Aquatic Botany* 4:103-126.

[28]RYDEN, J. C., SYERS, J. K. and HARRIS, R. F. (1973) Phosphorus in Runoff and Streams. *Advances in Agronomy* 25:1-45.

[29]SAWYER, C. N. (1974) Fertilisation of Lakes by Agricultural and Urban Drainage. *Journal of the New England Water Works Association* 61:109-127.

[30]SHAPIRO, J. (1978) The Need for More Biology in Lake Restoration. *University of Minnesota Limnological Research Center Contribution* no. 183. 17 pp.

[31]SHAPIRO, J., LAMMARRA, V. and LYNCH, M. (1975). Biomanipulation: An Ecosystem Approach to Lake Restoration. In "Symposium on Water Quality Management through Biological Control". (Brezonik, P. L. and Fox, J. L., Eds) (University of Florida: Gainesville) pp. 85-96.

[32]SMALLS, I. C. (1975) Nutrient Effects of Stormwaters on Lakes and Streams. In "Water in the Urban Environment: Proceedings of the Australian Water and Wastewater Association Summer School, February, 1975". Part A, Session 8, 46 pp.

[33]SPELDEWINDE, F. C. (1973) Water Pollution Control in Inland Australian Cities—Proposals for Canberra and Studies for Albury-Wodonga. In "Papers Presented to a Seminar on Pollution Problems of the River Murray, Wentworth, N.S.W., November, 1973". (Department of Decentralisation and Development: New South Wales) Section E, Paper A, 17 pp.

[34]STUMM, W. and LECKIE, J. O. (1970) Phosphate Exchange with Sediments; Its Role in the Productivity of Surface Waters. *Advances in Water Pollution Research: Proceedings of the Fifth International Conference on Water Pollution Research* 1970: III-26/1-III-26/16.

[35]VOLLENWEIDER, R. A. (1971) "Scientific Fundamentals of the Eutrophication of Lakes and Flowing Waters with Particular Reference to Nitrogen and Phosphorus as Factors in Eutrophication". (O.E.C.D.: Paris) 159 pp.

[36]VOLLENWEIDER, R. A. (1975) Input-Output Models with Special Reference to the Phosphorus Loading Concept in Limnology. *Schweizerische Zeitschrift für Hydrologie* 37:53-84.

[37]WESTLAKE, D. F. (1975) Macrophytes. In "River Ecology—Studies in Ecology". (Whitton, B.A., Ed.) (Blackwell Scientific: Oxford.) Vol. 2:106-128.

[38]WILLETT, I. R. (In preparation) Oxidation-Reduction Reactions in Soils. In "Soils, an Australian Viewpoint". (C.S.I.R.O. Division of Soils).

[39]WILLIAMS, W. D. and WAN, H. F. (1972) Some Distinctive Features of Australian Inland Waters. *Water Research* 6:819-836.

*[40]WILLIAMS, W. D. (Ed.) (1980) "An Ecological Basis for Water Resource Management." (Australian National University Press: Canberra) 417 pp.

*[41]WOOD, G. (1975) An Assessment of Eutrophication in Australian Inland Waters. *Australian Water Resources Council Technical Paper* no. 15, 238 pp.

*Key references and useful general reviews.

About the Author

Kathleen H. Bowmer is a graduate of the University of Nottingham (U.K.) School of Agriculture. Since 1969 Kathleen has been attached to the CSIRO Division of Irrigation Research at Griffith, where she is presently a Senior Research Scientist in the Aquatic Ecosystems Management Group.

AQUATIC PLANT MANAGEMENT IN NEW SOUTH WALES

with special emphasis on irrigation systems

In the last twenty years many methods of managing aquatic plants have been developed. They include chemical, mechanical, biological or habitat manipulation methods, or combinations of these techniques.

In some situations chemical control may be the only economical and practical method. In other situations there may be no need to take any action at all. Alternatively, fluctuation of water levels to limit the growth of aquatic plants may be all that is needed, or suitable competitive species may be cultivated to crowd out obstructive marginal or bank species.

When an aquatic plant management program does not achieve its aims, it is often because there is insufficient information on the biology and ecology of the plants, and on available management methods. Public reaction, political factors and lengthy bureaucratic procedures may alter, hinder or even prevent plant management programs.

There is a tendency among water managers and the public to classify all aquatic plants as "weeds". This is unfortunate, for it is just as natural for plants to grow in water as on land. Aquatic plants, like land plants, often do not cause problems and are valuable components of the ecosystem. However, in some situations, notably in rice fields, irrigation and drainage systems, and occasionally in recreational lakes, farm dams and wetlands, plants can be a problem; they obstruct the flow of water or reduce the use of the area for humans and wildlife.

About fifty species of aquatic plants, excluding algae, have been recorded in Australia as causing sufficient problem at one time or another to warrant control.

Some of the most troublesome plants in channels and streams are Water Couch *(Paspalum paspalodes)*; Paspalum *(Paspalum dilatatum)*; Cumbungi *(Typha* spp.*)*; Common Reed *(Phragmites australis)*; Water Lettuce *(Pistia stratiotes)*; Para Grass *(Brachiaria mutica)*; Ribbonweed *(Vallisneria gigantea)*; Pondweeds *(Potamogeton* spp.*)*; Water Milfoils *(Myriophyllum* spp.*)*; Elodea *(Elodea canadensis)*; Reed Sweetgrass *(Glyceria maxima)* and Arrowhead *(Sagittaria graminea)*.

In rice, Barnyard Grasses *(Echinochloa* spp.*)* and Dirty Dora *(Cyperus difformis)* if left uncontrolled can cause reductions in crop yields.

Other species, notably Water Hyacinth *(Eichhornia crassipes[7]);* Salvinia *(Salvinia molesta[12]);* Alligatorweed *(Alternanthera philoxeroides[11])* and Lagarosiphon *(Lagarosiphon major[9]),* although generally contained at present, are potentially very troublesome because of their capacity for rapid spread.

Is a Management Program Really Necessary?

Before deciding on a program, the following points need to be considered.

Make sure that there really is a problem and, if so, define it.

Consider whether the problem is likely to solve itself..

Clearly state the objectives of any management program.

Consider the alternative of doing nothing. Doing nothing can be positive management; it may be the most appropriate action.

Do not expect that the method used in one place will apply to another; every situation is different.

Select the appropriate control program and consider combining several methods.

Interference to an ecosystem by chemical or mechanical management or by habitat manipulation may result in a less stable system than previously existed. Once man has attempted some form of management, further action may be necessary to maintain the required level of plant control. It is often difficult to accept that the best action is to do nothing.

Stable or Artificial System?

The aquatic habitat is generally a difficult area in which to achieve satisfactory plant management. The submerged plants cannot be controlled by methods used on terrestrial plants.

Broadly, from a management viewpoint, the aquatic plant habitat can be divided into three "states".

1. Unstable—where the existing plants are undesirable and regular treatments are needed to maintain control.

2. Artificially stabilized—where management systems have been successfully introduced, e.g. biological agents and/or competitive species.

3. Stable—where a balanced plant population has established naturally and requires no maintenance.

The naturally stable state (No. 3) is by far the most desirable, and this is often achieved without human interference. An example of this is a channel carrying a dense marginal growth of Common Rush *(Juncus usitatus) (see photograph page 514)* and moderately dense Ribbonweed *(Vallisneria gigantea)* in the water. Mechanical or chemical removal of the Common Rush may result in an undesirable species such as Cumbungi *(Typha* spp.) growing in its place. Eradication of the Ribbonweed may bring a succession of more obstructive submerged plants. Such management would result in exchanging a stable plant population for an unstable one, and take many years to rectify.

Preventive Measures

Aquatic weed problems can often be prevented or at least limited. The most obvious preventive measure is to stop the introduction of aquatic weeds into farm channels, irrigation systems or wetlands. In addition, problems from naturally occurring plants can be minimised by proper channel design. The following points are directed at preventing aquatic weed problems in New South Wales.

1. Supply and Drainage Channels—Design Factors

(a) Channels with flat grades have usually more problems with water weeds. Slowly flowing water less than three metres deep aids the establishment of aquatic plants, particularly if the water is clear. Channels more than three metres deep with uniform grades and flows in excess of 0.3 metres per second seldom develop significant submerged weed problems. However, plant growth encroaching from the channel bank into the stream and below the waterline will still cause some obstruction, notable examples being Common Reed (*Phragmites australis*) and Cumbungi (*Typha* spp.).

It is important to properly survey and construct supply channels to uniform sloping grades to enable complete drainage when not in use. A fall of 1:10 000 is considered to be a minimum grade, but a more realistic minimum is 1:6000. Restricted flows from flatter grades are accentuated in small channels because of increased friction due to the high wetted perimeter in relation to cross-sectional area. Ideally, small drainage channels should have a uniform fall of about 1:2000. Depressions in a channel, where water ponds, provide an ideal habitat for water plants. If a suitable contour cannot be followed, channels that run through depressions require considerable earthwork to maintain grade uniformity.

(b) Channels must have sufficient dimension to provide adequate water flow to the anticipated agricultural development (Table 1 showing channel cross-section). The intended capacity should be carefully calculated, not guessed. Channels constructed to a size substantially larger than the necessary capacity have slower flows, provide a better habitat for the growth of waterplants, and may cause as many problems as channels with insufficient capacity.

(c) In some farm channels it is good practice to construct the bed with sufficient width to enable the operation of a rotary hoe or mechanical cultivator. Such a channel has a wide, flat bed; however, this has the disadvantage of providing a perfect site for the growth of emergent water weeds. The construction of a small secondary channel in the bed of larger channels will provide for rapid drainage, reduce the wetted area and discourage subsequent weed growth in periods of low flow.

TABLE 1

Dimensions of Channel for Various Flows

FLOW RATE (Ml/day)	A (m)	B(m)
2.5	0.53	1.00
5.0	0.63	1.00
12.0	0.70	2.00
12.0	0.65	2.50
20.0	0.77	2.50

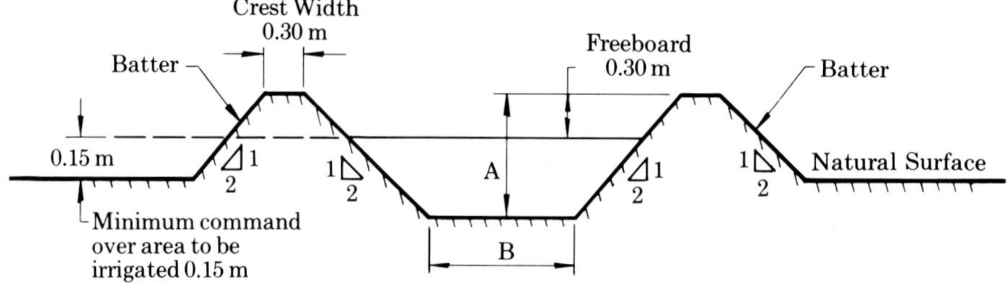

NOTE In all Cases: Crest Width 0.30 m.
Freeboard 0.30 m.
Min. Command 0.15 m.
Bank side slopes 1:2
Bed Grade 1:10 000
*Mannings *n* 0.035

*Mannings *n* is a roughness coefficient which makes allowance for the different surfaces over which the water is flowing. Mannings *n* can be increased and channel flow decreased by long dense plant growth.

(d) Ample room should be allowed for access to the channel to facilitate inspection and maintenance. Where possible design channels with access on both sides, and use ramps or grids, not gates, if cross fences are needed. Channels that require periodic mechanical clearing should have power lines positioned so that there is plenty of clearance for the safe operation of large machines. Increasing the height of the poles is one method of overcoming this problem.

(e) Channels should be fenced so that the area can be used for grazing by suitable animals. This will control many of the bank species, but where water is used for domestic purposes, contamination will result. To stabilise

banks and prevent damage from stock, plant species such as Kikuyu
(*Pennisetum clandestinum*) or Common Couch (*Cynodon dactylon*). (See
section "Plant Competition" on page 513).

(f) The number of cross fences should be limited. An obstruction by a
fence in the waterway may cause erosion, siltation and weed problems.

(g) Pipes are justifiable in some situations, despite their substantially
higher cost, as the exclusion of light eliminates weed growth. However, the
pipe capacities may diminish when fungal and bacterial growths develop
on the linings. Small bivalves may also obstruct flows in small pipes (see
pages 471-473). Concrete lining of small channels may not eliminate
problems, as ditch-bank plants like Paspalum (*Paspalum dilatatum*) trail
in the water, impeding flow. In large concrete-lined channels, flow is
impeded by filamentous algae attached to the concrete and submerged
weeds growing in silt.

(h) Diversion ponds to hold herbicide-contaminated water are useful to
enable the safe use of residual herbicides in supply channels and drains.
These ponds should be installed as far downstream as possible and
constructed so that channels can be drained completely. The block of
treated water can then be safely diverted and retained until the
contamination has been reduced by natural processes to a level which is
safe for return of water to the system. This approach should be
incorporated in the initial design of farm irrigation systems.

(i) Channels should be designed and built to enable rapid and complete
drainage (see also page 505 "Preventive Management—After
Construction").

2. Supply and Drainage Channels—Preventive Management After Construction

(a) Regular drying of the channel or fluctuating the water levels will
discourage most submerged weeds. This is the most common method of
aquatic weed control in New South Wales. Channels should be designed to
enable rapid and complete draining. Exposure of some submerged plants
to two days of drying in the summer may be as effective as many
knockdown-type herbicides. However, success with this technique
depends on drainage being complete. Alternative water supplies for stock
will make it possible to drain and dry the channel bed between irrigations,
thus limiting the growth of submerged plants. Slashing weeds in the
channel prior to irrigation will ensure good flows for a short period.
Supply channels should be completely drained in the winter months to
stress submerged weeds by drying. Drying is the most important factor. In
southern regions of New South Wales, frost on dormant aquatic plants
appears to have little effect on reducing regrowth the following season.

(b) Regular inspections should be carried out to see if any aquatic weeds
are present. Check the identity of any plants found in the channel.[1,14,18]
Early control measures may prevent a problem weed from getting out of

hand. Obstructive emergent plants like Water Couch (*Paspalum paspalodes*) and Cumbungi (*Typha* spp.) can be kept out of a channel by spot spraying with herbicides in the case of Water Couch or by hand-pulling where Cumbungi is invading. The control of plants on channel banks by residual and/or foliage-acting herbicides reduces weed problems within a crop. However, if channel banks are kept plant-free, erosion will occur on steep grades or loose soils. Suitable ground cover plants may be an alternative (*see page 513 of this section*).

(c) Once a supply or drainage channel has been constructed in areas where aquatic weeds are found, consideration should be given to a chemical treatment with a residual or soil-active herbicide to prevent the spread of water weeds into that system. However, treatment should be left until the channel has been used for a season, as any remedial earthworks on that channel will cause soil disturbance and negate the effect of the residual herbicide. Siltation also occurs, and water-borne soil may be deposited by slowly-flowing water, thereby covering the herbicide-treated soil with an overlay of untreated silt, and it is on these silt bars that water-plants tend to grow first.

3. Reservoirs, Lakes and Wetlands—Preventive Measures (see also "Water Level Management" on page 517)

It is difficult to isolate the factors which will cause aquatic plants and algae to be a problem in natural watercourses, lakes, reservoirs and marshes. Each situation requires separate evaluation; however, some explanations can be given for many existing problems and from these, measures can be taken to limit potential problems.

Large deep storages usually have less weed problems than small or shallow ones. Notable exceptions occur with free-floating species such as Salvinia *(Salvinia molesta)*[12], Water Hyacinth *(Eichhornia crassipes)* and Hornwort *(Ceratophyllum demersum)*[15] and rooted species such as Elodea *(Elodea canadensis)*[15] and Lagarosiphon *(Lagarosiphon major)*[9], which may break up and block hydro-electric inlet structures. Floating plants are usually only a problem where nutrient levels in water are high and wave action is limited. However, for many reasons, reducing nutrient levels in the water may not necessarily limit the growth of aquatic plants.

The growth of aquatic plants in lakes used for recreation and beautification, the stored water of weirs in rivers and creeks, and in town water storages, can be limited by the following measures:

(a) Structures could be installed to enable draining or water level fluctuations, providing there is an ample water supply. The frequency and magnitude of the water level change depends on the species causing problems and the degree of control needed. This measure does not prevent the growth of free-floating plants, and may increase the growth of marginal plants.

(b) Submerged rooted plants can be reduced in density by dredging existing storages or constructing artificial lakes, urban storages, marinas, etc., to depths that limit growth. In New South Wales, submerged growth is usually limited to a scattered population by water about four or five metres deep. Deepening the margin of recreational lakes may increase the hazard to non-swimmers.

(c) Coarse washed sand spread to a depth of about 200 millimetres at swimming and access points of recreational water will, in some situations, limit the growth of submerged plants. Dense growth elsewhere in the reservoir will probably cause no problem to swimmers, but may restrict fishing and hinder power boats and sailing craft.

(d) Earthen town-water storages should have small drainage channels constructed in the bed to enable rapid and complete drainage and drying to facilitate maintenance. Control of submerged plants is achieved when drying is sufficient.

(e) Plant growth may be slowed down by reducing inflows of nutrients to the water body. To be effective, nutrient inflows may need to be reduced for a long time. In intensive agricultural drainage basins, this may be impractical due to the many sources of nutrients *(see section on "Eutrophication" on page 491).*

Aquatic Plant Management in Flood Mitigation Drains of Coastal Areas

Flood mitigation drains have been extensively constructed in coastal regions of New South Wales, particularly in the North Coast, and it is often necessary to carry out measures to control aquatic plants in these drains.

General Description

Flood mitigation drains are designed to convey heavy local rainfall or floodwaters away from rural and urban areas; they are not designed to carry water at a uniform velocity.

During periods of flood, the velocity in the drain is variable because of the head differential at the headworks. The headworks are equipped with automatic flap floodgates.

On the coast, some drains are high level with the bed of the drain above the high tide mark and, because they are nearly always dry, aquatic plant growth is minimal. Other flood mitigation drains have beds which are below mean tide level. This type of drain is subject to heavy aquatic plant growth and requires regular maintenance. For example, in the Lower Clarence Valley of northern New South Wales, over 80 per cent of flood mitigation drains are below the low tide level. These drains are generally constructed with horizontal beds and batter slopes 1½:1 or 2:1. Because they are subject to high velocities during a flood recession (falling stage of a flood), it is important to have the batters well-grassed so as to avoid any scouring of the banks.

Aquatic Plant Control Methods

(a) When a drain has been mechanically cleaned to remove siltation (say once every 5-7 years), the drain may be treated with residual herbicide provided that contamination of streams is kept below maximum prescribed levels.

(b) Each succeeding year the drain is generally spot-sprayed (if required) as a routine maintenance operation.

(c) In drains where the bed is below low tide, submerged plants do not interfere greatly with the hydraulic efficiency of the drain. However, during periods of high flows, submerged plants can come adrift, pack up against the flood gates and clog up the whole system. For this reason, it is desirable to remove the plants.

(d) Saline water can be introduced into the drains by opening the flood gates, thereby controlling salt sensitive plants.

(i) Water Hyacinth *(Eichhornia crassipes)* is killed by sea water. A salt concentration in the order of 8000-10 000 ppm will also kill this plant after several weeks exposure.

(ii) *Salvinia molesta* is severely restricted in growth after several weeks of exposure to salinity levels around 4000 ppm.

(e) In many instances, a surface cover of Azolla (*Azolla* spp.) and/or Duckweed (*Spirodela* spp. and *Lemna* spp.) controls the rate of growth of submerged weeds, presumably by shading. In some situations, it may be desirable to encourage the growth of the floating plants.

Control of Aquatic Plants

Mechanical Methods

In Australia there has been relatively little development in the mechanical field, as herbicides generally provide more permanent control and are less expensive. However, control of aquatic weeds is entering a new phase where more emphasis is placed on biological control, habitat manipulation and mechanical control. At present only a few herbicides are registered for use in water and the restrictions on their use are becoming more stringent. These factors are increasing the need to develop more sophisticated mechanical weed control methods. However, mechanical control, unless well directed, may result in the elimination of desirable species and the encouragement of undesirable species.

Sophisticated harvesters are being developed overseas. One type throws the chopped submerged and floating vegetation in a stream of water onto the bank; another system cuts, loads and transfers the plants from the water to the bank, and lifts the vegetation onto trucks.[10] A further type draws the plant into a chamber and chops it into pieces small enough to prevent vegetative propagation. Buoyancy gases in the plant are exhausted, leaving a chopped debris on the lake or channel bed. A problem with this system is the detrimental effect of decaying debris on the water body and associated animals.

A number of weed-cutting boats have been developed overseas. These machines have hydraulically operated cutter bars designed in flat, U or inverted T shape. They can cut most species of submerged and emergent vegetation with a cutter bar that can be lowered down to about 1.5 metres into the water. The cut vegetation floats on the surface and is collected manually or mechanically with weed buckets.

A hand-held submarine saw, although labour intensive, is very effective in cutting submerged weeds in channels. The saw consists of a high-tensile steel ribbon with saw teeth on both edges. Weighted with lead at one metre intervals, the saw blade is drawn across the channel by operators on each bank.

Other devices used in channels include ships' anchor chains and arrow-cutters drawn by cables attached to tractors or A-frames towed by powerful boats. The disposal of huge quantities of cut material is a problem associated with all these mechanical methods.

Cutting emergent plants, particularly Common Reed *(Phragmites australis)* and Cumbungi *(Typha* spp.)* below the water line in autumn will severly retard plant growth, and with the latter will sometimes kill out the entire stand.

Wilder Weed Boat cutting *Typha* in Forest Creek near Deniliquin

A supply channel near Warren kept free of plants by mechanical scraping

Weeds on ditch banks and waterway margins are controlled efficiently with offset flail-type mowers, provided there is suitable access on the channel bank and the batter is well graded.

Sloper blades on graders are also used. The blade is angled down the batter, slicing bank and marginal vegetation and rolling it onto the top of the bank. This practice is common in channels supplying water to row crops in the United States and the cotton-growing areas of New South Wales.

V-shaped ditchers or scrapers with sloping blades that are drawn along the bed are commonly used to clean small irrigation and stock and domestic channels.

Dragline and excavating machines with weed or digging buckets attached are used to remove submerged and emergent weed growth. This method is a slow and expensive operation when compared with herbicides, and only a limited amount of channel cleaning is possible each year. Mechanical cleaning will aid the spread of plants by fragmentation, and in the case of Para Grass *(Brachiaria mutica)*, Common Reed *(Phragmites australis)* or Alligatorweed *(Alternanthera philoxeroides)*, may spread the plant into adjacent agricultural land. Another problem with excavation of channels is placement of the spoil and its effect on access tracks. Weed buckets are preferable to digging buckets because weeds and a little silt are all that is removed and the channel is not deepened unnecessarily. A foliage-applied translocated herbicide may be used on perennial weeds a few weeks before mechanical cleaning to reduce regrowth from roots.

Dragline with a weed bucket clearing cut Ribbonweed *(Vallisneria gigantea)* and Curly Pondweed *(Potamogeton crispus)* from Brays Dam Diversion Channel near Benerembah

The nutrients nitrogen and phosphorus accumulate in the tissue of aquatic plants, and the removal of this growth by mechanical means may be beneficial to a eutrophic water body. In lakes, harvesting will not have a significant impact on the nutrient status unless a large quantity of plants is removed in relation to the volume of the lake and the amount of incoming nutrients[13]. In stable natural systems these nutrients are removed via the food chain, mostly by fishes and birds. Commercial fishing is an efficient method of nutrient removal.

Efforts to utilize the cut plant material for food or fibre have generally failed, one of the few exceptions being the industry built up in Romania utilising Common Reed *(Phragmites australis)* for fibre.[15] At present the dry matter fibre and food yield of submerged and floating weeds are too low compared with terrestrial plants to allow for economical mechanical harvesting and transport. Most experiments to date have not concentrated sufficiently on incorporating the cut plant material into existing plant management practices. Aquatic plants consist largely of air and water. For example, freshly harvested Eurasian Water Milfoil *(Myriophyllum spicatum)* contains over 90 per cent water by weight and over 77 per cent air by volume.[10] If the weed is cut, large quantities of energy are consumed in lifting and transporting very heavy masses of plant material from the site. However, harvesting of plants allows for immediate use of the area and avoids deterioration in water quality from plant decay. Decay of submerged plants following a chemical treatment depletes oxygen supplies and releases nutrients for new plant growth.

Biological Control of Aquatic Weeds

Insects, Herbivorous Fish, European Carp

The use of insects to control serious aquatic weeds is described on page 474 of this book. Herbivorous fish, notably the White Amur or Chinese Grass Carp *(Ctenopharyngodon idella)*[11] and species of *Tilapia* are the subject of considerable research overseas. Their value in controlling aquatic weeds in Australia is not known. In southern Australia the European Carp, *(Cyprinus carpio)*, has had a substantial effect on submerged aquatic plants in some waterways, tearing the plants during feeding. Plague numbers of Carp in search of food disturb the silt in drains, channels and rivers, removing as much as 0.5 metres of the bed. As a result there have been cases of increased seepage. Large numbers of Carp turn water muddy and generally pollute the water. *(See "European Carp" on page 461)*.

Pathogens

Some fungi, bacteria, viruses, nematodes and mycoplasms are pathogens which may naturally attack plants. These pathogens usually do not infect animals or man, but are at times very destructive, and have great potential for reducing the vigour of aquatic plants. This field has received very little attention, but there is increasing interest by researchers in Australia and overseas.

Grazing Animals

A well-known form of weed control is the grazing of channel banks and lake shorelines by stock. Sheep and goats are very useful in keeping rank vegetation down; cattle are also useful, but because of their greater size and weight may cause damage to structures or break down channel banks. Contamination of the water from animal droppings limits this management technique.

Plant Competition

Ground-cover plants and replacement vegetation are other means by which long-term and inexpensive weed control can be achieved. Because this weed control measure offers long-term low-cost benefits, plant breeders are now turning their attention to the selection of ground cover variations chosen from existing useful species. A plant that is a weed in one situation may be very useful in another.

Planning Vegetation Replacement

If the newly constructed channels pass through undisturbed areas of native vegetation, it is best to wait and see whether natural plant populations are affected before conducting any extensive planting program. It is also necessary to guard against the spread of an introduced species into natural wetlands. Where channels deliver water to areas used for intensive agriculture, it is necessary to plan planting programs from an earlier stage. Channel banks can be sources of both water weeds and agricultural weeds so the establishment of a sward comprising species that will not be weeds in the irrigated crop is desirable.

The aim is to produce as complete a cover of low-growing plants as possible which can maintain itself with little active management and successfully compete with undesirable species. The use of fertilizers should be kept to a minimum and generally only sufficient used to ensure establishment. Fertilizers may promote excessive growth and could contaminate the water. If grazing is not practicable, then lawn species could be suitable. In either situation a mixed species sward is likely to be more vigorous and more resilient than a single species sward. The basic mix should include a rhizomatous or stoloniferous grass and a legume (e.g., species of *Trifolium*).

The available seed selection may not suit the locality where planting is planned. Check with the organisation marketing the seed before sowing, or propagate from pieces known to thrive locally.

Ground-Cover Plants for use Above the Waterline in New South Wales

If the area can be grazed, then a selection of the locally used pasture species would be appropriate. Also, consider the following species in addition to the pasture species or instead of the pasture species if grazing is not practicable:

Kikuyu *(Pennisetum clandestinum)* planted on a channel bank near Myall Park

Juncus usitatus forming a stable plant community on the margins of a supply channel near Bilbul. Removal of *Juncus* often results in a more troublesome species becoming established

514

Pennisetum clandestinum (Kikuyu)

A valuable competitive species for planting on channel banks. Generally, should only be used on channels with permanent flows and capacities above fifty megalitres per day. May become obstructive to water flow in farm channels, but easily controlled with herbicides and by grazing animals. Grows best in humid conditions but will thrive in the humid microclimate adjacent the water's edge in the dry south-west of New South Wales. Frost sensitive. Established from pieces or seed.

Cynodon dactylon (Common Couch)

Perhaps the most versatile species, it tolerates dry conditions, low fertility and high salinity levels better than most other species. Dies back in winter but resumes growth the following spring. Suitable for most areas of the State. Moderately resistant to glyphosate and atrazine, and low rates of these herbicides could encourage the spread of Couch. Vegetative propagation from local plants is often more successful than sowing seed.

Axonopus affinis (Carpet Grass)

A potentially useful species. Tolerates limited periods of waterlogging and low fertility levels. More useful in the wetter areas, but should be capable of growing in most areas of the State, competing successfuly against *Paspalum dilatatum.* Summer growing.

Sporobolus mitchellii (Rat's-tail Couch)

A native species which has great potential in the drier regions of the State. This species should survive under drier conditions and higher salinity levels than *Cynodon dactylon,* but produces less growth in winter.

Digitaria didactyla (Queensland Blue Couch)

A summer-growing native species which could be useful in coastal areas.

Agropyron repens (English Couch or Twitch)

A vigorous lawn species in areas with winter rainfall, but could become a problem in natural wetland areas or in areas adjacent to pastures. Probably more suited to channel banks adjacent to cultivated areas.

Juncus spp.

These species are difficult to plant but often occur naturally along the waterline. It is best not to disturb the smaller-growing species unless they are definitely blocking water flow. *Juncus* spp. often prevent more troublesome species from becoming established. Even the larger-growing *J. usitatus* appears to provide effective competition against *Typha* spp. and other more obstructive shoreline weeds.

Phylla nodiflora (Lippia)

A prostrate perennial herb which is a valuable lawn plant that also has potential for planting as a ground cover species adjacent to channels. Lippia thrives in the wetter parts of the State but could be worth including in seed mixes for other areas.

Dichondra repens (Kidney Weed)

Similar to *Phylla nodiflora* but rather hardier and may be useful in a wider range of conditions.

Trifolium arvense (Hare's-foot Clover); *T. dubium* (Suckling Clover); *T. campestre* (Hop Clover)

These are all annual clover species, listed in order of their decreasing ability to withstand moisture stress. One of these species is probably worth including in any seed mix for channel banks. Winter-spring growing.

Trifolium subterraneum (Sub Clover)

This is a more vigorous annual clover than the previously listed species. Winter-spring growing, it can maintain itself in areas of winter rainfall and under low nutrient levels.

Eragrostis curvula (African Love Grass)

A controversial pasture species which, under some conditions, can become a problem. Plantings should only be of palatable strains and should not be done without consulting local agricultural authorities. This species does seem to have some potential as a summer grower in the drier regions.

Pratia concolor and *P. darlingensis* (both Pratia)

Low-growing native perennials which will survive waterlogging. These species occur naturally on the heavy soil floodplains adjacent to New South Wales inland rivers and are potentially useful species for growing on channel banks.

Brachyachne convergens, B. ciliaris

Native species related to, and superficially similar to, *Cynodon dactylon*. Both *Brachyachne* species are worth trying in the drier, summer rainfall areas in the north of New South Wales.

Some tree species can be used to shade waterways and hence reduce waterplant growth. Care is needed, however, as the larger woody roots could damage concrete structures and the smaller, finer roots and trailing branches may restrict water flow. Trees may also prevent mechanical cleaning of the channel. *Salix* spp. (Willows), *Casuarina cunninghamiana* (River Sheoak), *Eucalyptus camaldulensis* (River Red Gum), *E. largiflorens* (Black Box), *E. microtheca* (Coolabah) and *E. robusta* (Swamp Mahogany) are worth trying in some circumstances.

Ground Cover Plants for Use Below the Waterline in New South Wales

Care is needed before introducing any aquatic species into irrigation or drainage systems. Any growth will reduce the flow of water through the channel and submerged growth should only be encouraged if the demand for water is substantially less than the design capacity. If this is the case then submerged growth could be encouraged to stabilize an unstable

channel bed or to replace species that are more obstructive or more
difficult to control.

The height of the submerged vegetation has a disproportionate effect on
the rate of water flow. For example, plants 10 centimetres high reduce
water flow by much more than twice the amount than do 5
centimetre-tall plants. The ideal submerged plant to encourage is a
uniformly low-growing perennial species that forms a turf, spreads by
vegetative means, can withstand periodic exposure to air, successfully
competes with other more obstructive species, but could still be controlled
comparatively cheaply.

Studies of submerged species have been very limited in Australia.
Species of the alga *Chara* have been found to compete with submerged
weeds in lakes. *Chara* species tend to taint water and are not particularly
suitable for farm dams or shallow, slowly flowing water.

Species of low-growing or dwarf spike rushes (e.g., *Eleocharis
coloradoensis, E. acicularis* and *E. parvula*), seem capable of competing
with a wide range of submerged aquatic plants in water up to seven metres
deep in North America. Research is continuing in California, U.S.A. on
methods of harvesting seed to enable sowing of these species on the beds of
irrigation channels. The Australian species which seem to be worth
investigating are the closely related *Eleocharis pusilla* and the slightly
larger-growing *E. atricha*. Again caution is necessary, as some species of
Eleocharis can become troublesome weeds on irrigated land.

Control by Environmental Modification

Water Level Management

Periodic lowering and raising of water levels may effectively reduce the
growth of undesirable plants.[13] This has been demonstrated in the
Tennessee Valley impoundments, where excess growth of Eurasian
Water Milfoil (*Myriophyllum spicatum*) has been managed with regular
changes in water level and herbicide treatments.[16] In shallow lakes and
marshes, fluctuating water levels may kill desirable species such as native
waterlilies and encourage bank and shoreline growth. This technique
must be used with caution and in the full knowledge of likely plant
population changes.

Draining the channel in winter or summer is a widely used technique to
control submerged species and is particularly effective if the bed is dried
out. The effect of freezing after draining varies with species, but is less
important than drying as a control measure *(see also section "Reservoirs
etc.—Preventive Measures" — page 506)*.

Shading—Reducing Light Penetration

The shading effect of bank trees such as Weeping Willow (*Salix
babylonica*) and the spreading of fibreglass shade cloth on the bed, are
control methods suitable for some species in small areas.

Dredging the lake, reservoir or channel to depths where light penetration reduces the density of submerged plants may be a suitable control measure in a few situations. However, it is difficult to generalize about this as the depths to which plants grow varies widely with the species; it is also affected by colonization from banks. Turbid and coloured water reduces light penetration and limits growth. For example, the very turbid water in stock dams or tanks constructed in soils containing bentonite clays is usually relatively free of submerged plants.

Other Suppressants

These include salinity, low temperatures, fast currents, wave action, and type of substrate; they may partially or totally limit the growth of water plants.

Chemical Control—Principles Applying to Irrigation Systems, Flood Mitigation Drains, Natural Watercourses, Lakes and Storages.

Generally, where labour and machinery costs are high and the work needs to be done quickly, chemicals are used to control weed problems. Because excessive herbicide contamination of water is always a risk, the task of selecting a herbicide and spraying should not be left to inexperienced operators.

Determining whether or not the problem requires treatment and the ramifications of a proposed treatment should be carefully evaluated.

Professional advice should be sought and Government regulations adhered to if chemicals are to be applied to water used by the public, or if the treated water is likely to flow away from the point of application and be used downstream.

There are five major considerations that follow a decision to chemically control aquatic weeds:

1. A herbicide registered for the purpose should be correctly applied.

2. Organisms other than the target plant must be protected; this includes the use of protective clothing and safety equipment by spray operators.

3. Levels of contamination of the herbicide in the water must not exceed those laid down by law.

4. The timing of the application should be close to optimum, using local knowledge of the susceptibility of the species.

5. Follow-up treatments should be planned to ensure that the benefit of the first treatment is not lost.

Application Techniques: Submerged Plants
Flowing Water: Factors Influencing Results

In flowing water the herbicide is usually injected into the stream at a point where turbulence will result in thorough mixing. The chemical then flows with the water, controlling weed growth as it makes contact with the plants. The slow-release pellet is another technique useful in flowing water. The herbicide is released from the pellet over a number of days. The distance over which control is achieved depends primarily on the persistence of the herbicide, and the water velocity. Other factors such as weed density, duration of application, turbidity, water temperature, water quality and concentration of the herbicide and susceptibility of target species, all affect the distance over which control is achieved.

The application should be planned so that a lethal quantity of herbicide reaches the most downstream section of the area to be treated. *Considerable local experience is often necessary to ensure an effective and economical treatment.* If necessary, booster treatments are made downstream at application rates that may be only one-half of the initial rate.

Before application it is important to identify the plants in the channel and the density of the growth. From this the correct amount of herbicide can be calculated. One method of assessing plant density and determining the species present is to cast a hook into the water at regular intervals along the channel.

Eriofloxine dye can be used for visual tracing of the herbicide in the flowing water. This can be done by adding 0.02 kilograms of dye for each one megalitre per day flow. If the dye is added about two hours after the injection of herbicide ceases, the centre of the block of dye will probably remain behind the herbicide block for at least a day. Before using a dye, check with the manufacturer that it is environmentally acceptable. Only dyes which are registered as tracers in water may be used. The block of water containing the herbicide will usually travel at about 0.7 times the velocity of an object floating on the water surface.

Injection into Flowing Water: Calculating Amount of Herbicide Required

To determine the quantity of herbicide required for a given treatment, it is necessary to know the channel discharge and the exposure to herbicide required to control the plants. This will depend on both the injection time and concentration. The formula $t/24 \times Q \times X$ gives the quantity of herbicide to be injected in kilograms of active ingredient where t is duration of injection in hours, Q is discharge in megalitres per day, and X is parts per million (active ingredient).

Where the injected material dissipates rapidly as it travels downstream (for example, acrolein), the quantity of herbicide injected will be considerably more than the quantity theoretically needed to control the plants.

For example, for acrolein in flowing water about half the material is lost for every four hours of travel time.[5] So if the lethal exposure of material required for a given species is K kilograms and the travel time to the furthest downstream point is four hours, the injection would need to be 2K kilograms. For a travel time of eight hours, 4K kilograms would be needed, etc.

Species differ considerably in their response to herbicides. For example, the quantity of acrolein required for control of Ribbonweed *(Vallisneria gigantea)* is about two ppm for two hours and for Floating Pondweed *(Potamogeton tricarinatus)* about 13 ppm for two hours.[5] Circumstantial evidence suggests that Elodea *(Elodea canadensis)* is far the most sensitive of these three species to the herbicide acrolein.

Static Water

Total Water Body Treatment: Calculation Based on Volume.

With this technique the entire water body is treated. A calculation based on volume is made and the herbicide is introduced at a concentration expressed in kilograms per megalitre or parts per million. The chemical is injected under the water surface by a trailing boom or sprayed over the water surface. This technique is being replaced with the bottom area treatment.

Bottom Area Treatment: Calculation Based on Area.

Recently it has often been found unnecessary to treat the entire water volume to control aquatic vegetation in static or very slowly flowing water. The area occupied by the weed is calculated and the chemical is placed on the bed or in bottom weed growth using granules, spray thickeners, invert emulsions, or polymers. The technique is most suited to lakes where bottom muds are shallow and there is very little water movement. The herbicide is mainly confined to the colder water at the bed of the lake and so fish may escape any effect of the chemical. However, some fauna which are part of the fish food chain may be destroyed by the treatment. This method is also suited to canals where the water can be retained in storage for a period, or in flood mitigation drains when they are not in use.

Both invert emulsions and liquid polymers can be used to sink herbicides onto submerged foliage. Invert emulsions can be thickened and weighted and incorporated with an aquatic herbicide in a specialised pump. Liquid polycarboxylate polymers are used successfully as herbicide carriers. These polymers are mixed with liquid or wettable powder formulations, to form a thick paste. In use they are handled in the same manner as an invert emulsion, being injected below the water surface and then breaking up to adhere to submerged aquatic plants. With a 2 per cent polymer solution, complete release of the herbicide occurs within about twenty-four hours. The polymeric material is degraded by micro-organisms and is not present after about the fourteenth day in static water.

Equipment and Application Techniques

Misting machines and backpack sprayers are useful in spraying small patches and inaccessible areas. The misting machine has the advantage of using very little diluent. Contamination of water with herbicides may also be reduced. However, misting machines are labour intensive and drift of fine droplets can sometimes damage adjacent crops.

More recently, controlled droplet applicators (CDA) have revolutionized low-volume application. There are a variety of models which are hand-held, light-weight units, comprising a 1 to 2.5 litre spray vat and a spinning disc connected to an electric motor powered by torch batteries. When correctly operated, uniform droplets of the herbicide are produced and these are of sufficient size to reduce drift. By using CDA with suitable herbicides, a better result is achieved when compared with high volume equipment. The capital cost of CDA equipment is small; also, operating costs are reduced by avoiding costly transportation of diluent, since application volumes are only a few litres per hectare. One disadvantage of CDA is that the spray is nearly invisible and water could be inadvertently contaminated. The nearly invisible droplets also increase the likelihood of spray operators being contaminated with concentrate.

An even more recent development is the rope or wick applicator. This simple method of applying herbicides has potential for ditch-bank, industrial and inter-row weed control. With one device the herbicide moves by gravity through a bank of ropes dragged or suspended from a bar. As the rope is drawn over the vegetation, concentrate from a tank, connected to the rope, is wiped onto the plants.

This technique enables the operator to control tall vegetation and leave low growing species untouched. Advantages over spray application include elimination of drift, reduced contamination of the spray operator with herbicides, and elimination of herbicide wastage on soil and other non-target objects.

Pumps that create an invert emulsion of the chemicals to be applied are a recent development in the United States of America.[6] The herbicide is incorporated in an oil phase and applied to ditch-bank foliage or floating plants, eliminating drift and reportedly reducing the application rate by as much as 50 per cent.

Aerial application of herbicides to emergent and floating plants is suitable where the areas are extensive or inaccessible. Fixed wing aircraft fitted with a Micronaire (R) boom can spray areas adjacent to crops with a greatly reduced chance of drift and crop damage. Helicopters may be

Rope applicators are
simple to operate and
reduce herbicide
contamination of water

A hovercraft applying a herbicide through a trailing boom to Azolla *(Azolla filiculoides)* and Hornwort *(Ceratophyllum demersum)* on a lake in a Brisbane park

useful in less accessible situations. Downdraft dispersion of the spray by the rotor may be a disadvantage if a narrow swathe is being treated, but this movement is advantageous where penetration into dense stands is necessary and the area to be treated is large.

When ditch-banks and water bodies are to be sprayed, tractor or truck mounted or trailed sprayers equipped with high pressure pumps are generally used. Where low pressure is needed, centrifugal pumps are satisfactory. Diaphragm pumps have largely replaced gear and roller pumps. Gear and roller pumps are not suited to abrasive or corrosive materials such as wettable powders or copper sulphate. Where spraying is continuous, the piston pump developed for use in orchards is very reliable. The piston pump generally resists corrosion and, although relatively expensive to repair, gives years of trouble-free performance.

Airboats and hovercraft and other amphibious craft are particularly useful to transport spray equipment in wetlands. Water surrounding the boat or craft is drawn into the pump. A venturi draws concentrated herbicide into the pump through an orifice disc, enabling operators on the craft to spray continuously in shallow or deep water.

Factors Influencing Results: Channel Weed Control

Because of the diversity of situations and the range of herbicides used to control aquatic plants, it is difficult to specify the many factors that will influence results. However, some generalisations can be made, and the following points are worth considering:

Application of Translocated Herbicides to Emergent Foliage

(a) Most of the major emergent aquatic weeds in New South Wales are more sensitive to herbicides applied in mid-summer and autumn than in the spring and early summer.

(b) Perennial plants usually require more than one application with most herbicides, and the follow-up sprays may be critical if eradication is required. Follow-up treatments are usually more closely spaced early in the growing season. The interval between applications is often not so critical in the autumn. A notable exception to this is the herbicide glyphosate. After the initial application with this chemical some plants are not receptive to closely spaced follow-up treatments.

(c) Prior to spraying, water should be drained from the channel. A less effective result may occur where a herbicide is applied to plants half covered with water. Depending on the herbicide, sufficient time should be allowed for the chemical to enter the plant before it is inundated again.

(d) Perennial weeds, such as Water Couch *(Paspalum paspalodes),* may encroach from outside the channel. If eradication is required, ensure that the whole infestation is fully sprayed.

Application of Residual or Soil Active Herbicides

(a) The spray unit should first be calibrated. Nozzle output is determined and then speed of travel is selected so that good coverage is achieved. Two or three practice runs are necessary to obtain an average speed. Next, practice spray an area of, say, 100 square metres, noting the time taken. Then calculate how much herbicide to add to the tank to achieve the correct rate per hectare.

(b) Where powder herbicide formulations are used, adequate agitation of the spray mix is usually necessary to ensure a uniform concentration.

(c) Some trees are sensitive to residual herbicides and will be damaged if roots extend into the treated zone. If it is essential to treat the area adjacent to trees, deep rip and sever the roots between the tree and the residual herbicide.

(d) Residual herbicides need to be incorporated into the soil. Depending on the herbicide, this is achieved by rainfall or physical incorporation or by holding the channel full of water for a few days after spraying. Good results can also be obtained by applying some herbicides with low solubility to static water.

Herbicides In Water – Reduction of Pollution and Assessing the Risk

Water may become contaminated with herbicides through runoff from surrounding agricultural and industrial land. Direct contamination usually results from herbicides used to control aquatic vegetation.

TABLE 2
SOME HERBICIDES USED IN THE AQUATIC ENVIRONMENT*–PERSISTENCE AND TOXICITY[2,4,5,17,19]

Chemical	Persistence In Water	TOXICITY**			Oral LD$_{50}$*** Rats mg/kg
		Mammals	Fish	Crops	
Acrolein	X	X	XXX	X	46
Xylene	X	X	XXX	X	1590
2,2-DPA	XX	X	X	XX	9330
Amitrole	XXX	X	X	XX	25 000
TCA (Sodium Salt)	XX	X	X	XXX	3320
Diuron	XXX	X	X	XXX	3400
Dichlobenil	XXX	X	X	XX	3160
Glyphosate	XX	X	X	X	4320
Copper Sulphate	XX	XX	XX	X	300
Diquat	X	X	X	X	231
AF 101 (Kerosene-Surfactant mixture)	X	X	X	X	Not available

KEY   X   Non-persistent in water (days) or non toxic

        XX   Moderately persistent in water (weeks) or very low to moderately toxic

        XXX Very persistent in water (months) or very toxic

*   Listing of the herbicide does not imply registration for use in aquatic situations in New South Wales.

**   Toxicity at concentrations normally found in water after direct injection, ditch bank spraying or soil application.

*** LD$_{50}$ The dose or amount of an active ingredient which, when taken by mouth or absorbed by the skin, kills half of the test animals; an expression of a compound's toxicity. LD$_{50}$ is generally expressed in mg/kg (milligrams per kilogram). It is commonly used to measure acute oral toxicity or acute dermal toxicity.

        (*Note:* The lower the LD$_{50}$ value, the more poisonous the pesticide.)

Regulations of Acts covering the use of herbicides have been framed to limit the harmful effects of herbicides on the environment. Regulations of the New South Wales "Pesticides Act, 1978" now make it an offence to use a herbicide in a manner other than specified on the label. The New South Wales "Clean Waters Act, 1970" limits the concentration of herbicides in water.

Licensing of applicators and the greater use of contractors is a further way of reducing the misuse of herbicides in aquatic weed control, as is a continuing educational program directed at spray operators and supervisors of herbicide programs.

Herbicides may affect aquatic ecosystems and fisheries and contaminate water used for agriculture, recreation and domestic purposes. The risk depends on the properties of the chemical and the amount entering the water (see Tables 2 & 3).

The level of contamination is affected by the life of the herbicide in water, the amount used, and the type of application.

TABLE 3

SOME IMPORTANT AQUATIC HERBICIDES—THEIR LIFE IN SOIL AND UPTAKE BY PLANTS*[17,18]

	Life in Soil (at normal rates of application)	Uptake by plants through	
		Soil	Leaves
Glyphosate	XX**	—	✔✔
Dichlobenil	XX	✔✔	—
Amitrole	XX	✔	✔✔
2,2-DPA	XX	✔	✔✔
TCA	XX	✔✔	✔
Diuron	XXX	✔✔	✔

KEY
- X   Non persistent (Life of hours or days)
- XX   Moderately persistent (weeks)
- XXX   Persistent (more than 3 months)
- *   This table is a broad guide only
- **   No bioactivity when applied to soil
- —   No uptake
- ✔   Very limited uptake
- ✔✔   Taken up

Types of application include:
(a) rope application to marginal vegetation;
(b) spraying bank and marginal vegetation with CDA, low or high volume equipment;
(c) spraying floating vegetation by ground rigs or from the air;
(d) injection into flowing or static water; granules spread into static water;
(e) spraying the channel bed.

> (a) will result in from nil to a few per cent of the herbicide applied entering the water;
> (b) usually results in up to about 25 per cent of the herbicide entering the water;
> (c) and (d) will result in up to 100 per cent of the herbicide entering the water;
> (e) will result in from a few per cent to nearly all entering the water, depending on incorporation in the soil, degradation, the type of herbicide and the interval between application and period of flushing.

Maximum residue limits of herbicides in water used for domestic and irrigation purposes must not be exceeded. However, there is a diversity of crops and irrigation techniques, and crop tolerance levels are not always available. For some sensitive crops it is desirable that there is no contamination of irrigation water. If data is not available, the herbicide must not be used.

Calculating Contamination Levels

Static Water

Where levels have been set, the following formula allows calculation of the potential degree of contamination.

- Determine the amount of active ingredient in the herbicide.
- Determine the volume of the water.

If static water in a channel: Cross section area × length.

If static water in a lake: Average depth × average width × average length *or* surface area × average depth.

Obtain the concentration in kilograms per megalitre or parts per million (ppm) active ingredient as follows:

$$\text{Concentration} = \frac{\text{kilograms active ingredient}}{\text{megalitres water}}$$

Example

A herbicide contains 360 grams (0.36 kilograms) active ingredient per litre, and if, say, ten litres were used in a volume equalling one megalitre, the calculation would be:

$$\frac{0.36\,(\text{kilograms}) \times 10\,(\text{litres})}{1\,(\text{megalitre})} = \quad 3.6 \text{ kilograms per megalitre or } 3.6 \text{ ppm in the water}$$

This assumes 100 per cent of the application enters the water and there is no loss from degradation or absorption.

A further calculation is necessary to allow for the percentage of the spray entering the water.

For example, if 25 per cent of the herbicide enters the water, multiply by 25/100 which means, for the example above, the herbicide concentration in the water is 0.9 ppm. The breakdown rate of the herbicide is a further consideration when calculating the level of contamination.

Flowing Water

For an injection of herbicide into flowing water the same principles apply, but the injection time must also be taken into account in calculating the volume of contaminated water.

Example

For an injection lasting 3 hours into a flow of 1 megalitre per day, the volume of contaminated water is 3/24 megalitres and the concentration in the water from an injection of 10 litres of herbicide containing 0.36 kilograms of active ingredient per litre would be:

$$\frac{0.36\,(\text{kilograms}) \times 10\,(\text{litres})}{\dfrac{3}{24}\,(\text{megalitres})} = 28.8 \text{ kilograms per megalitre or ppm}$$

If the same quantity of material were injected over a day the concentration would be

$$\frac{0.36 \times 10}{1} = 3.6 \text{ ppm}$$

Action that will *reduce* the risk and level of contamination includes:
>strict adherence to label instructions;
>directing herbicide onto the target plant and away from water;
>timing application to periods of maximum flow and thus greater dilution of the herbicide, or to periods when there is no water present;
>using herbicides when the size of the infestation is small; for example early in the growing season or after flood has removed problem vegetation or a dry season has diminished growth, less herbicide will be necessary;
>restricting application to a period when the plant is most susceptible so that less herbicide will be necessary;
>selecting a herbicide with a short life in preference to one with a long life so that contamination will be less in the long term.

Public Liability and Common Law

Herbicide users have a legal responsibility.

The fact that a herbicide is registered does not automatically make it safe to use in every situation.

Equally, the strict observance of label instructions is in itself no guarantee that non-target organisms are safe from damage.

In a 1978 court action* concerning herbicide crop damage, a New South Wales Supreme Court Judge said:

"Every person . . . must take *reasonable care* to avoid acts or omissions which he . . . can reasonably foresee, or should reasonably foresee, would be likely to cause harm or loss to others."

The critical words are *reasonable care.*

"The defendant does not guarantee that its activities will not harm the property of others. It is not required to be infallible or perfect but it must take reasonable care in the circumstances to prevent foreseeable harm by what it does or it fails to do."

"The more inherently dangerous a thing or operation is, whether to persons or to their property, the higher in general is the care required in order to fulfil the duty which the law imposes of taking reasonable care in the circumstances."

It is the responsibility of every herbicide user to take reasonable care, and this could include informing neighbours of intended treatments which may lead to damage. Because water flows through boundaries, a "neighbour" may be someone living a hundred kilometres downstream.

**Savage and Others v Water Resources Commission, 1978.*

References

[1] ASTON, Helen I. (1973) "Aquatic Plants of Australia." (Melbourne University Press: Melbourne) 368 pp.

[2] BILL, S. M. and GRAHAM, W. A. E. (1970) "Chemical Weed Control in Irrigation Channels and Drains." 3rd ed. (State Rivers and Water Supply Commission: Victoria) 33 pp.

[3] BOWMER, Kathleen, H. (1979) Management of Aquatic Weeds in Australian Irrigation Systems. *Proceedings 7th Asian Pacific Weed Science Society Conference, Sydney* pp. 219-222.

[4] BOWMER, Kathleen H. and SAINTY, G. R. (1978) Herbicide Residues in the Murrumbidgee and Coleambally Irrigation Areas, Australia. *Proceedings 5th International Symposium on Aquatic Weeds, Amsterdam,* pp. 163-170.

[5] BOWMER, Kathleen H. and SAINTY, G. R. (1977) Management of Aquatic Plants with Acrolein. *Journal of Aquatic Plant Management* 15: 40-46

[6] BURKHALTER, A. P., CURTIS, L. M., LAZOR, R. L., BEACH, M. L. and HUDSON, J. C. (Eds)(1973) "Aquatic Weed Identification and Control Manual." (Bureau of Aquatic Plant Research and Control: Tallahassee, Florida) 100 pp.

[7] DUNK, W. P. (1974) The Environmental Implications of Weed Control in Aquatic Situations. *Proceedings Weed Society of New South Wales* 4: 28-33.

[8] DUNST, R. C. *et al.* (1974) Survey of Lake Rehabilitation Techniques and Experiences. *Wisconsin Department of Natural Resources Technical Bulletin* no. 75. 182 pp.

[9] GRAHAM, W. A. E. (1976) "Aquatic Weeds: Observations on their Control in New Zealand." (State Rivers and Water Supply Commission: Victoria) 149 pp.

[10] KOEGEL, R. G., BRUHN, H. D. and LIVERMORE, D. F. (1972) Improving Surface Water Conditions through Control and Disposal of Aquatic Vegetation. *University of Wisconsin Water Resources Center Technical Report* OWRR B-018-Wis. 45 pp.

[11] MITCHELL, D. S. (1978) "Aquatic Weeds in Australian Inland Waters." (Australian Government Publishing Service: Canberra) 189 pp.

[12] MITCHELL, D. S. (1979) "The Incidence and Management of *Salvinia molesta* in Papua New Guinea". (Office of Environment & Conservation: Papua New Guinea) 51 pp.

[13] NICHOLS, S. A. (1974) Mechanical and Habitat Manipulation for Aquatic Plant Management. *Wisconsin Department of Natural Resources Technical Bulletin* no. 77. 36 pp.

[14] SAINTY, G. R. (1973) "Aquatic Plants Identification Guide." (Water Resources Commission: Sydney) 110 pp.

[15] SCULTHORPE, C. D. (1967) "The Biology of Aquatic Vascular Plants." (Edward Arnold: London) 610 pp.

[16] SMITH, G. E. (1971) Resumé of Studies and Control of Eurasian Watermilfoil (*Myriophyllum spicatum*) in the Tennessee Valley from 1960 through 1969). *Hyacinth Control Journal* 9: 23-25.

[17]SWARBRICK, J. T. (1979) "The Australian Weed Control Handbook." 4th ed. (Plant Press: Toowoomba) 341 pp.

[18]VICTORIA. *State Rivers and Water Supply Commission. Water Research Section* (1976) "Aquatic Weed Identification." 12 pp.

[19]WEED SCIENCE SOCIETY OF AMERICA (1979) "Herbicide Handbook." 4th ed. (113 North Neil Street, Champaign, Illinois, U.S.A.) 479 pp.

The Authors of this Book

Geoff Sainty is the Senior Field Officer specialising in aquatic plant management for the Water Resources Commission, New South Wales. He was born and raised in Sydney, graduated from Wagga Agricultural College in 1956 and worked for five years as an Extension Officer with the New South Wales Department of Agriculture. First became interested in waterplants in 1962 and in 1973 was granted a Churchill Fellowship to study the subject in America. In that same year he started work on this book. In 1980 Geoff was awarded a Diploma in Extension from the Hawkesbury Agricultural College

Surrey Jacobs is a research scientist at the National Herbarium of New South Wales. Graduated B.Sc.Agr. from Sydney University in 1967 and received his Ph.D from the same University in 1974. Has since specialised in the study of Australian grasses and chenopods, becoming involved with waterplants in the last few years.

GLOSSARY

Abbreviations

cm — centimetre
kg — kilogram
m — metre
mg — milligram
L — litre
ml — millilitre
ML — megalitre
ppm — parts per million
sp. — species (one)
spp. — species (more than one)
ssp. — subspecies
var. — variety

A

Abaxial: The side or face of an organ facing away from the plant axis (e.g., the lower surface of a leaf).

Abiotic: Without life.

Absorption: Incorporation of molecules *into* a plant, animal, soil or a solid.

Achene: A dry indehiscent one-seeded fruit formed from a superior ovary with one carpel.

Actinomorphic: Regular or radially symmetrical.

Acuminate: Tapering into a narrow, drawn-out tip.

Adaxial: The side or face of an organ facing toward the plant axis (e.g. the upper surface of a leaf).

Adjuvant: A substance added to a mixture to assist the action of the active component.

Adsorption: The holding or loose-binding of a gas or a dissolved or a suspended substance in the form of a surface film of molecules on the *surface* of a solid.

Aerobic: Supplied with air (particularly oxygen) or requiring air.

Alga (pl. Algae): One of a group of plants including "seaweeds".

Amorphous: Non-crystalline, with no well-defined shape or form.

Anabranch: A branch of a river which re-enters or alternates with the main stream.

Anaerobic: Not supplied with air (particularly oxygen) or not requiring air.

Annual: A plant completing its life cycle, from germination to fruiting, then dying, in one year.

Anther: The part of the stamen producing the pollen, usually borne on a filament.

Antheridium (Antheridia): The sac-like male reproductive organ.

Anthesis: The time when pollen is shed.

Antrorse: Pointing or bent upwards.

Apiculate: With a small abrupt point at the apex.

Archegonium (Archegonia): The female reproductive organ, usually flask-shaped.

Aril: A fleshy outgrowth from the point of attachment or funicle of certain seeds.

Articulate: Having obvious joints: separation may or may not occur at these points.

Auricle: A rounded to pointed appendage at the base of a leaf.

Auriculate: Having auricles.

Awn (Awned): A bristle-like appendage: may be straight, bent or twisted.

Axil: The upper angle between the axis (stem usually) and any organ which arises from it.

Axillary: Arising from the axil.

B

Bacteria: A large group of unicellular or filamentous organisms without a well defined nucleus, and surrounded by a thin wall.

Batter: The slope of a wall, terrace or bank from the perpendicular.

Beak: A pointed projection.

Bed Grade: The ratio of fall to length, i.e., it represents the slope of the channel.

Bentonite Clay: A clay used in filters; formed by the decomposition of volcanic glass under water; consists largely of montmorillonite.

Berry: A fleshy indehiscent fruit without any hard layer and with usually more than one seed.

Biennial: A plant which completes its life cycle from germination to seed set and death in more than one but less than two years.

Bifacial: Having two faces or surfaces.

Biotype: One individual of a population composed of organisms which are genotypically identical.

Bisexual: Having both fertile male and female organs in the same flower; hermaphrodite.

Boom: A length of pipe or tubing connecting two or more nozzles and thereby allowing chemical treatments to be applied over a wider area.

Borrow-pit: The site from which fill material is removed from outside a construction zone.

Bract: A leaf-like structure or scale subtending an inflorescence or flower.

Bracteate: Having bracts.

Bracteole: A small bract on the pedicel or sometimes on the calyx of the flower but not subtending it.

C

Caespitose: Tufted, with the leaves all basal and ± erect.

Callus: The usually sharp basal projection of the floret of some grasses.

Calyx: The sepals collectively, usually used when the sepals are fused to some degree.

Capitate: Enlarged at the tip (e.g. the head of a pin); formed like a head.

Capsule: A dry dehiscent fruit of two or more fused carpels.

Carotenoid: Yellow, orange, brown or red tetraterpene (40 C atoms) plant pigments. They consist of two vitamin A units joined at the hydroxl end of the molecule.

Carpel: A unit of the female part of the flower (gynoecium) in which one or more ovules are enclosed; usually divisible into stigma, style and ovary. Carpels may be free or variously fused.

Carpodium: Sterile female flower produced by species of *Typha*. A partially developed club-shaped carpel.

Carpophore: The persistent central axis found in some fruits.

Cartilaginous: Hard and tough (like cartilage).

Caruncle: A usually spongy outgrowth from the seed coat near the hilum.

Cataphylls: The first formed, reduced leaves at the base of the shoots of monocotyledons (especially grasses).

Catkin: A spike or spike-like inflorescence of unisexual flowers and often with scale-like bracts.

Cauline: Attached to the stem.

Cell: (i) a locule or compartment of the anther
(ii) the basic unit of plant tissue, surrounded by a cell wall.

Chartaceous: Papery.

Chasmogamous: Flowers which open, exposing stigmas and anthers; such flowers are often potentially cross-pollinated.

Chloroplast: A plastid or cell organelle containing chlorophyll; the site of photosynthesis.

Chromosome: One of the rod-like bodies containing genes in the cell nucleus.

Ciliate: Fringed with hairs (cilium, cilia).

Ciliolate: Fringed with small hairs.

Cleistogamous: Flowers which remain closed; such flowers are either self-pollinated or apomictic.

Colloid: A substance in which the particle size varies in liquid between that of true molecules to that of coarse suspensions (10^{-7}-10^{-5}cm).

Connate: Fused to one or more organs of the same floral whorl.

Convection Currents: Currents in liquid or gas set in motion by variation in density caused by variations in temperature.

Convolute: Rolled so that the margins overlap.

Cordate: Heart-shaped in outline, particularly in reference to the base of a leaf.

Coriaceous: Leathery.

Corm: A short, swollen, upright, underground stem formed annually, often in the stem base, usually storing food reserves and surrounded by protective leaf bases.

Corolla: The collective term for a whorl of petals; usually used when the petals are variously united.

Corona: A ring of tissue, crown-like.

Cotyledon: A seed-leaf of the embryo of seed-producing plants.

Crenate: Having rounded, regular lobes, applied to margins.

Crenulate: Minutely crenate.

Cryptomonads: Algae belonging to the family Cryptophyceae.

Culm: An aerial stem of grasses or sedges, terminating in an inflorescence.

Cultivar: A variety or race that originated and persists under cultivation, not necessarily referable to a botanical species.

Cuspidate: Having a cusp or sharp point.

Cuticle: Non-cellular waxy layer on most external surfaces of plants.

Cytoplasm: A semi-transparent semi-fluid substance of complex chemical composition forming the matrix around cell organelles.

D

Deciduous: Not persistent, falling off.

Decumbent: The lower portions on or $\pm$ parallel to the ground but the upper portions ascending.

Dentate: Having regular (symmetrical) pointed teeth; applied to margins.

Denticulate: Minutely dentate.

Desorption: The removal of a substance from the surface at which it is absorbed.

Destratification: The break-down and intermixing of strata or layers.

Dichotomous: Dividing into two approximately equal branches.

Dicotyledons (Dicots): A group of flowering plants (angiosperms) whose embryos usually have 2 cotyledons.

Diffusion: The passive molecular inter-penetration or mixing of two or more liquids or gases without chemical combination.

Digitate: With 2 or more simple branches from a short vertical axis; finger-like.

Diluent: The substance (usually a fluid) that serves to increase the proportion of liquid (usually inert) in a mixture.

Dimorphic: Occurring in two different forms.

Dioecious: Having unisexual flowers borne on separate plants, i.e. having male and female plants.

Disc Floret: The small flower or floret (usually tubular) borne on the central portion of the head in some Compositae (Asteraceae).

Dissolution: Separation into parts or constituent substances.

Distichous: Arranged in two opposite rows.

Diurnal: Completed in one day, active during daylight hours.

Divaricate: Very divergent, spreading or straggling.

Dorsal: The back of, or attached to the back of an organ.

E

Elliptic: A symmetrical oval outline with the length greater than the width.

Embryo: A young plant within the seed or developing seed.

Emergent: Growing or protruding above the water surface, as opposed to floating or submerged.

Enzyme: A protein in living cells that increases the efficiency of biochemical reactions; cf adjuvant.

Epiphyte: A plant growing on other than a soil or water substrate, e.g. another plant, rockfaces, etc.

Erose: Appearing eroded or gnawed, irregularly lobed or divided.

Espatheate: Without a spathe.

Eutrophic: Conditions conducive to supplying high nutrients. In aquatic areas, usually associated with low oxygen levels.

F

Fascicles: Bundles or clusters (fascicled).

Filament: The stalk of a stamen.

Filiform: Long and thin; almost thread-like.

Flagellum (Flagella): A small whip-like organ.

Free Board: Difference between crest level and water level; provides a safety margin for variation in flows.

Frond: The leaf-like organ of a fern.

Fucoxanthin: A brown carotenoid pigment, common in some algae

Fulvous: Covered with brown hairs.

Fungus (Fungi): A major group of living organisms without chlorophyll and differing from plants in a number of other fundamental characters.

G

Gametophyte: The haploid generation producing the sexual reproduction organs in algae, bryophytes and ferns.

Gel: A jelly-like material formed by the coagulation of a colloidal liquid.

Gene: An hereditary factor which either alone or in conjunction with other genes, produces a character in an organism.

Gene Pool: Pertaining to the sum of genetic variation (i.e. the range of genes) possessed by a population or a taxon.

Genetics: The study of variation and heredity.

Geniculate: Bent abruptly, like a knee.

Gibbous: With a rounded, swollen, projection.

Glabrous: Without hairs.

Glaucous: Covered with a bluish-white waxy or powdery coating.

Globose: Globular, spherical.

Glume: (i) one of usually 2 sterile bracts at the base of the spikelet in the Gramineae.
(ii) the bract subtending a flower in the spikelet of Cyperaceae.

Gynophore: The stalk of a superior ovary.

H

Hard Water: Water containing soluble salts of calcium and magnesium or, less commonly, iron.

Hastate: Triangular with spreading basal lobes.

Herb: A plant which does not produce a woody stem.

Hermaphrodite: Having both male and female sex organs in the same flower; bisexual.

Heteromorphous: Of more than one type or morphology.

Heterostylous: Producing styles of more than one type, e.g. some short, some long, some intermediate.

Hilum: The scar remaining on the seed at the point of attachment.

Hirsute: Covered with spreading hairs.

Hispid: Densely covered with stiff, usually short, hairs or bristles.

Homostylous: Producing styles of only one type.

Hyaline: Translucent, thin.

Hybrid: A plant resulting from a cross of plants from two different species.

Hydrated: Combining with water.

Hydrolysis: Decomposition of water into ions, H^+ and OH^-.

Hydrosoil: The bed of a waterbody or stream.

Hypanthium: The elongated tissue above an inferior ovary bearing the perianth and stamens at its summit; common in Hydrocharitaceae.

Hypogonous: Referring to stamens or also to the petals and sepals which are attached to the receptacle below the gynoecium.

I

Inferior Ovary: An ovary which is situated below the point of attachment of the calyx.

Inflorescence: A group of flowers arising from a common axis (or stem).

Inorganic: Used for chemical substances that do not contain two or more carbon atoms bonded directly to each other.

Internode: The portion of a stem between two adjacent nodes.

Invert Emulsion (Invert System): The suspension of minute water droplets in a continuous oil phase.

Involucre: One or more whorls of bracts surrounding an inflorescence or flower.

Involute: Rolled inwards or upwards.

Ion: A charged atom, molecule or radicle.

K

Keel: A ridge, like the keel of a boat.

Knock-Down Herbicide: Equivalent to a "contact" herbicide, i.e. one that kills plant material that it comes in contact with, it is not translocated and its field life is too short (hours or days) to be regarded as residual.

L

Laciniate: Divided into narrow strips or lobes.

Lamina: Leaf-blade; the expanded portion of a leaf or frond.

Lanceolate: Approximately 3-8 times as long as broad, tapering at both ends and broadest below the middle.

Leaflet: The leaf-like unit of a compound leaf.

Lemma: The lower of two bracts enclosing the flower of a grass.

Lenticular: Lens-shaped.

Ligulate: Having a ligule.

Ligule: A membranous flap or row of hairs; especially used for the structure at the junction of sheath and blade in grasses and for the corolla of ray florets in the Compositae (Asteraceae).

Linear: Narrow and parallel sided (except for apex), many times longer than broad.

Loculus (Locule): A compartment; used in reference to ovaries or anthers especially.

Lodicule: One of the reduced perianth of the grass flower. Usually 1-3 per flower and inserted below the stamens and ovary.

M

Macrophyte: Literally "big plant", used to describe water plants other than microscopic algae, sometimes apparently used to exclude filamentous algae or even all algae except for members of the Characeae.

Macrospore: Used interchangeably with megaspore (hence also macrosporangium, macrosporocarp).

Manning's *n* Coefficient: A roughness coefficient which makes allowances for the different surfaces over which the water flows.

Megalitre: 1 000 000 litres; 1000 cubic metres of water.

Megaspore: The name given to the larger type of spores if the plant produces spores of two distinct sizes; megaspores germinate to form megagametophytes which produce female sex organs.

Membrane: A thin sheet-like structure, usually fibrous and of a complex structure, enclosing or lining all or part of an organ or organelle.

Methanol: Methyl alcohol, the structurally simplest alcohol with one carbon atom (CH_3OH).

Microspore: The name given to the smaller type of spore if the plant produced spores of two distinct sizes; microspores germinate to form microgametophytes which produce male sex organs or spores.

Microsporocarp: A sporocarp containing microsporangia which produce microspores.

Misting (Mist Spraying): A method of spraying where the concentrated spray is atomized into a high velocity air stream, the air acting as a diluent.

Monocotyledons, Monocots: A group of flowering plants (angiosperms) whose embryo usually has only one cotyledon.

Monoecious: Having unisexual male and female flowers both growing on the one plant.

Mucro: A short projection (usually terminal or sub-terminal).

Mucronate: Having a mucro.

Mycoplasm: A non-mycelial vegetative stage of a fungus in which the fungal plasmodium or amorphous body becomes indistinguishably fused with the cell contents to form a mycoplasm.

N

Nectary: A specialised gland which secretes nectar.

Nematodes: Small thread-like worms belonging to the class *Nematoda*.

Node: The area of the stem from which a leaf, branch or bract arises.

Nucleolus: A homogeneous spherical body or organelle within a cell nucleus.

Nucleus: A cell organelle, bounded by a nuclear membrane and containing a complicated system of proteins, nucleoli and chromosomes.

Nut: A dry, indehiscent hard or bony 1-seeded fruit formed from 2 or more carpels.

O

Ochrea (Ochreous) (also sometimes as ocrea): Stipular sheath formed by the fusion of two stipules (especially Polygonaceae).

Oogonium: The female reproductive organ producing oospores in the algae and some other cryptogams.

Orbicular: Circular or almost so.

Organelle: Membrane-bound structures found within a cell, eg. chloroplasts, mitochondria.

Organic: Used to describe chemicals which have two or more carbon atoms directly bonded together.

Ovary: The lower portion of one or more fused carpels enclosing one or more ovules and which, after fertilisation, develops into the fruit.

Ovate: A rounded shape with the length greater than, but less than twice as long as the width with the greatest width below the middle; resembling the longitudinal section of a hen's egg.

Ovule: The megasporangium (or female structure) together with the integuments of a seed plant. After fertilisation the ovule develops into the seed.

Oxidation: Combining with oxygen, or the process in which electrons are lost from an atom or ion.

P

Palea: The upper of two bracts enclosing the flower of a grass.

Palmate: Divided or lobed so that all lobes meet near the point of attachment to the petiole. Palm-like or hand-like.

Panicle (paniculate): Strictly a raceme of racemes but used to describe any open compound inflorescence.

Papilla (Papillae): A small elongated protuberance, usually obtuse and not hardened.

Papillose: Covered with papillae.

Pappus: A whorl of hairs or scales on the top of fruit ("seeds") in many Compositae.

Paramylon: An insoluble carbohydrate related to starch, with the same empirical formula as starch but does not respond to the usual tests for starch.

Parasitic: Living on or in and deriving nourishment from another organism (the host).

Pathogen: A disease-producing organism.

Pedicel: The stalk of a single flower in an inflorescence.

Peduncle: The stalk of an inflorescence or of a solitary flower.

Peltate: (Of a leaf) having the petiole attached on the undersurface and near the middle of the leaf.

Perennial: Plants which complete their life cycle over more than two years.

Perianth: The calyx and corolla of a flower but mostly used when either of these two whorls are not clearly differentiated or when only one or more than two whorls are present.

Pericarp: The wall of a fruit, developed from the ovary wall after fertilisation.

Petal: One of the segments of the inner of the two whorls of flattened structures (perianth) usually surrounding the sexual organs of a flower.

Petiole (Petiolate): The stalk of a leaf.

pH: The symbol of a scale used to measure the acidity of a solution. pH 7.0 is neutral, pH 7 to 1 indicates increasing acidity and pH 7 to 14 indicates increasing alkalinity.

Photosynthesis: The conversion of carbon dioxide and water into carbohydrates on exposure to light.

Phycobiliprotein: Pigments found in some algae. The phycobilin pigments are always associated with proteins.

Phytoplankton: One or few celled small free-floating plants.

Pinna: A leaflet or unit formed on the primary division of a pinnate leaf.

Pinnate Leaf: A compound leaf with leaflets arranged on opposite sides of an elongated axis.

Pinnatifid: Divided into lobes on both sides, about halfway to the midrib.

Pinnatisect: Divided into lobes on both sides almost to the midrib.

Pinnule: The segment formed by the secondary division of a bipinnate leaf; a segment of a divided pinna.

Plankton: A collective name for one to few celled small free-floating organisms.

Plumose: Like a feather with a central elongated axis and finer hairs arising from it.

Pneumatophore: A modified root with a different structure from "normal" roots and adapted to allow gas exchange in roots growing under reducing conditions (especially mangroves).

Pollen: The powdery grains or threads produced by the anthers of flowering plants (angiosperms) or microsporangia of gymnosperms. Pollen grains initiate fertilisation when lodging on the stigmas.

Polymorphic: Of many different forms (usually more than 3).

Protandrous: The male parts of the flower (i.e. the stamens) maturing before the female parts (stigmas).

Prothallus: The structure produced by the germinated asexual spores produced by the sporophyte, also known as a protonema.

Pseudolamina: Literally "false-blade" used in some cases for theoretically secondarily expanded blades.

Pubescent: Covered with short soft hairs.

R

Raceme: An inflorescence with a single axis and stalked flowers with the youngest flowers at the apex.

Radical (of leaves): Arising from the base of the stem, forming a rosette.

Radicle: The rudimentary root of the embryo.

Ray Floret: A small flower or floret with a single petal-like "ray" or lobe, these florets occur on the circumference of the head in some Compositae (Asteraceae).

Receptacle: The parts of the axis which bears the floral parts.

Recurved: Curved backwards or downwards.

Redox: Abbreviation for oxidation-reduction, which is used as an indicator to measure the ease of loss (oxidation) or gain (reduction) of electrons.

Reduction: The process in which one or more electrons are added to an atom or an ion.

Repand: Having an undulating margin with the plane of the undulations at right angles to the plane of the blade.

Residual Herbicide: A pesticide remaining effective in the environment for a comparatively long time (usually weeks or months).

Respiration: Absorption by living organisms of oxygen from the air (or water) and the release of carbon dioxide.

Retinaculum (Retinaculae): A connecting band or means of retention. Used for the bract-like appendages near the margins of the spadix in Zosteraceae.

Retrorse: Bent backwards or downwards.

Revolute: Rolled backwards from the margins on to the under surface.

Rhachilla: The axis of a grass spikelet.

Rhachis: The axis of a compound leaf or of an inflorescence.

Rhizome: An underground stem.

S

Sagittate: Shaped like the head of an arrow (of Roman times) with a point and two acute backward pointing lobes.

Scaberulous: Slightly rough to the touch.

Scabrous: Rough to the touch; with short, stiff emergences.

Scarious: Dry and membranous.

Secund: Arranged along one side only.

Sepal: One of the outer of the two whorls of flattened structures generally surrounding a flower.

Septate: Having septa.

Septum (Septa): A partition.

Serrate: Notched on the margin with asymmetrical forward pointing teeth.

Serrulate: Minutely notched.

Sessile: Without a stalk.

Setose: Covered with stiff hairs or bristles (seta).

Silicon: A non-metallic element, one of the most abundant, commonly occurring combined with oxygen as silica.

Siliqua: A dry dehiscent fruit formed from a superior ovary of 2 carpels and with 2 parietal placentas (placentas attached to the wall) connected by a false septum. A characteristic fruit of the Brassicaceae (Crucifereae).

Sinuate: With a deep wavy margin.

Sinus: A depression or cleft between adjacent lobes.

Spadix (Spadices): A spicate inflorescence with a fleshy axis, usually surrounded by a spathe.

Spathe: A large bract which more or less encloses sheath-like the inflorescence, or partial inflorescence it subtends.

Spatheate: Subtended by a spathe.

Spathulate: Spoon-shaped; enlarged and rounded towards the summit.

Spike: A racemose inflorescence of sessile flowers.

Spikelet: A small, reduced spike; usually applied to grasses and sedges.

Spinulose: Covered with small spines.

Sporangium (Sporangia): An organ which produces spores.

Spore: A unicellular or few-celled asexual or sexual reproductive unit not containing an embryo.

Sporocarp: A compact mass of sporangia surrounded by sterile tissue forming a ± globular dispersal unit.

Sporophyte: The diploid generation; the part of the plant which produces asexual spores.

Stamen: The male part of a flower which produces the pollen and consists of an anther and filament.

Staminode: A sterile stamen, often morphologically differentiated from the fertile stamen.

Stigma: The part of the style adapted for the retention and germination of pollen grains.

Stipe: A small stalk.

Stipulate: Having stipules.

Stipule: Appendage, usually in pairs, at the base of the petiole.

Stolon: A lateral stem growing on or above ground and rooting at the nodes.

Stoloniferous: Producing stolons.

Stratification: The formation by natural processes of strata or layers. Used in water to describe the formation of non-mixing layers.

Striate: With fine longitudinal markings.

Style: The part of the carpel above the ovary and bearing (usually terminating in) the stigma.

Sub- (as a prefix): Almost or approaching.

Subulate: Narrow and gradually tapering to a point.

Sudd: A mass of floating vegetation; originated from floating islands of *Cyperus papyrus* on the White Nile.

Superior Ovary: One which is attached to the receptable above the point of insertion of the calyx.

Surfactant: A material which increases emulsifiability, spreading, wetting, dispersibility or other surface-modifying properties.

T

Terete: Cylindrical or almost cylindrical.

Ternate: Arranged in threes.

Thallus (Thalli): The vegetative part of a plant body when it is not differentiated into stems and leaves.

Translocation: The movement of substances in solution inside the vascular system of the plant; with herbicides, referring to the transference of the spray from the point of contact throughout the plant via the vascular system.

Translucent: Allowing the passage of diffuse light.

Trichotomous: Dividing into three approximately equal branches.

Trigonous: With three angles.

Tripinnatifid: Three times pinnatifid; i.e., the lobes on the lobes of the lobes.

Triquetrous: With 3-angles in cross-section.

Truncate: Terminating abruptly; base or apex nearly straight across.

Tuber: The swollen end of an underground stem containing food reserves.

Tuberculate: With swellings or nodules; warty.

Turion: A specialized shoot with numerous shortened internodes and often with reduced leaves; produced by some water plants as over-wintering vegetative propagules.

U

Umbel: A racemose inflorescence in which all the pedicels arise at the tip of the peduncle and the flowers lie at the same level.

Undulate: Wavy, corrugated.

Unisexual: Each flower being either male or female only.

Utricle: A membranous or hardened covering, possibly modified floral parts, surrounding the ovary of some Cyperaceae and Gramineae.

V

Vacuole: A fluid-filled organelle found in many cells.

Vascular: Used to describe the conducting system of the plant; comprised of xylem and phloem.

Veins: The vascular bundles, usually visible externally.

Venation: The arrangement of the veins or vascular bundles of a leaf.

Ventral: The front of, or attached to the front of, an organ.

Venturi Nozzle: A small diameter nozzle fitted axially in a large diameter pipe and uses Venturi's principle to draw fluid through the larger pipe.

Villous: Covered with long, soft, weak hairs.

Virus: Minute parasitic organisms, smaller than bacteria and with characteristic protein structures.

W

Whorl: Three or more organs or appendages arising from one level on the axis.

Z

Zooplankton: One to few celled small, free-floating, animals.

Zygomorphic: Irregular or radially asymmetrical.

Zygote: The cell resulting from the union of two reproductive cells or gametes.

LEAF MARGINS

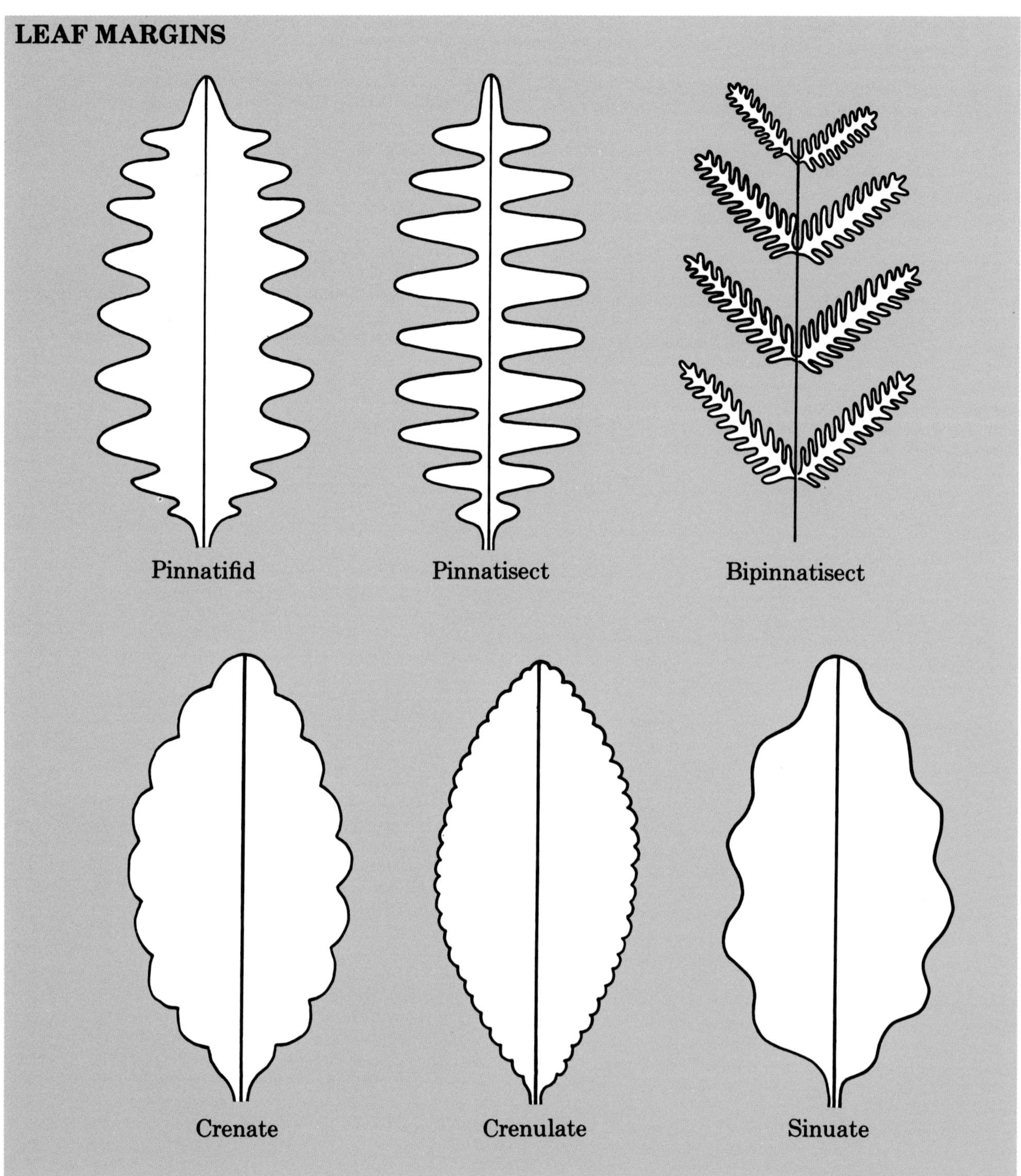

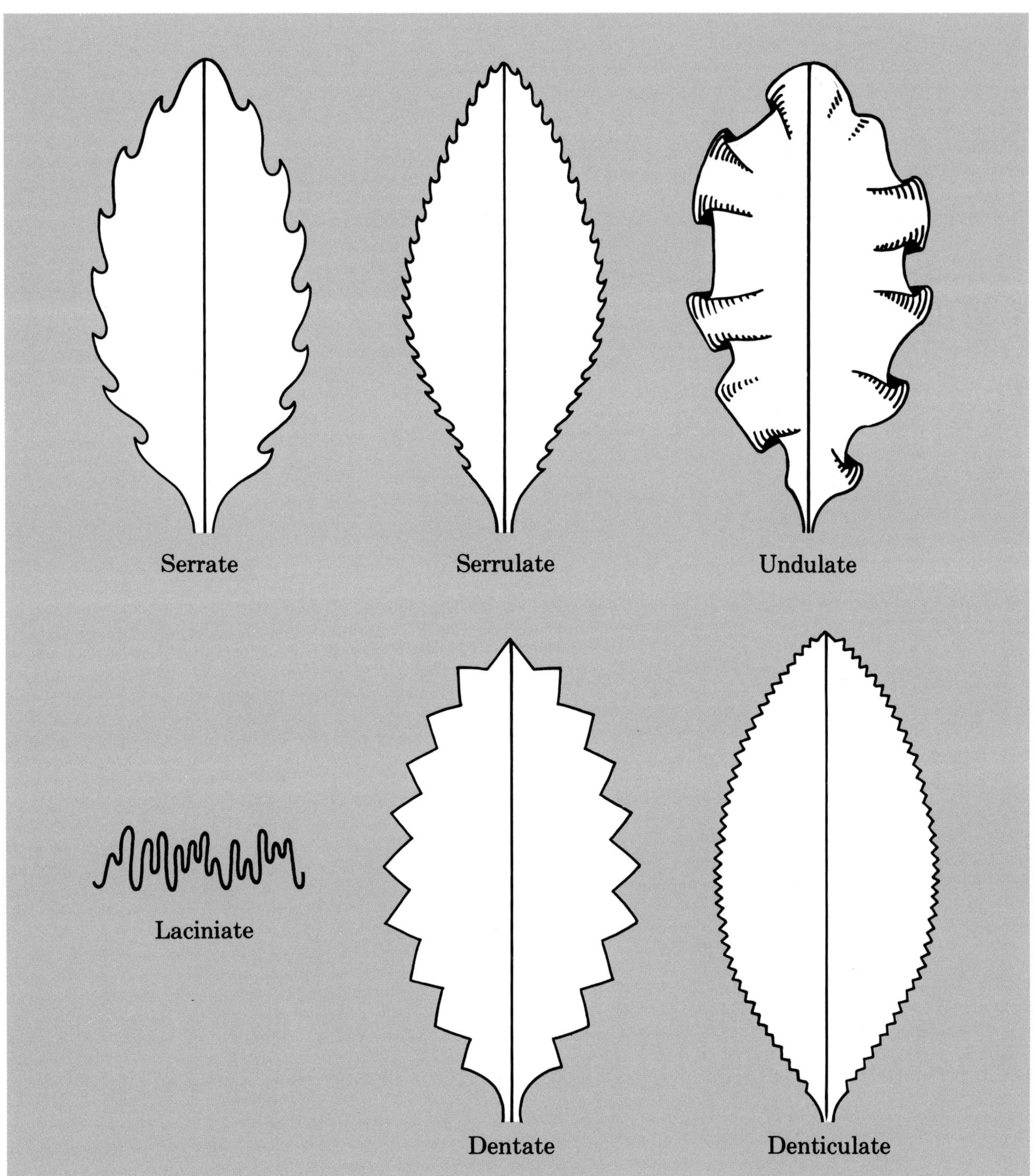

Serrate
Serrulate
Undulate
Laciniate
Dentate
Denticulate

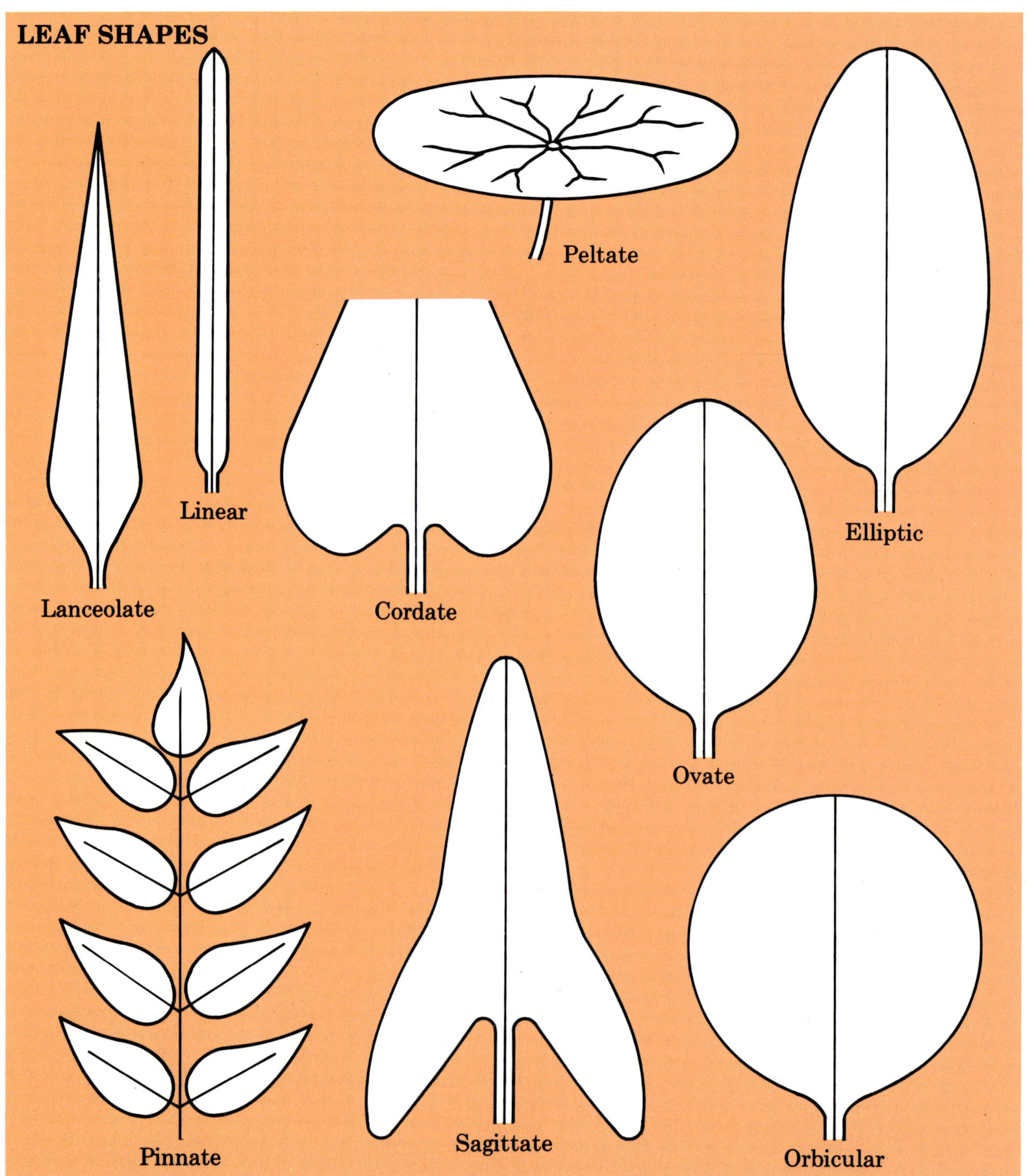

LEAF SHAPES
Lanceolate
Linear
Peltate
Elliptic
Cordate
Ovate
Pinnate
Sagittate
Orbicular

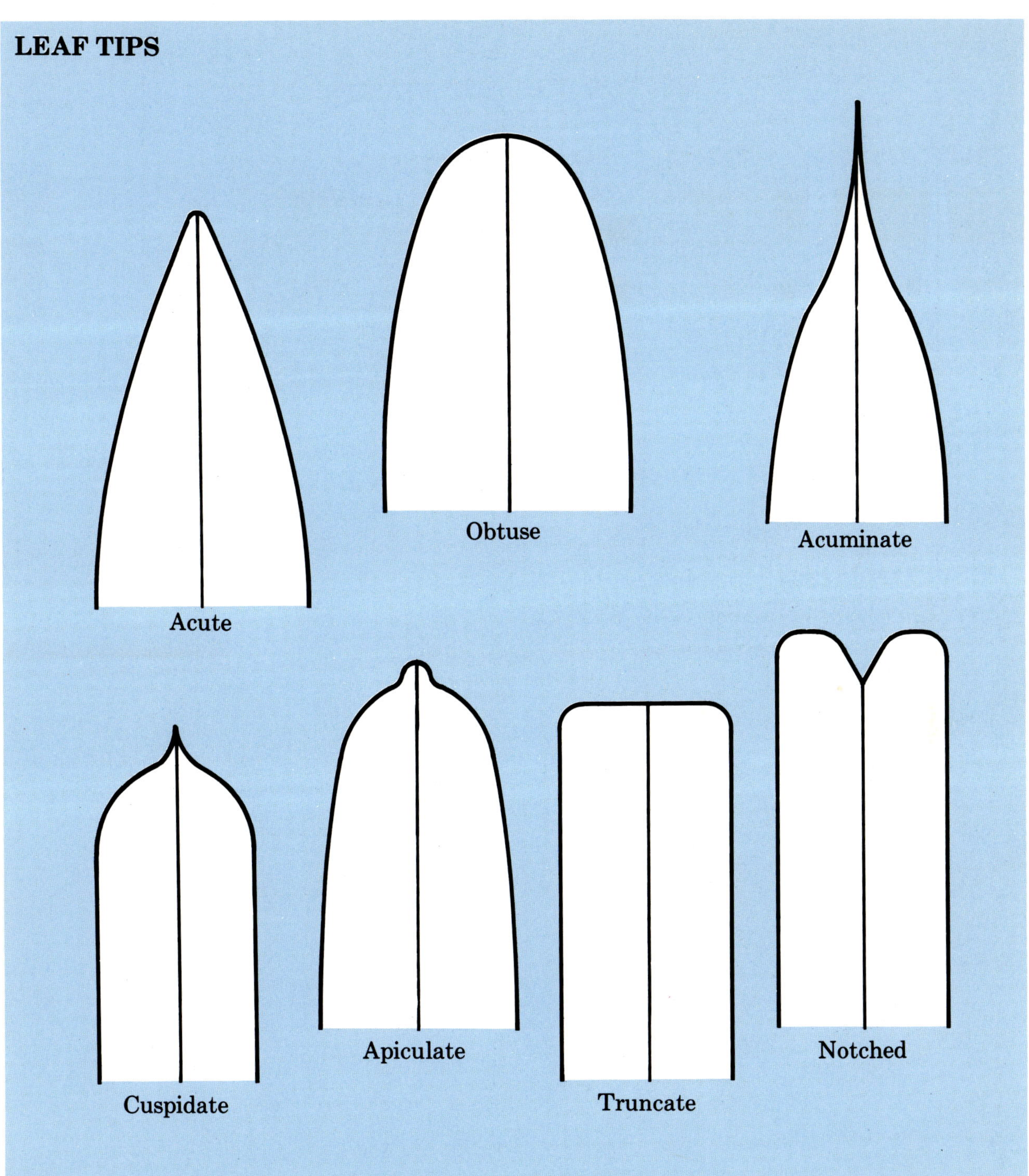

Acute
Obtuse
Acuminate
Cuspidate
Apiculate
Truncate
Notched

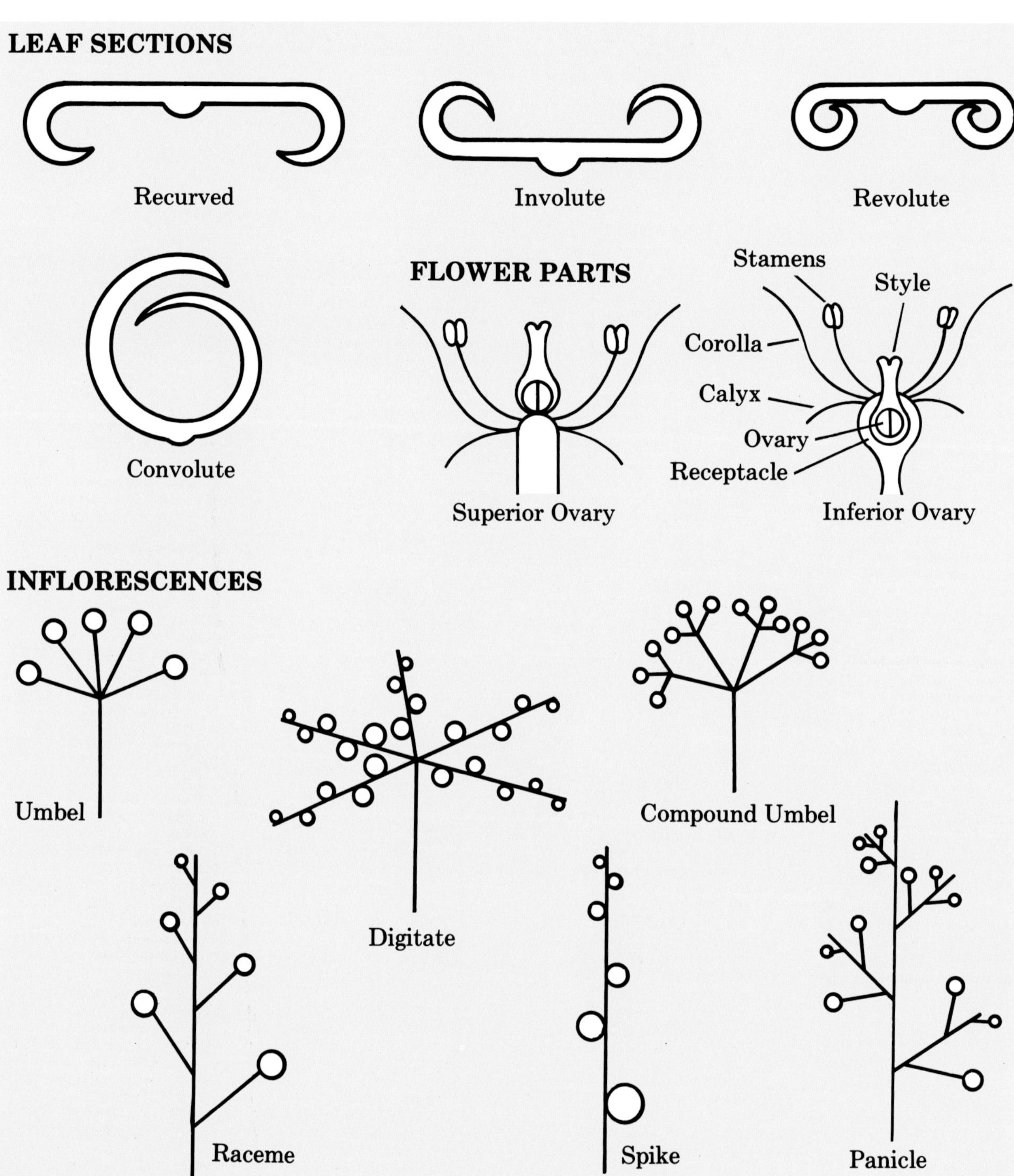

LEAF SECTIONS
Recurved
Involute
Revolute
Convolute
FLOWER PARTS
Stamens
Style
Corolla
Calyx
Ovary
Receptacle
Superior Ovary
Inferior Ovary
INFLORESCENCES
Umbel
Digitate
Compound Umbel
Raceme
Spike
Panicle

INDEX

WATER RESOURCES COMMISSION
NEW SOUTH WALES